Advances in Aquatic Ecology

— Volume 8 —

The Editors

Dr. V.B. Sakhare is Head of Post Graduate Department of Zoology, Yogeshwari Mahavidyalaya, Ambajogai. He has 14 years' experience as an outstanding teacher and researcher. He is recipient of fellowship of Indian Association of Aquatic Biologists, Hyderabad. He has done pioneering work in the field of Reservoir Fisheries and Limnology. Dr. Sakhare has successfully organized *National Conference on Emerging Trends in Fisheries and Aquaculture (ETFA-2012)*, *National Conference on Current Perspectives in Limnology (NCCPL-2009)* and *Regional Workshop* on *Water Quality Assessment (Implications in potability, productivity and pollution control)*.

Dr. Sakhare has been editing an international journal '*Ecology and Fisheries*' (ISSN 0974-6323). He is member of Editorial Advisory Board of *Journal of Science Information* (ISSN 2229-5836) and *E-international Scientific Research Journal (EISRJ)* published by BCTA Inc. Philippines (ISSN 2094-1749). Dr. Sakhare has authored/edited few books such as '*Applied Fisheries*', '*Reservoir Fisheries and Limnology*', '*Reservoir Fisheries and Ecology: A Literary survey*', '*Methodology for Water Analysis*', '*Aquatic Ecology*', '*Aquatic Biology and Aquaculture*', '*Inland Fisheries*', '*Applied Ecology*', '*Perspectives in Ecology*', and '*Advances in Aquatic Ecology (Vols. I–VIII)*.

Dr. Sakhare has supervised a research project funded by University Grants Commission, New Delhi and he is a recognized post graduate teacher and research guide of Dr. Babasaheb Ambedkar Marathwada University, Aurangabad, Solapur University, Solapur and J.J.T. University, Rajasthan. One student has completed Ph.D. under his guidance and three students are working for doctoral degree. He has published 30 research articles and reviews in peer reviewed journals and about 70 marathi articles in newspapers and magazines.

Dr. Sakhare has chaired a number of sessions of different seminars/symposia. He has been invited to different colleges/institutes to deliver lectures on different topics in aquatic ecology and reservoir fisheries.

Dr. B. Vasanthkumar is working as Head and Associate Professor in the Department of Zoology, Government Degree Arts and Science College, Karwar. He was awarded with Ph.D. from Gulbarga University. He has published more than 30 research papers and 30 popular articles in different aspects of ecology and environment. He has also published 11 text books for under graduate students of Karnataka University and guided 4 students for M.Phil. Dr. Vasanthkumar is also co-author of a reference books like '*Aquatic Ecosystem and its management*' and '*Applied Ecology*'. Presently he is working with Major Research Project sanctioned by University Grants Commission, New Delhi.

Advances in Aquatic Ecology

— Volume 8 —

Editors

Dr. Vishwas B. Sakhare

Head,
Post Graduate Department of Zoology
Yogeshwari Mahavidyalaya,
Ambajogai – 431 517
Maharashtra
INDIA

&

Dr. B. Vasanthkumar

Head
Department of Zoology
Government Degree Arts and Science College,
Karwar – 581 301
Karnataka
INDIA

2014

Daya Publishing House®

A Division of

Astral International Pvt. Ltd.

New Delhi – 110 002

Published by : **Daya Publishing House®**
A Division of
Astral International Pvt. Ltd.
– ISO 9001:2008 Certified Company –
4760-61/23, Ansari Road, Darya Ganj
New Delhi-110 002
Ph. 011-43549197, 23278134
E-mail: info@astralint.com
Website: www.astralint.com

Laser Typesetting : **Classic Computer Services**, Delhi - 110 035

Printed at : **Thomson Press India Limited**

PRINTED IN INDIA

Preface

We are delighted to write about the eight volume of *Advances in Aquatic Ecology*. This volume is the compilation of esteemed articles of internationally acknowledged experts in the field of aquatic ecology with the intention of providing a sufficient depth of the subject to satisfy the need of a level which will be comprehensive and interesting. It is an assemblage of up to date information of rapid advances and developments taking place in the field of aquatic ecology. With its application oriented and interdisciplinary approach, we hope that the students, teachers, researchers, scientists, policy makers and environmental lawyers in India and abroad will find this volume much more useful. The articles in the book have been contributed by eminent scientists/academicians active in the areas of aquatic ecology.

We express our sincere gratitude to Dr. S. T. Khursale, President, Yogeshwari Education Society, Ambajogai who has been source of constant inspiration. We are especially thankful to Dr. R. B. Chavan, Principal, Yogeshwari Mahavidyalaya, Ambajogai, Prof. Siddaramu, Principal, Government Arts and Science College, Karwar, Prof. K. Vijaykumar, Department of Zoology, Gulbarga University, Gulbarga and Dr. P. K. Joshi, Dnyanopasak Mahavidyalaya, Parbhani for the encouragement.

Our special thanks and appreciation goes to experts and research workers whose contributions have enriched this volume. We are thankful to C. Stella of Alagappa University, Thondi Campus, Thondi, H. M. Ashashree of Shivamogga, Poonam Bhadja of Rajkot, Dr. S. P. Chavan of Swami Ramanand Teerth Marathwada University, Nanded, Dr. C. Sarla and Dr. P. Deepthi of Centre for Water Resources, Jawaharlal Nehru Technological University, Kukatpally, Dr. Patricio De los Ríos Escalante of Catholic Univeristy, Chile, Dr. R. P. Mali and Dr. A. R. Jagtap of Yeshwant Mahavidyalaya, Nanded, Dr. S. A. Khabade of D. K. A. S. C. College, Ichalkaranji, L. P. Lanka of Devchand College, Arjunnagar, Poonam Lata of Government Dungar College, Bikaner, A. N. Lonkar of Nutan Adarsh Arts, Commerce and Smt. M. H. Wegad Science College, Umrer, R. Uma Maheswari of Arulmigu Palaniandavar Arts College for Women, Palani, M. B. Mule of Dr. Babasaheb Ambedkar Marathwada University, Aurangabad, Dr. J. L. Rathod and U. G. Naik of Karnatak University Post Graduate Centre,

Karwar, V. Ravi of Centre and Advanced Study in Marine Biology, Annamalai University, Parangipettai, Dr. B. Laxma Reddy of Kakatiya University, Warangal, J. B. Solanki of Junagadh Agricultural University, Veraval, Ankush Sharma of Government Dungar College, Bikaner.

We thank our publisher Shri Anil Mittal of Daya Publishing House, Delhi for taking pains in bringing out the book.

Finally, we will always remain a debtor to all my well-wishers for their blessings, without which this volume would not have come into existence.

Dr. V.B. Sakhare

Dr. B. Vasanthkumar

Contents

List of Contributors

Ashashree, H.M.
Department of Zoology, Sahyadri Science College (Autonomous), Kuvmepu University, Shivamogga – 577 203

Balachandar, S.
Centre and Advanced Study in Marine Biology, Annamalai University, Parangipettai – 608 502

Begum, Nafeesa
Department of Botany, Sahyadri Science College (Autonomous), Kuvempu University, Shivamogga – 577 203

Benarjee, G.
Fisheries Research Laboratory, Department of Zoology, Kakatiya University, Warangal – 506 009

Bhadja, Poonam
Department of Biosciences, Saurashtra University, Rajkot – 360 005

Bharathi, R.
Centre and Advanced Study in Marine Biology, Annamalai University, Parangipettai – 608 502

Bhosale, B.J.
Department of Environmental Science, Dr. Babasaheb Ambedkar Marathwada University, Aurangabad – 431 004

Chavan, S.P.
School of Life sciences, Swami Ramanand Teerth Marathwada University Campus, Nanded – 431 605

Chellaiyan, D.
Department of Animal Science, Bharathidashan University, Thiruchirapallai

Deepthi, P.
Centre for Water Resources, IST, Jawaharlal Nehru Technological University, Kukatpally – 500 085

Dhananjaya, S.G.
Department of Zoology, Government Science College, Davangere University, Chitadurga, Karnataka,

Dhapate, A.S.
Sub-Regional Office, Maharashtra Pollution Control Board, Kolhapur

Dodia, A.R.
College of Fisheries Science, Junagadh Agricultural University, Veraval – 362 265

Escalante, Patricio De los Ríos
Laboratorio de Ecología Aplicada y Biodiversidad, Escuela de Ciencias Ambientales, Facultad de Recursos Naturales, Universidad Católica de Temuco, Casilla 15 – D, Temuco, Chile

Hashmi, Seema
Department of Microbiology, Milliya College, Beed.

Hena, M.K. Abu
Department of Animal Science and Fishery, Faculty of Agriculture and Food Sciences, Universiti Putra Malaysia Bintulu Sarawak Campus, Nyabau Road, Post Box No. 396, 97008 Bintulu, Sarawak, Malaysia

Jagtap, A.R.
Post Graduate Department of Zoology, Yeshwant Mahavidyalaya, Nanded

Jayadev, A.
Post Graduate Department of Environmental Sciences, All Saints' College, Thiruvananthapuram

Jeevan, E.N.
Department of Zoology, Sahyadri Science College (Autonomous), Kuvmepu University, Shivamogga – 577 203

Kalaiarasi
Department of Oceanography and Coastal Area Studies, Alagappa University, Thondi Campus – 623 409

Kamble, S.P.
Department of Zoology, D.K.A.S.C. College, Ichalkaranji, District Kolhapur, Maharashtra

Karennawar, M.H.
Department of Zoology, Raje Ramrao Mahavidyalaya, Jath

Kaur, Harbhajan
Post Graduate Department of Zoology, Government Dungar College, Bikaner – 334 003

Khabade, S.A.,
Department of Zoology, D.K.A.S.C. College, Ichalkaranji, District Kolhapur, Maharashtra

Kundu, Rahul
Department of Biosciences, Saurashtra University, Rajkot – 360 005

Lanka, L.P.
Department of Zoology, Devchand College, Arjunnagar, District Kolhapur, Maharashtra

Lata, Poonam
Post Graduate Department of Zoology, Government Dungar College, Bikaner – 334 003

Late, A.M.
Department of Environmental Science, Dr. Babasaheb Ambedkar Marathwada University, Aurangabad – 431 004

Lonkar, A.N.
Department of Zoology, Nutan Adarsh Arts, Commerce and Smt. M.H. Wegad Science College, Umrer – 441 203

Mahadevan, G.
Centre and Advanced Study in Marine Biology, Annamalai University, Parangipettai – 608 502

Maheswari, R. Uma
P.G. Department of Zoology, Arulmigu Palaniandavar Arts College for Women, Palani

Maknikar, Bhagwaan
Sub-Regional Office, Maharashtra Pollution Control Board, Latur

Mali, R.P.
Post Graduate Department of Zoology, Yeshwant Mahavidyalaya, Nanded

Mule, M.B.
Department of Environmental Science, Dr. Babasaheb Ambedkar Marathwada University, Aurangabad

Murgunde, Rekha
Department of Zoology, D.K.A.S.C. College, Ichalkaranji, District Kolhapur, Maharashtra

Naik, K.L.
Department of Zoology, Sahyadri Science College (Autonomous), Kuvmepu University, Shivamogga – 577 203

Naik, T. Vasantha
Department of Botany, D.R.M. Science College, Davangere University, Davangere – 577 066

Niture, S.D.
Department of Zoology, Shivaji Mahavidyalaya, Udgir – 413 517

Parmar, H.V.
College of Fisheries Science, Junagadh Agricultural University, Veraval – 362 265

Patil, Archana
Department of Zoology, D.K.A.S.C. College, Ichalkaranji, District Kolhapur, Maharashtra

Peje, B.S.
College of Fisheries Science, Junagadh Agricultural University, Veraval – 362 265

Priya, A.
Department of Marine sciences and Technology, Madras Christian college – Chennai

Rahman, M.M.
Institute of Oceanography and Maritime Studies, International Islamic University Malaysia, Jalan Sultan Ahmed Shah, Bandar Indera Mahkota – 25200, Kuantan, Pahang, Malaysia

Rajkumar, M.
CAS in Marine Biology, Faculty of Marine Sciences, Annamalai University, Parangipettai – 608 502

Rathore, N.S.
Post Graduate Department of Zoology, Government Dungar College, Bikaner – 334 003

Ravi, V.
Centre and Advanced Study in Marine Biology, Annamalai University, Parangipettai – 608 502

Ravichandran, R.
Department of Oceanography and Coastal Area studies, Alagappa University, Thondi campus, Thondi – 623409

Reddy, B. Laxma
Fisheries Research Laboratory, Department of Zoology, Kakatiya University, Warangal – 506 009

Sakhare, V. B.
Director, Post Graduate Studies, Yogeshwari Mahavidyalaya, Ambajogai – 431 517

Saleem, Quazi
Department of Zoology, Milliya College, Beed

Sarala, C.
Centre for Water Resources, IST, Jawaharlal Nehru Technological University, Kukatpally – 500 085

Saravanakumar, A.
CAS in Marine Biology, Faculty of Marine Sciences, Annamalai University, Parangipettai – 608 502

Sayeswara, H.A.
Department of Zoology, Sahyadri Science College (Autonomous), Kuvmepu University, Shivamogga – 577 203

Serebiah, J. Sesh
Marine Studies and Coastal Resource Management, Madras Christian College, Tambaram, Chennai – 600 059

Shaikh, F.I.
Department of Zoology, Milliya College, Beed

Sharma, Ankush
Post Graduate Department of Zoology, Government Dungar College, Bikaner – 334 003

Solanki, J.B.
College of Fisheries Science, Junagadh Agricultural University, Veraval – 362 265

Stella, C.
Department of Oceanography and Coastal Area Studies, Alagappa University, Thondi Campus, Thondi – 623 409

Sumanthrappa, D.B.
Department of Zoology, Sahyadri Science College (Autonomous), Kuvmepu University, Shivamogga – 577 203

Sumathi, C. Latha
Department of Oceanography and Coastal Area studies, Alagappa University, Thondi campus, Thondi – 623 409

Thanga, V.S.G.
Department of Environmental Sciences, University of Kerala, Thiruvananthapuram

Thayalan, M.
Department of Oceanography and Coastal Area Studies, Algappa University, Thondi Campus – 623 409

Vaghela, Ashokkumar
Department of Biosciences, Saurashtra University, Rajkot – 360 005

Vasanthkumar, B.
Government Arts and Science College, Karwar – 581 301

Veeramani, T.
Centre and Advanced Study in Marine Biology, Annamalai University, Parangipettai – 608 502

Chapter 1

Biodiversity and Population Ecology of Intertidal Macrofaunal Assemblage at Rocky Coasts of Kathiawar Peninsula (India) Off Arabian Sea

☆ *Poonam Bhadja, Ashokkumar Vaghela and Rahul Kundu*

ABSTRACT

The marine fauna is rich and varied. The coastline encompasses almost all types of intertidal habitat. Each local habitat reflects prevailing environmental factors and is further characterized by its biota. Studies on population dynamics are very essential in understanding the effects of various factors governing these aspects in different invertebrate species inhabiting the intertidal zone. The present study deals with the biodiversity and man-made pressure on the coastal health as well wealth of the rocky intertidal macrofauna in four different stations along the Saurashtra coastline. For spatial analysis Saurashtra coastline divided in to four sampling site, *viz.* Dwarka, Mangrol, Veraval and Kodinar. These shores were selected on the basis of their strategic locations, existing industries, infrastructural facilities for the likelihood of being developed as industrial zones, different anthropogenic activities along the entire coastal area of Saurashtra peninsula. The intertidal zone of each sampling sites were surveyed regularly on monthly basis and all the macrofauna and flora encountered were recorded. Population density and abundance values of macrofaunal groups at four different sites were significantly influenced by space and time. The mean density and abundance of each group in the present study was high during most of the time at Dwarka than the other sampling sites because of the selected sites are often greatly complicated by cracks and crevices, rocks pools and variations in slop and rock type. The population density and abundance of intertidal macrofauna was more during winter and low during summer season. It has been observed that winter and post-monsoon seasons were most favorable for most of the animal groups while summer season was the least favorable condition for these free moving animals that probably migrate towards deeper areas of the intertidal to avoid exposure. It has been observed that during summer season macrofaunal population was high at Dwarka followed by Mangrol and Veraval. At the sampling site Kodinar population density and abundance have been the least.

Keywords: *Anthropogenic impact, Arabian Sea, Kathiawar peninsula, Macrofaunal diversity, Population ecology, Rocky shore.*

Introduction

The Ocean is great reservoir of life, but most studies on biological diversity relate to terrestrial systems and thus, the knowledge of marine biodiversity lags far behind than that of the land. The study of organisms which deals with marine habitat is known as marine ecology. The marine fauna is rich and varied. The coastline encompasses almost all types of intertidal habitat. Each local habitat reflects prevailing environmental factors and is further characterized by its biota. The coastal zone comprising, land mass, intertidal areas and inshore waters, is characterized by ecosystem diversity. Rocky shores are one of the most easily accessible marine habitats and are a transition zone between the land and sea. Many animals and plants in the rocky intertidal such as barnacles, limpets and seaweeds, are sessile as adults and can be quickly and easily surveyed using non-destructive methods. The rocky intertidal has a long history as a test area for important concepts in ecology and a good knowledge of processes and species biology.

Studies on population dynamics are very essential in understanding the effects of various factors governing these aspects in different invertebrate species inhabiting the intertidal zone. A population is a collection of individuals of the same species that live together in a region. Population ecology is the study of populations (especially population abundance) and how they change over time. Crucial to this study are the various interactions between a population and its resources. A population can decline because it lacks resources or it can decline because it is prey to another species that is increasing in numbers. Populations are limited by their resources in their capacity to grow; the maximum population abundance (for a given species) an environment can sustain is called the carrying capacity. As a population approaches its carrying capacity, overcrowding means that there are fewer resources for the individuals in the population and this result in a reduction in the birth rate. A population with these features is said to be density dependent. Of course most populations are density dependent to some extent, but some grow (almost) exponentially and these are, in effect, density independent. Ecological models that focus on a single species and the relevant carrying capacity are single species models. Alternatively, multi-species or community models focus on the interactions of specific species.

The coastal environment is being altered at ever-increasing rates, often without looking ahead at future consequences. This is due to a multitude of human activities. The Indian Ocean is the third largest of the four major oceans. It covers an area of 74 million sq. km which comprises about 20 per cent of the total area of water in the world. The ocean has been and remains the frontier of intercontinental trade. A large number of countries, India, are increasingly dependent on the Indian Ocean for their foreign trade and fishing industries. The future is expected to make the sea lanes of the indian ocean important not only to India, but also to the littoral states of the Indian ocean that are dependent on the ocean for fishing, shipping and transportation. Recently study suggested that about 70 per cent of the total sea transport is ferried through Indian waters (Anon, 2003). Moreover, Asia's largest ship breaking yard Alang and fish landing site Veraval is also located along the west coast of India. Thus, the long coastal belt of India, which is known to be rich in fishery and mineral resources, is therefore, at high risk of a serious ecological imbalanced. India has a vast coastline of 7500 km along the mainland in addition to that of the Andaman and Nicobar Islands in the Bay of Bengal and the Lakshadweep Island in the Arabian Sea (Nayak, 2005). The Indian coast has a large variety of sensitive ecosystems like lagoons, sand dunes, coral reefs, mangroves, sea grass beds and wetlands (Ingole, 2005). These coastal habitats are considered highly productive in terms of biological production. Some of these areas have been act as spawning and nursery ground for commercially important fishes, molluscs, crustaceans and various other species that constitute the coastal fisheries.

The coastal stretches of Gujarat have several industries, which are based on salt as raw material. The salt pan activities not only provide livelihood for large number of unskilled workers but also provides raw material for several such chemical industries. Various industries on the Gujarat coastline like the Birla factory in Porbandar, GHCL in Sutrapara, LNT and Cement factories in Kodinar, Bhavnagar and Jaffrabad and various other Gujarat chemical industries on the Saurashtra-Kachchh coastline have been dumping millions of liters of industrial effluents and toxic wastes into the coastal waters every day, as they have no treatment plants. For example, there are four soda ash industries in Saurashtra producing more than 60 percent of the country's total production. One factory or another at one time releases effluents with more ammonia than maximum permissible limits. Veraval is one of the important port cities located the western coast of Gujarat. In addition to this, load of pollution generated from the operation of boats and vessels, the domestic wastewater generated from the Veraval town is being discharged into the fishing harbour area without collection/treatment and also the effluent generated from the 45 fish processing industries located in the nearby GIDC also being discharged in to this fishing harbour area. All these sources of pollution from different spheres are contributing the load of pollution in fishing harbour area and subsequently contaminating the nearby coastal waters. An additional, about 14 fish processing industries are also located in Mangrol and Dwarka.

The present study was undertaken to set up an innovative trend of monitoring of the human-nature interaction and its effect on the natural system to set up the openings of the future study on this tract at this area. In this context, a detailed study on the Saurashtra coastline, one of the biggest one in India desired a detailed monitoring to work out the present status of the ecosystem, the threats mounting and impending, natural resistance and adaptation in response to the pressure and a possible negotiation to the neutralize the harsh condition to offer a better tomorrow. The present study deals with the biodiversity and man-made pressure on the coastal health of the rocky intertidal macrofauna in four different stations along the Saurashtra coastline. With a view to assess the status of the few key species of intertidal mollusca, the heavy metal contamination of the coasts and the interaction between the fauna and anthropogenic activities were investigated. The Western coastal belt of India, these days is considerably being exploited heavily by various kinds of Industries. This study revealed how this is affecting the ecosystem of this area.

Material and Methods

Study Area

India has a coastline of 7517 km (total 4700 mile), which is bounded by the Indian Ocean on the south, the Arabian Sea on the west and the Bay of Bengal on the east. Among that, Gujarat is situated on the north-western part of peninsular India. It has about 1,650 km longest coastline; that accounts for 22 per cent of the total coastline available to the country, with a continental shelf of 164,200 km^2 (35.3 per cent of the country) and an Exclusive Economic Zone (EEZ) of 214,000 km^2 (9.9 per cent of the country). Gujarat coastline consist 28 per cent sandy beach, 21 per cent rocky coast, 29 per cent muddy flats, 22 per cent marshy coast. Further, the coastal zone of Gujarat state can broadly be divided into three major geographical parts, two major gulfs namely Gulf of Khambhat and Gulf of Kachchh and the Saurashtra coastline, each one with its own distinctive characteristic and diverse geo-environmental features, which embrace diverse coastal habitats as well as biota of ecological significance. Saurashtra is a region located south-western part of Gujarat which occupies a total coastal stretch of 865 km. Saurashtra is a part of an arid peninsula called as Kathiawar. On an average, it receives 500 mm rainfall annually. The south coast of Saurashtra from Dwarka-Kodinar segment stretches for about 250 km. with smooth and straight sandy of rocky-sandy beaches. The beaches are usually calcareous and dominated by bio-clasts, the consolidated ancient equivalent of these biogenic sands are famous milliolite rocks. The milliolite underline the beach sands and occur as cliffs, wave cut platforms and submerged dunes, all along the shoreline indicating quaternary sea level fluctuations (Stanley, 2004). The development of the Saurashtra region was driven solely by the trading possibilities offered by its long coastline and ports. It would appear that today the problems consequent to the high degree of industrialization along the Saurashtra coastline are being addressed with even more industrialization. There are two developed fishing harbours with allied at Veraval and Mangrol, which face 62 per cent fish production out of the total fish production. The industrial groups that have grater dominance are cement; food industry and the existing port with facilitate import or export of fish and fishery products, fertilizers, salt, cement, soda ash and lime stone etc.

The whole of south Saurashtra coast was surveyed extensively from physical and biological point of view for their coast characteristics. The main reason behind the selection of the Saurashtra coastline was the difference exists between the four sampling sites which different level of interferences with the human community. It would give a clear indication of the impact of different anthropogenic pressure on marine ecosystem keeping all the physicochemical parameters constant. The other point of view for selection of this region was differences between locations, in terms of slop, substratum type and the length of the intertidal zone.

Sampling Stations

Before the selection of study sites, locations of the sampling sites were selected according to a preliminary survey of the coastline in view of different anthropogenic pressure on coastal area. Now a day's especially Saurashtra coast is being hot-spot for various mega industries, fishery related opportunities and further, more tourism is also one of the related problems on the coastal zone of Saurashtra peninsula. For spatial analysis Saurashtra coastline divided in to four sampling site, *viz.* Dwarka, Mangrol, Veraval and Kodinar (Figure 1.1). These shores were selected on the basis of their strategic locations, existing industries, infrastructural facilities for the likelihood of being developed as industrial zones, different anthropogenic activities along the entire coastal area of Saurashtra peninsula. The Saurashtra coastline is basically rocky-sandy, being rockier in the east and west, sandier in the central part and more rocky-muddy in the far eastern part. It is remarkable by having milliolite limestone formation along the coastline.

Dwarka (22° 13′ N, 68° 58′ E) is situated on the west coast of India and a major pilgrim town owing to that it is also be a tourist place on the coastal area. It is nearest about Okha port, which is well known as entry point of the Gulf of Kachchh and to western India, and around 175 km west of Veraval and around the northernmost corner of the Kathiawar peninsula. There are small scales fishing industries also available. The local community mainly depends on the tourism and fishing related opportunities. The total length of the sampling site was about 1.5 km.

Mangrol (21° 07′N, 70° 07′E) is a small hamlet and important harbour around 50 km west of Veraval with predominantly fisherman population. There are many small scale fisheries industries

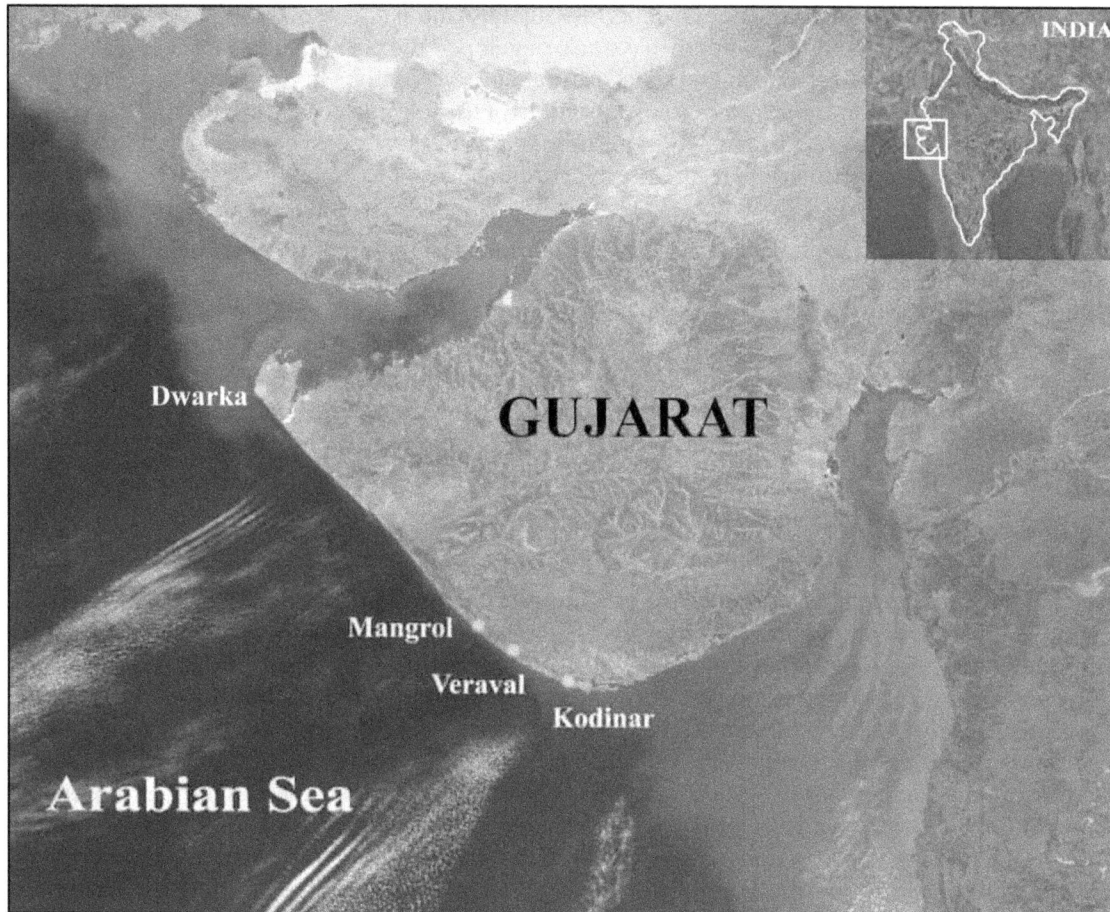

Figure 1.1: Sampling Locations along the Saurashtra Coastline

located along the coastline and it exports the fishery related products to many other countries. The Mangrol port is having a small but proper landing place inclusive of all infrastructure facilities such as storage of catch, ice factories, repairing of boats and engines etc. The total length of the shoreline of this area is about 2 km. The local communities which live nearest about the coastline mainly depend on the fishing related opportunities and changes from time to time.

Veraval (21° 35′ N, 69° 36′ E) is one of the largest fish landing site of India situated around 35 km east of Mangrol, surrounded by a large chemical factory, a medium scale cement factory, number of small to medium scale industries and fish processing units. It involves port activities like transport, boat manufacture and receive waste from different sources. In addition to that, the area, being one of the most developed spot from industrialization point of view is a hot spot for both heavy and small scale chemical industries. The area favours the fish processing industries too due to its proximity to the landing center and easy supply of the raw materials. The total length of the shore area is about 3 km.

Kodinar (20°41′ N, 70° 46′ E) is a small town, situating southeastern part of Junagadh district, on the southern coastal region of Saurashtra peninsula. Tourists find ample places to visit in Kodinar and its nearby region. Very near to the costline, the Gujarat Ambuja Cement Group has established its flagship cement factory and the company have also developed the port of Muldwarka. It is also well-known for the sugar factory, which is situated near about the coastline and also minute level of fish catching unit located near the selected site. Total length of the selected coastal stretch of the sampling site is about 1.5 km. The coastal area between Kodinar and Veraval is fast emerging as an industrial hot-spot and few mega industries are already in operation.

The selected locations are situated at South Saurashtra coastline off Arabian Sea, which are significantly rocky with irregular patches of sand or mud. The rocky portion is generally formed of rocks of miliolite and laterite stone. Extensive limestone deposits are seen to occur in the coastal areas of Gujarat. The intertidal rocks of Saurashtra shoreline are calcareous sand stone. The substratum type is varies at the Saurashtra coastline. The intertidal belt is interspersed with many tide pools, puddles, crevices and small channels. The upper portion of the intertidal belt is generally covered with an admixture of silt and sand mixed with pieces of broken shells. The intertidal belt is intersected by many tide pools. Since the pools are natural ones, the shape and size are not precisely same. The upper intertidal pools have light accumulation of sand settled over the rocky base.

Methodological Approach

Study was intended to conduct the spatial as well as temporal variations of rocky intertidal macrofauna, seawater quality and anthropogenic impact along the Saurashtra coastline. In this regards, four study sites; from Saurashtra coastline were considered as various anthropogenic impacts with different magnitude of human disturbances as seawater quality. Community stress, if any, discriminated by population ecology, community structure and various statistical methods to the recognized anthropogenic disturbances on intertidal macroinvertebrate assemblages and water quality from Saurashtra coastline, Arabian Sea. The study focused on intertidal macrofaunal groups existing on rocky intertidal shores of selected locations. The study examined the differences in density, abundance of all macro-invertebrate groups between four study localities. For the study of community structure and distribution of various macrofaunal groups on intertidal belt, six groups such as porifera, coelenterata, annelida, arthropoda, mollusca and echinodermata were recognized. These groups are significant on the coastal belt for the detailed study of coastal biodiversity. This investigation was undertaken due to the different anthropogenic pressures affects the intertidal community, which experienced by the coastal ecosystem.

Sampling Methods for Macrofaunal Diversity

For Quantitative Analysis

The intertidal zone of each sampling sites were surveyed regularly on monthly basis and all the macrofauna and flora encountered were recorded. Extensive photography was employed for the identification of the animal species with the identification keys, literature available in the form of books, journals, reports and with extensive use of internet. The complete study was conducted in a non-destructive manner in which the organisms were not at all disturbed and in some cases if disturbed, it was limited to the bare minimum, let alone killing any. Once the organisms were identified, during the successive surveys just the record of the encounter was made. However, few algal samples were collected and stored immediately in 10 per cent formaldehyde. They were then brought to the laboratory and washed in running tap water, and then it was subjected for temporary herbarium preparation. During the study, all sampling sites were frequently surveyed at regular intervals during the lowest tides. All intertidal macrofauna and algae observed were recorded properly and later classified systematically. Thus animals under various phyla were recorded and checklist was prepared.

For Qualitative Analysis

The structural attributes of the intertidal fauna were studies by transect method (Misra, 1968). Belt and Foot transect methods in were used for generating the data on the selected belt and criss-cross direction was followed to cover the maximum exposed area on the intertidal belt. The surveys were made at the lowest tides of the months. Sampling used to be started with the start of the low tide and attempts were made to finish two sites within the stipulated duration of about 4 hours. Quadrates of $0.25 \ m^2$ were laid while following an oblique direction covering maximum area at almost regular occurrence. At least 10 quadrates were laid vertically across the complete intertidal area from upper littoral to lower littoral zone for recording the attributes.

Data Analysis

Population Ecology

Among the ecological attributes, seasonal variations in the population density and abundance of major phylum in each sampling stations were calculated (Misra, 1968). The collected data of ecological attributes were calculated by below formula were treated as raw data from which the total density and total abundance values were calculated.

$$\text{Density} = \frac{\text{Total number of individuals recorded from the sample plot}}{\text{Total number of sample plot studied}}$$

$$\text{Abundance} = \frac{\text{Total number of individuals recorded}}{\text{Total number of sample plot where the individuals occurred}}$$

Statistical Analyses

The collected monthly data were presented as seasonally for the seasonal approach like winter, summer, monsoon and post-monsoon. The obtained data were initially subjected to various descriptive statistical analyses like mean and standard deviation. The obtained data were further subjected to different statistical analyses for their cumulative acceptability (Sokal and Rohlf, 1969). Significance of spatial and temporal variations was compared by using single factor ANOVA. More advanced analyses like Regression and Correlation Coefficients analyses were also performed to find out relationship between various water quality parameters within a sampling site and to assess the influence of seawater quality parameters on the macrofaunal community structure (Southwood, 1978).

Results and Discussion

The present study was undertaken to set up an innovative trend of monitoring of the human-nature interaction and its effect on the natural system to set up the openings of the future study on this tract at this area. In this context, a detailed study on the Saurashtra coast line, one of the biggest one in India desired a detailed monitoring to work out the present status of the ecosystem, the threats mounting and impending, natural resistance and adaptation in response to the pressure and a possible negotiation to the neutralize the harsh condition to offer a better tomorrow. The present study deals with the biodiversity and man-made pressure on the coastal health as well wealth of the rocky intertidal macrofauna in four different stations along the Saurashtra coastline. With a view to assess the status of the intertidal macrofauna, the physico-chemical characteristic of the coast and the interaction between the fauna and anthropogenic activities were investigated. The western coastal belt of India, these days is considerably being exploited heavily by various kinds of Industries. This study revealed how this is affecting the ecosystem of this area.

Macrofaunal Diversity

The selected sampling locations along the South Saurashtra coastline off Arabian Sea are predominantly rocky with some patches of sand which is represented various macrofauna and flora species. All the sampling stations have been surveyed regularly and all the species of macrofauna that occurring along the entire intertidal area has been recorded and checklist was prepared. In the present study, a total of 120 intertidal macrofaunal species represented by seven diverse phyla such as porifera, coelenterata, platyhelminthes, annelida, arthropoda, mollusca and echinodermata were recorded, of which mollusca, coelenterata and arthropoda was most prominent groups for the major macrofaunal population of selected intertidal zone. In the present study, amongst four sites, Dwarka showed more macrofaunal diversity on the intertidal belt than the other sites. A clear dominance was observed between the sampling site of Veraval and Kodinar based on the macrofaunal diversity.

The macro-invertebrates showed fluctuations among different sampling sites of Saurashtra coast. There has been a renaissance of taxonomy and related subjects such as abundance and distribution of species as biodiversity in the last two decades. This has resulted from growing awareness that ecological, economic and livelihood securities of mankind are inseparably linked with the maintenance of the diversity of the biological components in land, water and atmospheric environments. For the variety of reasons, macrofauna are extremely important in the functioning of coastal system, from a logistic standpoint that they make a good study specimen, because they are abundant, readily surveyed and taxonomically rich. Macrofaunal groups like coelenterate (Patel, 1978, 1988; Pillai and Patel, 1988; Deshmukhe, *et al.*, 2000) and mollusca (Misra and Kundu, 2005; Vaghela *et al.*, 2010) were studied along the Saurashtra coast. Spatial variation in macrofaunal diversity phylum porifera represents six species of sponges throughout the study period (Table 1.1). It was observed during the present study that in the case of macrofauna, the sponge population was less because sponge is very delicate and damaged by fishermen and other people. Some species such as *Halichondria panacea*, *Microciona* sp., *Oscareila lobularis* and *Tethya* sp. were recorded at all the selected sites, while *Leucosolenia punctata* and *Grantia* sp. were observed only at Dwarka and Mangrol respectively.

Sponge population was relatively high at Dwarka than the other sapling sites. Among the various intertidal faunal groups, porifera has an evolutionary history of about 570 million years and so far, 486 species have been described in India (Thomas, 1998). The Gulf of Kachchh has the highest diversity about 25 species of sponges (Venkataraman and Wafar, 2005). Sponges were seen mainly in middle and lower littoral zone, somewhat present in upper littoral but not in dried area. Sponge diversity was relatively less during summer season at all the sampling sites. Some species such as *Halichondria panacea* were common and highly diverse along the intertidal zone of all the selected sampling sites.

A total of 17 species were recorded in the phylum coelenterata, in case of six coral species were observed from intertidal region at four sampling site. Among them, three species *Goniastraeapectinata*, *Hydnophora exesa* and *Montipora folisa* were observed at all the sampling sites, while *Favia favulus* and *Goniopora* sp. were recorded at Dwarka and Mangrol respectively. It has been found in group coelenterata that the variation in species was high in lower littoral zone and minimum was in upper littoral zone. This may be due to the fact that lower littoral zone was least exposed zone and upper littoral zone was the maximum exposed one, providing the habitat for only some selected species of this sessile group. Among the corals from the intertidal region three species (*Goniastraea pectinata*, *Hydnophora exesa* and *Montipora folisa*) were recorded from all the sampling sites. However, all these coral species have a patchy distribution along the rocky intertidal coast (Raghunathan, *et al.*, 2004). The occurrence of the corals in the intertidal zone is restricted between middle littoral and lower littoral zones. Species like *Portis lutea* and *Favia favulus* were recorded mostly in rock pools. The *Zoanthus* population is quite good here. In the pools of rocky beach, sea anemones were found. However, the lower zone consists of big boulders usually covered with *Zoanthus*.

Table 1.1: Checklist of the Intertidal Macrofauna Recorded at various Sampling Stations

	Occurrence in Sampling Sites					*Occurrence in Sampling Sites*			
	D	M	V	K		D	M	V	K
Porifera					*Beguina variegata*	+	+	−	−
Grantia sp.	−	+	−	−	*Berthellina citrina*	+	−	−	−
Halichondria panicea	+	+	+	+	*Brusa granularis*	+	+	+	+
Leucosolenia punctata	+	−	−	−	*Cantharus spirallis*	+	+	+	+
Microciona sp.	+	+	+	+	*Cantharus undosus*	+	+	+	+
Oscareila lobularis	+	+	+	+	*Cellana radiata*	+	+	+	+
Tethya sp.	+	+	+	+	*Cerithium columna*	+	+	+	+
Coelenterata					*Cerithium morus*	−	−	+	−
Favia favulus	+	−	−	−	*Cerithium scabridum*	+	+	+	+
Goniastraea pectinata	+	+	+	+	*Chiton peregrinus*	+	+	+	+
Goniopora sp.	−	+	−	−	*Conus miliaris*	+	+	+	+
Hydnophora exesa	+	+	+	+	*Conus cumnigii*	+	−	+	+
Montipora folisa	+	+	+	+	*Conus figulinus*	+	+	+	+
Portis lutea	+	+	+	−	*Cronia subnodulosa*	+	+	+	+
Anthopleura sp.	+	+	+	+	*Cyprea lynx*	+	+	+	+
Auralia aurita	+	−	+	−	*Cyprea ocellata*	−	+	−	+
Isaurus tuberculata	−	+	+	−	*Engina zea*	+	+	+	+
Metridium sp.	+	+	+	+	*Euchelus asper*	+	+	+	+
Palythoa tuberculosa	+	+	+	+	*Hexaplex cichoreus*	−	+	−	−
Physalia physalia	+	+	+	+	*Janthina globosa*	+	−	−	+
Porpita porpita	+	+	+	+	*Mancinella bufo*	+	+	+	+
Protopalythoa vestitus	+	+	+	+	*Mitra ambigua*	+	+	+	+
Utricina sp.	−	+	+	−	*Mitras cutulata*	+	+	+	+
Vellella vellella	+	−	−	−	*Monodonta australis*	+	+	+	+
Zoanthus sociatus	+	+	+	+	*Murex bruneus*	+	+	+	+
Platyhelminthes					*Murex ternispina*	+	+	+	+
Pseudoceros indicus	+	−	−	−	*Mytilus* sp.	+	−	−	−
Pseudobiceros stellae	+	−	+	−	*Nassarius distortus*	+	+	+	−
Pseudoceros susanae	+	+	+	+	*Nerita albicilla*	+	+	+	+
Annelida					*Nerita chamaeleon*	−	−	+	−
Baseodicus hemprichii	+	+	+	+	*Octopus vulgaris*	+	+	+	+
Chetopterus chetopterus	+	+	+	+	*Oliva oliva*	+	+	+	+
Eulalia viridis	+	+	+	+	*Onchidium verruculatum*	+	+	+	+

Contd...

Table 1.1–*Contd...*

	Occurrence in Sampling Sites					Occurrence in Sampling Sites			
	D	M	V	K		D	M	V	K
Eurythoa complanata	+	+	–	–	*Paphiaala-papilionis*	+	–	–	+
Hetronereis	+	+	+	+	*Perpura panama*	+	+	+	+
Nereis pelagica	+	+	+	+	*Pyrene flava*	–	–	+	+
Sabella pavonica	+	+	+	+	*Rhinoclavis sinensis*	+	+	+	+
Serpula vermicularis	+	–	–	–	*Siphoneria siphoneria*	+	+	+	+
Arthropoda					*Sunetta donacia*	+	+	+	+
Atergatis sanguinolentus	+	+	+	+	*Thais lacera*	+	–	–	–
Balanu samphitrite	+	+	+	+	*Thais rugosa*	+	+	+	+
Cancer pagurus	+	–	+	+	*Tibia insuladchorab*	+	+	+	+
Carcinus means	+	+	+	+	*Trachicardium flavum*	+	+	–	–
Clibanarius nathi	+	+	+	+	*Trochus radiatus*	+	+	+	+
Clibanarius zebra	+	+	+	+	*Turbo brunnes*	–	+	–	–
Pachygrapsus crassipes	+	+	+	+	*Turbo cornetus*	+	+	+	+
Pagurus longicarpus	+	+	+	+	*Turbo intercostalis*	+	+	+	+
Palaemon serratus	+	+	+	+	*Venus reticulate*	+	+	+	+
Pilumnus hirtellus	+	+	+	+	*Xancus pyrum*	–	+	–	–
Pinaeus indicus	+	–	+	+	**Echinodermata**				
Pinaeus monodon	+	+	+	–	*Antedon* sp.	+	–	+	+
Portunus granulatus	+	–	–	+	*Asterina gibbosa*	+	–	–	–
Portunus pelagicius	–	+	–	+	*Clymeaster* sp.	+	+	–	–
Squilla squilla	–	+	–	–	*Echinus* sp.	+	–	+	–
Mollusca					*Ophioderma brevispinum*	+	+	+	+
Aplysiao culifera	+	+	+	+	*Strongylocentrorus* sp.	+	+	–	–
Austrea stellata	+	+	+	+					

Platyhelminthes group comprised three species, such as *Pseudoceros indicus*, *Pseudoceros stellae* and *Pseudoceros susanae* which were present in tide pools with the existence of water during low tide and associated with algae. Among this species *Pseudoceros susanae* was recorded at all the sampling sites while *Pseudoceros stellae* observed at Dwarka and Veraval. *Pseudoceros indicus* was found only at Dwarka sampling site during post-monsoon season.

The animals of annelids phylum were present all time at the intertidal area of selected sites, it represents eight species during the study period. *Nereis pelagica* and *Baseodicus hemprichii* were observed almost throughout the study period. *Baseodicus hemprichii* was recorded maximum time at Dwarka; however, it found mostly in middle littoral zone at all the sampling site. On the other hand, *Eulalia viridis* was observed mostly at Kodinar during the study. *Nereis pelagica* and *Hetronereis* sp. was present in sandy portion and under the rock in pools and due to its nature of burrowing, rarely came out in the open. The annelids were mainly seen in middle and lower littoral zone but rarely showed in upper littoral zone during the study. *Chetopterus chetopterus*, *Serpula vermicularis* and *Sabella pavonica* were mostly found in lower littoral zone and attached with rocks. *Serpula vermicularis* was found to be attached with rock and observed only at Dwarka through the study period.

The phylum arthropoda was well represented in all the sampling sites throughout the study. A total of 15 species with the *Balanas amphitrite*, *Atergatis sanguinolentus* and various species of hermit crabs like *Clibanarius zebra* and *Clibanarius nathi* as dominant at the intertidal region of selected locations. Arthropoda covers all three littoral zone of the intertidal area. *Pilumnus hirtellus*, *Balanas amphitrite* and *Carcinus means* was recorded commonly at the entire littoral zone of all the sampling sites throughout the study. *Clibanarius zebra* and *Clibanarius nathi* were present in deserted shell of gastropod molluscs in all three zones. The arthropoda group is prefers to be in association with intertidal algae at upper and middle littoral zone, especially in the pools and puddles. Arthropoda feeds on the algae as well as zooplankton, thus, vigorous tidal activity of the lower littoral zone might not be a suitable place for them.

Phylum mollusca were highly occurred and most prominent group than any other phylum in the selected intertidal area of study. About 65 species of mollusca were recorded in the selected sites throughout the study period (Table 1.1). The intertidal areas with rocky and partly sand substrate provide the habitat preferred by the molluscs under study. The most abundant species in all the sampling sites as mollusca the gastropods with the prominent species like *Turbo coronetus*, *Turbo intercostalis*, *Trochus radiatus*, *Nerita albicella*, *Cellena radiata*, *Rhinoclavis sinensis* and *Mancinella bufo* etc. The vertical upper littoral was uniformly covered by small sized *Cellanaradiata* and juveniles of *Cerithiums cabridum*. *Chiton peregrinus* prefer the small pools of the upper and middle intertidal zone. *Turbo coronetus*, a dweller of the upper littoral zone occupies the plane substratum. *Cantharus spiralis* and *Cantharus undosus* were found in quite good number.

The shell-fewer molluscs *Onchidium verruculatum* and *Berthellina citrina* were also present in the selected sites of the study area. *Berthellina citrina* was observed only at Dwarka during winter season. Among the entire animal recorded of phylum mollusca, many species were found in mostly upper littoral zone, while in middle littoral zone all species were present except *Octopus vulgaris* which were present mainly in lower littoral zone. Group Mollusca showed more or less similar trend in upper and middle littoral zones. This trend may be due to the fact that the mollusca mainly feed on the marine algae and thus, always associated with intertidal seaweeds. Mollusca have been recorded 3379 species along the Indian coast from the marine habitat (Subba Rao, *et al.*, 1991, 2004; Subba Rao and Mantri, 2006). Among that eight species of oysters, two species of mussels, 17 species of clams, six species of pearl oysters, four species of giant clams and other gastropods such as *Trochus, Turbo* as well as 15 species of cephalopods are exploited from the Indian marine region (Venkataraman and Wafar, 2005). The phylum mollusca is a large assemblage of animals having diverse shapes, sizes, habits and occupy different habitats and have received more attention because of their aesthetic and gastronomic appeals (SubbaRao, 1977). Molluscan community structure are effective indicators of overall ecosystem health and species diversity (Rittschof and McClellan-Green, 2005), making them ideal study organisms of conservation and biodiversity study.

At all the sampling site six species of phylum echinodermata have been recorded during study period. The members of this phylum were totally absent in upper littoral zone but abundantly occurred in middle and lower littoral zone. Particularly echinoderms were commonly observed hidden in crevices, small caves and between algal cover. All six species were recorded at Dwarka, while three species at Mangrol and Veraval. *Ophioderma brevispinum* was found at all the sampling site during the study period (Table 1.1). The species diversity of this phylum was comparatively less at all the selected intertidal region of the study area because *Asterina gibbosa* and *Ophioderma brevispinum* inhabiting deep water but also present in intertidal zone. On the whole, it appears that in general, this area rich in macrofauna and algae. Most of the species of this phylum were recorded during post-monsoon and winter season at the selected sampling sites.

The results of present investigation suggested that higher dominance of intertidal invertebrate species in the middle and lower littoral zone, as compared to the upper littoral, due to organisms of intertidal zone preferred a healthier environment. Misra and Kundu, (2005) reported, the marine animals along the intertidal have to protect themselves against high salinity, desiccation and against the predators. This they achieve through taking shelter under the thick cover of the seaweeds which grow better on the middle and lower littoral zone. Seaweeds on the upper littoral zone get desiccated out during the period of emergence, and therefore the animals cannot get shelter much at upper littoral zone. Therefore, with the advent of the disappearances of seaweed along the upper and middle littoral, the animal migrates towards the lower littoral as a safer habitat (Misra, 2004; Ramoliya, *et al.,* 2007; Vaghela, *et al.,* 2010). The lower littoral zone of the entire Saurashtra coastline is quite different in nature from other coasts. The rough edge of the lower littoral zone creates very strong wave force, thus, generating strong tearing force. That's hampers the settlement of the algae and feeble footed molluscs. However, the zone seemed to be the best suited one for intertidal organisms. It is clear that there was greater degree of similarity between the middle and lower littoral levels of the intertidal than upper and middle littoral levels with respect to their biota. During the period of emergence *i.e.,* at low tides, the first part to get emerged is the upper littoral zone and the last emerged is the lower littoral zone. This results in a maximum exposure time of the upper littoral and minimum exposure time of the lower littoral. Different organisms adapted to these environments take shelter at a suitable tidal level. So far as the seaweeds are concerned, these inhabiting the upper littoral get dried out and die first. On the contrary, organisms inhabiting the lower littoral level are the least exposed to the ambient environment and they get here more stable habitat as compared to the upper and middle littoral levels. Intertidal macrofaunal community is characterized by temporal and spatial changes in the population and the vertical zonation is the most obvious distribution pattern of hard substrate communities (Witman and Dayton, 2001). The macrofaunal invertebrate were found to have a linkage with vegetation through food web (Bell, 1979). The flora and fauna present on intertidal rock platforms currently show relatively little variation between locations, with seasonal fluctuations. Most sites were dominated by the molluscan group than the other groups in the present investigation.

Seasonal variations were moderately erratic at selected locations, maximum macrofaunal occurrence during winter and post-monsoon. However, all the earlier finding and the present study confirms one important point and that is, maximum seaweed growth occurs during the winter months when the seawater temperature shows a minimum couple with maximum dissolved oxygen content (Vaghela, *et al.,* 2010). The intertidal harbours many microhabitats like tide pools, small puddles, crevices, and small channels. Thus, the spore lings of seaweeds germinate, settle and grow at a particular microhabitat. In these four sites, among all the invertebrate macrofaunal groups, mollusca constituted highest number of species and seasonal occurrence. While most of the groups exhibited obviously discontinuous seasonal occurrence. Certain macrofauna groups like platyhelminthes and Echinodermata were appeared only once or twice during the study. The association of animals on the bases that the algae provided them protection from extreme high and low temperature and their dislodgement by wave action. Further reason for their algal association may be that they also feed on spores, filaments or detritus matter of these algae as evident by their food content (Misra, 2004). From the results of the association of different species of fauna with the flora on the upper littoral zone, it is apparent that *Cellana radiata* seldom associated with *Ulva lactuca*, *Chiton* with *Ulva lactuca* and *Chaetomorpha antennina*. On the other hand, it appears that the *Trochus hanleyanus* did not show any specific affinity to any seaweed species. Thus, it is discernible that this species does not have any specific choice with particular algae and may be using the assemblage as source of food. At the middle littoral zone, *Conus miliris* showed affinity with *Ulva lactuca*. *Nereis* and *Eurythoa complanata* have affinity to association with algae. In this zone *Trochus radiates* and *Turbo* sp. also present and associated with algae too for feeding and breeding reasons. At the lower littoral zone Chitonperegrinus associated with *Sargassum swartzii* and *Ulva lactuca*, *Cyprea* sp. with *Sargassums wartzii*, *Ulvalactuca* and *Gracilaria corticata*. While *Aplysia benedicti* associated with *Sargassum* sp. and *Ulvalactuca*. In general, it appears that, all the animal species were well associated with particularly two seaweed species *Ulva lactuca* and *Sargassum* sp. While, the gastropods *Trochus radiatus* and *Monodonta australis* associated with almost all species of algae.

Similarity Index between Sampling Site

The macrofaunal species occurrence was subjected to similar fauna of the intertidal belt found at various sampling sites. The Sorenson's index of similarity allowed to compares the species composition between sampling sites. The result showed values varied from 0.813 to 0.855 (Table 1.2). The similarity index of four sites indicates that Veraval and Kodinar sites are more similar throughout the study period than Veraval and Dwarka (0.848), and Kodinar and Dwarka (0.846). However, during the study period similarity index of Mangrol and Kodinar (0.826) was closer to that of Mangrol and Veraval (0.831). The lowest value of similarity found between Mangrol and Dwarka (0.813), while between other sampling sites result suggest moderate standing macrofaunal species occurrence during the study. Each of these four sites, intertidal macrofaunal species showed high degree of similarity as all the indices values were above 0.80.

Table 1.2: Sorenson's Index of Similarity for Occurrence of Intertidal Macrofauna at the Four Sampling Sites

	Dwarka	Mangrol	Veraval	Kodinar
Dwarka	1.000			
Mangrol	0.813	1.000		
Veraval	0.848	0.831	1.000	
Kodinar	0.846	0.826	0.855	1.000

Population Ecology

Seasonal density and abundance values of macrofauna at four different sites are given in Table 1.3. In general, the abundance of intertidal macrofauna was more during winter and low during summer season comparatively. Among those mollusca was dominated at all the sites followed by coelenterata and arthropoda. Dwarka and Mangrol recorded highest density during winter and monsoon. The population density at sampling site Veraval and Kodinar was found to be varied from 0.02 to 2.80, 0.02 to 1.68 no/0.25 m^2 respectively (Table 1.3). The mean density of each phylum was significantly influenced by space and seasons. In the present study, the density value was low during most of the time at Kodinar, Veraval and Mangrol than the Dwarka comparatively. At Dwarka macrofaunal population showed a higher density as well as abundance values than the other sites during summer season. Density and abundance was found less at sampling site Kodinar during the most of season. At the sampling site Kodinar maximum density of porifera recorded was 0.30 no/0.25 m² and at Veraval, it was 0.25 no/0.25 m² during monsoon season. The abundance varied greatly among the sampling sites, the highest abundance was found in case of phylum mollusca at site Dwarka (35.75 no/0.25 m²) during winterseason, where large number of gastropods species found that time. Mean abundance value of annelids at Dwarka and Mangrol was 2.33 no/0.25m² followed by Kodinar (1.83 no/0.25m²) during summer (Table 1.3).

Table 1.3: Seasonal Mean Density and Abundance Values of Various Groups in Each Sampling Sites during the Study Period

Macrofaunal Group	Density				Abundance			
	W	S	M	PM	W	S	M	PM
Dwarka								
Porifera	0.32	0.15	0.35	0.30	3.17	1.50	3.50	3.00
Coelenterata	2.85	1.62	2.28	2.67	25.78	15.92	20.83	26.17
Annelida	0.45	0.23	0.30	0.28	3.83	2.33	2.58	2.67
Arthropoda	1.92	1.53	1.73	1.75	14.31	11.75	13.00	14.33
Mollusca	4.10	3.17	3.40	3.15	35.75	29.42	29.28	29.50
Echinodermata	0.05	0.00	0.05	0.07	0.50	0.00	0.50	0.50
Mangrol								
Porifera	0.33	0.08	0.35	0.32	3.33	0.83	3.50	3.17
Coelenterata	1.65	0.88	0.95	1.65	14.67	8.83	8.92	13.25
Annelida	0.35	0.28	0.23	0.30	2.42	2.33	1.83	2.83
Arthropoda	1.38	0.80	1.10	0.97	11.58	7.25	9.58	9.00
Mollusca	3.18	2.38	2.48	2.40	29.25	22.08	22.39	22.00
Echinodermata	0.02	0.03	0.02	0.03	0.17	0.33	0.17	0.33
Veraval								
Porifera	0.22	0.00	0.25	0.15	2.17	0.00	2.50	1.50
Coelenterata	1.78	1.03	1.23	1.45	16.28	9.75	10.33	13.67
Annelida	0.28	0.15	0.13	0.23	2.25	1.50	1.33	1.92
Arthropoda	0.77	0.37	0.52	0.67	6.75	3.50	4.75	6.50
Mollusca	2.80	1.75	1.93	2.07	25.67	16.58	18.92	19.00
Echinodermata	0.03	0.00	0.03	0.02	0.33	0.00	0.33	0.17
Kodinar								
Porifera	0.25	0.12	0.30	0.25	2.50	1.17	3.00	2.50
Coelenterata	1.42	1.08	0.92	1.45	13.47	10.42	8.92	14.25
Annelida	0.37	0.20	0.18	0.18	3.17	1.83	1.67	1.67
Arthropoda	0.73	0.68	0.60	0.52	6.33	5.92	5.42	4.67
Mollusca	1.68	1.33	1.37	1.33	16.25	12.75	13.08	12.92
Echinodermata	0.02	0.00	0.00	0.00	0.17	0.00	0.00	0.00

W: Winter; S: Summer; M: Monsoon and PM: Post-monsoon.

In the intertidal zone of the selected sampling sites, the density and abundance values of the coelenterata showed no meaningful variations however decreased during summer season. Coelenterata was dominated at all the sampling sites followed by arthropoda. Annelid was represented highest density value (0.45 no/0.25m^2) at Dwarka during winter season. Echinodermata showed lowest density and abundance values during winter season with 0.02 and 0.17 no/0.25m^2 respectively (Table 1.3). The abundance value of mollusca at Veraval site ranged from 16.58 no/0.25m^2 to 25.67 no/0.25m^2 during summer and winter season respectively. The least value of density and abundance found during summer and maximum during winter season. In the present study significant spatial variations in coelenterata, arthropods and molluscan groups were well supported by ANOVA test (Table 1.4). Population density and abundance values of macrofaunal groups at four different sites were significantly influenced by space and time. The mean density and abundance of each group in the present study was high during most of the time at Dwarka than the other sampling sites because of the selected sites are often greatly complicated by cracks and crevices, rocks pools and variations in slop and rock type. Cracks and crevices provide protection from waves and from desiccation and will increase species richness of a shores and abundance of some species.

Table 1.4: Results of ANOVA for Density and Abundance of Major Macrofaunal Groups in the Selected Sampling Sites

Phylum	Density	Abundance
Porifera	1.2396	1.2396
Coelenterata	7.0755*	8.8941*
Annelida	1.8958	2.6520
Arthropoda	36.3633*	29.8796*
Mollusca	19.7270*	20.4185*
Echinodermata	3.0370	2.9111

* Significant at $p = 0.05$.

The population density and abundance of intertidal macrofauna was more during winter and low during summer season. It has been observed that winter and post-monsoon seasons were most favourable for most of the animal groups while summer season was the least favourable condition for these free moving animals that probably migrate towards deeper areas of the intertidal to avoid exposure (Vaghela, *et al.,* 2010). The density and abundance values were slightly decreased during monsoon to post-monsoon season then it was increased up to winter season in case of porifera group. It has been also observed that during summer season macrofaunal population was high at Dwarka followed by Mangrol and Veraval. At the sampling site Kodinar population density and abundance have been least values of macrofaunal group. Overall the least value of the population density and abundance found during summer and maximum during winter season. In the present study significant spatial variations in coelenterata, arthropoda and mollusca were supported by ANOVA test which may be due to the variations in general abiotic parameters and difference population of these species between the sampling sites. Other animal groups did not show significant spatial variation either in population density or abundance (Table 1.4).

Conclusion

Population density and abundance values of macrofaunal groups at four different sites were significantly influenced by space and time. The mean density and abundance of each group in the present study was high during most of the time at Dwarka than the other sampling sites because of the selected sites are often greatly complicated by cracks and crevices, rocks pools and variations in slop and rock type. The population density and abundance of intertidal macrofauna was more during winter and low during summer season. It has been observed that winter and post-monsoon seasons were most favourable for most of the animal groups while summer season was the least favourable condition for these free moving animals that probably migrate towards deeper areas of the intertidal to avoid exposure. It has been observed that during summer season macrofaunal population was high at Dwarka followed by Mangrol and Veraval. At the sampling site Kodinar population density and abundance have been the least.

References

Anon, 2003. Fan Shells, Family Pinnidae. Online Guide to Check Jawa. http://www. wildsingapore. com/chekjawa/text/s411. htm (26/9/2005).

Bell, S.S., 1979. Short and long term variation in a high marsh meiofauna community. *Estuarine, Coastal and Marine Science*, 9: 331–350.

Deshmukhe, G., Ramamoorthy, K. and Sen Gupta, R., 2000. On the coral reefs of the Gulf of Kachchh. *Current Science*, 79(2): 160–162.

Ingole, B.S., 2005. Indian ocean coast: Coastal ecology. In: *Encyclopedia of Coastal Science*. (Ed.) Maurice L. Schwartz. Western Washington University, WA, USA. Spinger Publisher, Netherlands, p. 546–554.

Misra, R., 1968. *Ecology Workbook*. Oxford and IBH Publishing Co., Calcutta, India.

Misra, S. and Kundu, R., 2005. Seasonal variations in population dynamics of key intertidal molluscs at two contrasting locations. *Aquatic Ecology*, 39: 315–324.

Misra, S., 2004. Studies on the biodiversity and the impact of anthropogenic pressure on the ecology and biology of certain intertidal macrofauna. *Ph.D. Thesis*, Saurashtra University, Rajkot.

Nayak, G.N., 2005. Indian Ocean coast, coastal geomorphology.In: *Encyclopedia of Coastal Science*. (Ed.) Maurice L. Schwartz. Western Washington University, WA, USA. Spinger Publisher, Netherlands, p. 554–557.

Patel, M.I., 1978. Generic diversity of Scleractinians around Poshitra Point, Gulf of Kutch. *Indian Journal of Marine Sciences*, 7: 30–32.

Patel, M.I., 1988. Patchy corals of the Gulf of Kutch. In: *Proceedings of the Symposium on Endangered Marine Animals and Marine Parks*, (Ed.) E.G. Silas. Marine Biological Association of India, Cochin, p. 411– 413.

Pillai, C.S.G. and Patel, M.I., 1988. Scleractinian corals from the Gulf of Kutch. *Journal of Marine Biological Association of India*, 30: 54–74.

Raghunathan, C., Sen Gupta, R., Wangikar, U. and Lakhmapurkar, J., 2004. A record of live corals along the Saurashtra coast of Gujarat, Arabian Sea. *Current Science*, 87(8): 1131–1138.

Ramoliya, J., Kamdar, A. and Kundu, R., 2007. Movement and bioaccumulation of chromium in an artificial freshwater ecosystem. *Indian Journal of Experimental Biology*, 45: 475–479.

Rittschof, D. and McClellan-Green P., 2005. Mollusks as multidisciplinary models in enviroment toxicology. *Marine Pollution Bull.*, 50: 369–373.

Sokal, R.R. and Rohlf, F.J., 1969. *Biometry*. WH Freeman and Company, San Francisco.

Southwood, T.R.E., 1978. *Ecological methods*. Chapman and Hall, London.

Stanley, O.D., 2004. Wetland ecosystems and coastal habitat diversity in Gujarat, India. Review. *Journal of Coastal Development*, 7(2): 49–64.

SubbaRao, N.V., 1977. On the collection of Strombidae (Mollusca : Gastropoda) from Bay of Bengal, Arabian sea and Western Indian Ocean with some new records 1. Genus Lambis, Terebellum, Tibia and Rimella. *Journal of Marine biological Association of India*, 19: 21–34.

SubbaRao, N.V., Surya Rao K.V. and Maitra, S., 1991. Marine molluscs. State Fauna series 1, Part 3. Fauna of Orissa. Kolkata: Zoological Survey of India, pp. 1–175.

SubbaRao, P.V. and Mantri, V.A., 2006. Indian seaweed resources and sustainable utilization: Scenario at the dawn of a new century. *Current Science*, 91(2): 164–174.

SubbaRao, P.V., Eswaran, K. and Ganesan, M., 2004. Cultivation of agarophytes in India.

Thomas, P. A., 1998. Porifera, pp: 27–36. In: *Faunal Diversity of India*, (Eds.) J.R.B. Alfred, A.K. Das and A.K. Sanyal. ENVIS Centre, Zoological Survey of India, Calcutta, pp. 497.

Vaghela, A., Bhadja, P., Ramoliya, J., Patel, N. and Kundu, R., 2010. Seasonal variations in the water quality, diversity and population ecology of intertidal macrofauna at an industrially influenced coast. *Water Science and Technology*, 61(6): 1505–1514.

Venkataraman, K. and Wafar, M., 2005. Coastal and marine biodiversity of India. *Indian Journal of Marine Sciences*, 34(1): 57–75.

Witman, J.D. and Dayton, P.K., 2001. Rocky subtidal communities. In: *Marine Community Ecology*, (Eds.) M.D. Bertness *et al.* Sinauer Associates, Sunderland, pp. 339–366.

Chapter 2

Trace Metal Analysis in the Worm Eel, *Myrophis platyrhynchus* Breder, 1927 from Parangipettai, Southeast Coast of India

☆ *T. Veeramani, V. Ravi, R. Bharathi, G. Mahadevan*
and S. Balachandar

ABSTRACT

The objective of the present study was to determine the trace metal accumulation in worm eel, *Myrophis platyrhynchus* caught from Parangipettai coastal waters and were analyzed trace metals (Cd, Cr, Cu, Fe, Mn, Ni, Pb, Zn, and Hg) accumulations in tissues of the worm eel. Among the nine metals, the order of accumulation was as: Fe > Cu > Mn > Zn > Pb > Cr > Ni > Cd and Hg. The higher concentration of metals was found to be iron and copper but lower concentration was mercury.

Keywords: Trace metals, Worm eel, Muscle, M. platyrhynchus.

Introduction

Impact of pollution in the sea changes water chemistry and upset the ecological balance in the ocean; eventually it can disrupt the productivity of the ocean (Mirnov, 1978). Among the many pollutants, attention must be focused on heavy metals because of their environmental persistence, toxicity at low concentration and ability to incorporate into food chain of aquatic organism (Negilski, 1976; Harte *et al.*, 1991). Trace elements occur in minute concentration in biological system. They may exert beneficial or harmful effect on plant, animal and human life depending upon the concentration (Forstner and Wittman, 1981). The source of metals can be taken up by the fish from water, food, sediment and suspended particulate material (Harderson and Wratten, 1998). However, the presence of metals in high concentration in water or sediment does not involve the direct toxicological risk to fish, especially in the absence of significant bioaccumulation. It is known that bioaccumulation is to a large extent mediated by abiotic and biotic factors that influence metal uptake (Rajotte *et al.*, 2003).

Metal also act as constituent of oxygen carriers, the best known are hemoglobin (Fe) hemocyanin (Cu) (Bryan, 1980). Most heavy metals whether essential or not are potentially toxic to living organisms at higher concentration (Bryan, 1976). The worm eel (*M. platyrhynchus*) (Anguilliformes: Ophicthidae) inhabits as a demersal in nature and their food items along with the organic matters may be subjected to accumulate the trace metals. Hence the present study is aimed to find out the levels of trace metals (Cd, Cr, Cu, Fe, Mn, Ni, Pb, Zn, and Hg) accumulations in tissues of the worm eel from the study area.

Materials and Methods

Fish samples of *M. platyrhynchus* were collected from Mudasalodai landing centre during January to December 2010. The specimen (30 – 50 cm TL of similar maturity stage) was properly cleaned and the total length, total weight and sex and maturity stages were determined. In the present study, the trace metals such as Cadmium (Cd), Chromium (Cr), Copper (Cu), Iron (Fe), Manganese (Mn), Nickel (Ni), Lead (Pb), Zinc (Zn) and Mercury (Hg) were evaluated. For trace metal analysis, a portion of the muscle from the widest part of the body (devoid of bones) after removal of the skin was taken carefully from male and females separately and used for trace metal analysis. The analysis of trace metals was carried out using the method suggested by Alam *et al.* (2002). For this, all the reagents used are analytical grade. The sample was digested with concentrated nitric acid. Dissected samples were transferred to a clean beaker. Then 10 ml of concentrated nitric acid was added and the sample was heated using a hot plate for near dryness. Finally 2 ml of 1N HNO_3 was added to the residue and the solution evaporated again on the hot plate, continuing until every sample was completely digested. After cooling, further 10ml of 1N HNO_3 was added. The solution was then diluted and filtered through a 0.45 µm nitrocellulose membrane filter. Determination of the elements in all samples was carried out by ICP- AES (Optima 2100 DV, Perkin-Elmer, USA).

Results and Discussion

A total of nine trace metals such as Cadmium (Cd), Chromium (Cr), Copper (Cu), Iron (Fe), Manganese (Mn), Nickel (Ni), Lead (Pb), Zinc (Zn) and Mercury (Hg) were analyzed. Among the nine metals the order of accumulation Fe> Cu > Mn > Zn > Pb > Cr > Ni > Cd and Hg were estimated. The higher concentration metals are iron and copper but lower concentration was mercury. The statistical analysis of trace metal concentrations are shown in Table 2.1 and Figure 2.1.

In the present investigation, the value obtained for *M. platyrhynchus* was in the trace metal concentration of male as: the concentration of Cd 0.486 µg/g – 1.587 µg/g, Cr 3.89 µg/g – 8.32 µg/g, Cu 10.32 µg/g – 26.96 µg/g, Fe 300.84 µg/g – 662.31 µg/g, Mn 7.92 µg/g – 24.51 µg/g, Ni 1.02 µg/g – 2.41 µg/g, Pb 3.52 µg/g – 7.36 µg/g, Zn 13.26 µg/g – 58.16 and Hg 0.36 µg/g – 1.41 µg/g were recorded. In female, the concentration of Cd 0.452 µg/g – 1.45 µg/g, Cr 3.21 µg/g – 7.125 µg/g, Fe 314.23 µg/g – 635.21 µg/g, Mn 6.521 µg/g – 22.14, Ni 1.01 µg/g – 2.21 µg/g, Pb 3.47 µg/g – 7.74 µg/g, Zn 17.13 µg/g – 59.32 and Hg 0.054 µg/g – 1.98 µg/g were noticed.

The highest and lowest value of trace metal concentration in male 'Cd' during October (1.587 (µg/g) and July (0.483 (µg/g), 'Cr' November (8.3 µg/g) and July (3.89 µg/g), 'Cu' November (26.96 µg/g) and June (10.32 µg/g), 'Fe' November (662.31 µg/g) and June (300.84 µg/g), 'Mn' November (24.51 µg/g) and June (7.92 µg/g), 'Ni' November 2.41 µg/g and 1.02 µg/g during June, 'Pb' November (7.36 µg/g) and August (3.52 µg/g), 'Zn' November and July (58.16 µg/g) and (13.26 µg/g) respectively,

Hg during November and June (1.41 µg/g) and (0.36 µg/g) respectively. The maximum and minimum value of trace metal concentration in female during the months are, 'Cd' during October (1.45 (µg/g) and July (0.452 (µg/g), 'Cr' October (7.125 µg/g) and April (3.21 µg/g) respectively, 'Cu' November 24.13 µg/g and July 10.23 µg/g, 'Fe' October (635.21 µg/g) and June (314.23 µg/g), 'Mn' November (22.14 µg/g) and June (6.521 µg/g), 'Ni' October (2.21 µg/g) and June (1.01 µg/g), 'Pb' November (7.74 µg/g) and July (3.47 µg/g) respectively, 'Zn' October (59.32 µg/g) and April (17.13 µg/g), 'Hg' November (1.98 µg/g) and July (0.054 µg/g) respectively.

Table 2.1: Analysis of Sum, Average and Variance between the Trace Metal Concentrations in the Muscle of Male and Female of *M. platyrhynchus*

Metals	Sex					
	Male (µg/g)			Female (µg/g)		
	Sum	Average	Variance	Sum	Average	Variance
Cd	9.98	0.907273	0.105205	9.01	0.819091	0.087283
Cr	62.08	5.643636	2.010565	58.364	5.305818	1.275909
Cu	183.46	16.67818	21.5477	187.115	17.01045	19.36826
Fe	5562.36	505.6691	14530.79	5396.22	490.5655	12530.01
Mn	167.76	15.25091	32.33903	169.241	15.38555	28.99563
Ni	17.72	1.610909	0.192949	16.9	1.536364	0.186905
Pb	58.75	5.340909	1.769349	58.971	5.361	2.032089
Zn	366.49	33.31727	168.3489	384.93	34.99364	226.1589
Hg	8.98	0.816364	0.112845	8.553	0.777545	0.269754

The statistical analysis of correlation was followed by SPSS software. The correlation was made between the trace metals. In the male species, the 'Cd' correlated with Fe and the significant level showed at 0.05 levels, other metals were non-correlated. Cadmium and Chromium (Cr) values are significant with Cu, Fe, Mn, Ni, Pb, Zn and Hg the results shows positive correlation at the significant level 0.01, 'Cu' correlated with Fe, Mn, Ni, Pb, Zn and Hg and the significant level 0.01 was recorded, 'Fe' correlated with Mn, Ni, Pb, and Hg with significant level 0.01 and Zn significant level 0.05 was reported, 'Mn' correlated with Ni, Pb, Zn and Hg with positive correlation and significant level recorded 0.01 level, 'Pb' correlated with Zn and Hg significant level 0.01, Zn correlated with Hg and the significant level 0.01, according to the correlation 99 per cent of metals significant were recorded.

In female 'Cd' correlated with Cr, Cu, Fe, Ni and Zn and the significant level 0.01, 'Cr' indicated insignificant level and without correlated with other elements, 'Cu' correlated with Fe, Mn, Ni, Pb, Zn and Hg and significant level 0.01. 'Fe' correlated with Mn, Ni and Zn and the significant level 0.01 and Pb, Hg was correlated with 0.05 significant levels. 'Mn' correlated Pb and Zn with 0.01 significant level and Ni, Hg significant with 0.05 level, 'Ni' correlated with Zn and Hg and significant level 0.01, 'Pb' correlated with Zn and Hg with 0.05 significant level, 'Zn' correlated with Hg and the significant level 0.05 was reported.

According to the present result, the maximum concentration was found during August to December in both sexes, and minimum concentration was noticed during April - July, based on the concentration variation in the peak trace metal concentration which was reported during November in male, and in female during October and November respectively, less concentration reported during June and July

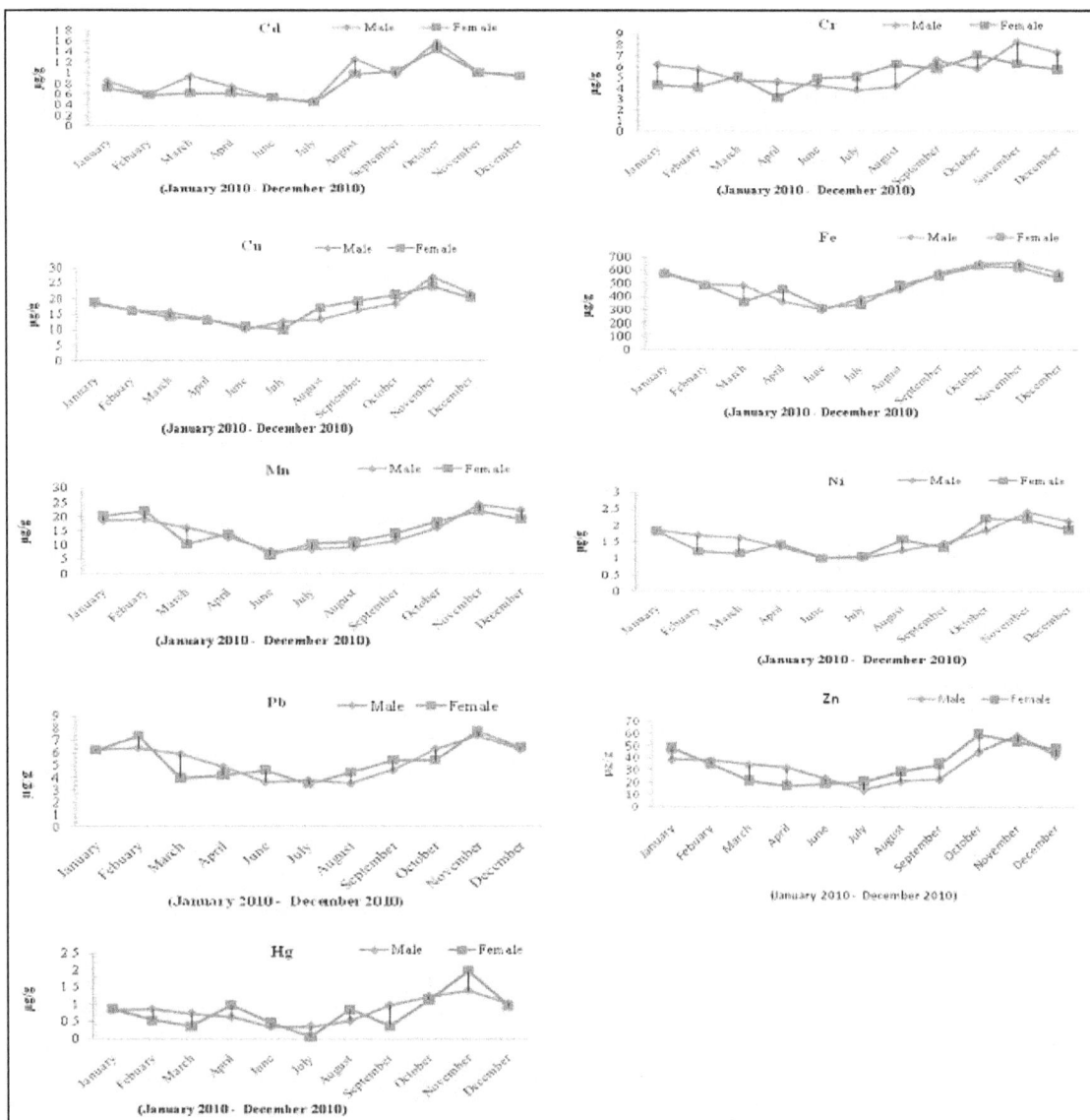

Figure 2.1: Concentration of Cadmium, Chromium, Copper, Iron, Manganese, Nickel, Lead, Zinc and Mercury in Male and Female of *M. platyrhynchus*.

in male and female April, June and July, the trace metal accumulations were differed depending upon the food and feeding, maturation and spawning factors, the availability of food rich in the season of monsoon due to heavy rain, river water inflow and domestic and industrial wastages were accumulated in the shore region, based on the heavy nutrient rich and food chain stimulate the trace metal accumulation and more in monsoon season. Low concentrations of trace metals were reported in the month of June and July may be the reasons for less feeding activity in dry season (summer season) because lack of river or domestic, industrial wastage inflow, availability of foods were less than monsoon season.

In female fishes, concentration of trace metals are varied between the months and seasons, main reason indicate maturation and spawning activity. The maturation period and feeding activities were less than other seasons. After spawning feeding activity are more, in the present study and the result shows the maximum concentration of trace metals found during October and minimum during April, June, and July. In these three months the concentrations are not much varied were noted. The fluctuation and sudden changes of trace metal were found more in female than male, monsoon season showed gradual increasing of trace metal accumulation that is based on the food and feeding activity.

Accumulation of trace metals are caused by several natural and anthropogenic activities, increasing industrialization along with the violation of effluent disposal norms that has caused heavy contamination of water bodies. Fish and other aquatic biota in the vicinity of industrial areas is a good indicator for gauging the level of pollution. Heavy metals like Cadmium (Cd), Chromium (Cr), Zinc (Zn), Copper (Cu), Lead (Pb), Nickel (Ni), and Mercury (Hg) have proven to be persistent pollutants.

The trace metals analyzed in the present study has been reported as being commonly bioaccumulated (*e.g.*, Miramand *et al.*, 1998; Ni *et al.*, 2000) and in fishes, metal levels generally diminish from the bottom to the top of the food chain (Mance, 1987). The present study showed little evidence for such phenomena, (*i.e.*, bioaccumulation). With the exception of Zn, Fe, Cd and other trace metals, the present results are in contrast with that of Fernandes *et al.* (1994). Concentration of trace-metals was decreased with age and size of the organisms because contamination does not always happen through ingestion but through contact across a permeable membrane such as the gill (Mance, 1987). Though present in traces and being lipophilic, heavy metals tend to bioaccumulate and biomagnifying. Bioactive metals play important role in metabolism, thus in physiology and pathology of fish, Zn, Cu and Mn function as a cofactor in several enzyme systems and Fe directly involved with haemoglobin formation in fish blood described by Bury *et al.*, 2003.

Non-essential metals namely, Pb and Cd and essential metal namely Zn were quite high and continues input could risk the ecosystem. Continuous accumulation in fish muscle could transport metals to human through diet and fishes are safe to be taken for daily diet since all metals detected in fishes were below the permissible limit (Shuhaimi Othman, 2010). Concentration of metal in fish body tissue from the surrounding medium or food, either by absorption or ingestion is known as bioaccumulation (Forstner, and Wittmann, 1981), based on the health the fish can regulate metal concentration to a certain limit after which bioaccumulation occur (Heath, 1987), in marine or freshwater environment the higher level metal accumulation in fish tissue than compared with water and the higher metal bioaccumulation could be unsafe for human consumption, the highest level metal general ration $Zn_2^+ > Cu_2^+ > Pb_2^+ > Cd_2^+$ (Mohammed Al-Kahtani, 2009) the same ration of trace metal was accumulated in tissue of eelfishes. Lipid accumulation in eelfish for increased growth phase to provide sufficient energy to enable migration, gonad maturation and spawning, but the accumulations of cadmium affect the lipid storage in yellow eelfish in order to evaluate the possible contribution of this pollutant to the reported decline of European eel populations. Eels were exposed to dissolved cadmium at nominal concentrations of 0 and 5g L^{-1} for 1 month (Fabien Pierron *et al.*, 2007). Present study may be useful to monitor bioaccumulation of fish tissue and control of trace metal accumulation.

The metal concentration due to the assessment of marine contamination and derived from natural origins rather than anthropogenic activity, in marine fish and bivalves contain low concentration of Hg and very higher concentration of Cd according to the food chain and bioaccumulation (Stephen de

Mora *et al.* (2004)). Copper is the most common biocide in antifouling paints and this leaches to marine environment through boats and impact would be on surrounding organisms through food chain and bioaccumulation. Concentration of Fe, Mn, Mg, Ni in five fishes were studied from Parangipettai coastal waters and result of Fe and Mn highest level than Ni and Mg, (Lakshmanan *et al.*, 2010) based on cited author result the present study focused result Fe and Mn higher concentration than other metals.

Jyotsna Singh *et al.* (2008) described the Cd concentration in freshwater fish tissue the range value <5 ppm, when compared present result of worm eel Cd concentration was very low rank and maximum 1.5 ppm were reported. Urena *et al.* (2007) reported the *Anguilla anguilla* species heavy metals ratio. The concentration of heavy metals was maximum Pb (0.67 µg/g), Cu (5.0 µg/g), Zn (12.97 µg/g), Cd (0.08 µg/g), Mn (0.1 µg/g), Hg (0.08 µg/g), and Fe (3.72 µg/g) were found and compared with the present heavy metals are similar ratio, according to above mentioned result the eelfish had less concentration were noticed. Based on the other fishes, the worm eel contain similar concentration of trace-metal.

Hg concentration in the diet were determined by eight marine fish tissue and the Mercury levels in food ranged from 0.08 to 0.32 ppm, dry weight, less concentration was detriment by Magalhae *et al.* (2007), the same concentration found in eelfish trace metal estimation. Testuro, Agusa *et al.* (2005) described the trace elements in marine fish and risk assessment, higher concentration of trace mental risk for human being and daily intakes of Cr, Mn, Cu, Zn, Cd and other accumulated fish effect the people especially Hg effect pregnant women because fetus is more susceptible to Hg toxicity. According to above mentioned trace metal accumulation, the present study result of trace metal ratio Fe > Cu > Mn > Zn > Pb > Cr > Ni > Cd and Hg indicated the concentration of trace metal originated from pollution and accumulated through food chain, higher concentration of trace metal effect the regular fish consumer.

Acknowledgement

Authors are thankful to the Director, CAS in Marine Biology and authorities of Annamalai University for the facilities provided.

References

Agusa, Tetsuro, Kunito, Takashi, Yasunaga, Genta, Iwata, Hisato, Subramanian, Annamalai, Ismail, Ahmad and Tanabe, Shinsuke 2005. Concentrations of trace elements in marine fish and its risk assessment in Malaysia, *Marine Poll. Bull.*, 51: 896–911.

Alam, M.G.M., Tanaka, A.G., Allinson, L.J.B., Stagni, F. and Snow, E., 2002. A comparison of trace element concentration in cultured and wild carp, *Cyprinus carpio* of lake Kasumigaura, Japan. *Ecotoxicol. Environ. Saf.*, 53: 348–354.

Bryan, G.W., 1976. Some aspects of heavy metal tolerance by aquatic organisms. In: *Effects of Pollutants on Aquatic Organisms*, (Ed.) A.P.M. Lockwood. Cambridge University Press, Cambridge, UK.

Bryan, G.W., 1980. Recent trends in research on heavy metal concentration in the sea. *Helgolanda Measurement*, 32: 6–25.

Bury, N.R., Walker, P.A. and Glover, C.N., 2003. Nutritive metal uptake in teleost fish. *The J. Exper. Biol.*, 206: 1123.

Fabien Pierron, Magalie Baudrimont, Angelique Bossy, Jean-Paul Bourdineaud, Daniel Brethes, Pierre Elie and Jean-Charles Massabuau, 2007. Impairment of lipid storage by cadmium in the European eel (*Anguilla anguilla*). *Aqu. Toxi.*, 81: 304–311.

Fernandes, H.M., Bidone, E.D., Veiga, L.H.S. and Patchineelam, S.R., 1994. Heavy metal pollution assessment in the coastal lagoons of Jacapaguá, Rio de Janeiro, Brazil. *Envir. Pollut.*, 85: 259–264.

Forstner, U. and Wittman, G.W.T., 1981. *Metal Pollution in the Aquatic Environment*. Springer-Verlag, Berlin, Heidelberg, Germany, p. 272–486.

Harderson, S. and Wratten, S.D., 1998. The effect of carbaryl exposure of the penultimate larval instars of *Xathocnemis zealandica* on emergence and fluctuating asymmetry. *Ecotoxicol.*, 7: 297–304.

Harte, J., Holdren, C., Schneider, R. and Shirley, C., 1991. *Toxics A to Z: A Guide to Everyday Pollution Hazards*. University of California Press, Oxford, England, p. 478.

Heath, A.G., 1987. *Water Pollution and Fish Physiology*. FL, CRC Press. Boca Raton, pp 245.

Lakshmanan, R., Kesavan, K., Sakkarvarthi, K. and Rajagopal, S., 2010. Concentration of (Fe, Mn, Mg, Ni) in five species of fishes caught in Parangipettai coastal waters. *International Journal of Chemtech Research*, 2(1): 622–623.

Mance, G., 1987. *Pollution Threat of Heavy Metals in Aquatic Environments*. Elsevier, New York.

Magalhaes, Maria C., Costa, Valentina, Menezes, Gui M., Pinho, Mario R., Santos, Ricardo S., Monteiro, Luýs R.Z., 2007. Intra- and inter-specific variability in total and methylmercury bioaccumulation by eight marine fish species from the Azores. *Marine Poll. Bull.*, 54: 1654–1662.

Miramand, P., Fichet, D., Bentley, D., Guary, J.C. and Caurant, F., 1998. Concentrations em métaux lourds (Cd, Cu, Pb, Zn) observées le long du gradient de salinité dans le réseau trophique de l'estuaire de la Seine. *Comptes-Rendus de l'Académie des Sciences de Paris, Sciences de la Terre et des Planètes*, 327: 259–264.

Mirnov, 1978. In the first American-Soviet symposium on the biological effect of pollution on marine organism (EPA) 600/9–78–007.

Mohammed A. Al-Kahtani, 2009. Accumulation of heavy metals in Tilapia fish (*Oreochromis niloticus*) from Al-Khadoud Spring, Al-Hassa, Saudi Arabia. *American J. Appl. Sci.*, 6(12): 2024–2029.

Mora, Stephen de, Fowler, Scott W., Wyse, Eric and Azemard, Sabine, 2004. Distribution of heavy metals in marine bivalves, fish and coastal sediments in Gulf and Gulf of Oman; *Marine Pollut. Bull.*, 49: 410–424.

Negilski, D.S., 1976. Acute toxicity of Zinc, Cd and Chromium to the marine fishes, Yellow eye mullet (*Aldrichetta porsteri* C. and V.) and small mouthed hardy headed (*Antherinasona microstoma*). *Aust. J. Mar. Fr. Res.*, 1: 137–149.

Ni, I.H., Wang, W.X. and Tam, Y.K., 2000. Transfert of Cd, Cr, and Zn from zooplankton prey to mudskipper *Periophthalmus cantonensis* and glassy *Ambassis urotaenia* fishes. *Marine Ecol. Prog. Seri.*, 194: 203–210.

Rajotte, J. and Couture, P., 2003. Indicator of chronic metal tress in wild yellow perch from metal contaminated environments. In: *Conf. Present. Mining and Envir., 28th Annual Meeting*.

Shuhaimi Othman, M., 2010. Heavy concentrations in sediments and fishes from lake Chini, Pahang, Malaysia. *J. Biol. Sci.*, 10(2): 93–100.

Singh, Jyotsna, Kant, K., Sharma, H.B. and Rana, K.S., 2008. Bioaccumulation of cadmium in tissues of *Cirrihna mrigala* and *Catla catla*. *Asian J. Exp. Sci.*, 22(3): 411–414.

Urena, R., Peri, S., Ramo, J. Del and Torreblanca, A., 2007. Metal and metallothionein content in tissues from wild and farmed *Anguilla anguilla* at commercial size. *Env. Inter.*, 33: 532–539.

Chapter 3

Impact of Uncontrolled and Illegal Fishing Practices on Biodiversity of Gulf of Mannar

☆ *R. Uma Maheswari*

ABSTRACT

Biological diversity means the variability among living organisms from all sources including interalia, terrestrial, marine and other aquatic. India has rich biological diversity and one of the 12 diverse counters of the world, with only 2.5 per cent of land area, India accounts for 7.8 per cent of the recorded species at global level. India is also rich in traditional and indigenous knowledge both coded as well as informal. Among 5 the Asian countries, India is perhaps the only country that has a long record of inventories of coastal and marine biodiversity dating back to at least two centuries. Much of the world's wealth of biodiversity is found in highly diverse marine and coastal habitats. Fishing is the major activity in the Gulf of Mannar and nearly 225 fishing village are located along the Gulf of Mannar coast. Several types of net operated in this region cause over exploitation of marine resources especially finfishes and shellfishes. The productivity of the benthic organisms at the base of food webs leading to food fishes is seriously impacted by bottom trawling, as is the survival of their juveniles when deprived of the biogenic bottom structure destroyed by that form of fishing. The demand for marine fisheries has increased many folds as the oceans have been exploited profusely, using highly developed fishing vessels and gears. Overfishing has led to over-exploitation of some marine species, which has in turn led to a progressive change in the composition of the global catch to species of lower economic value. As the resources are severely affected due to fishing practice the conservative measures should be implemented to save the biodiversity. While over fishing could be a major reason for declining in fish stocks, increase in sea temperature, decrease in freshwater flow into the sea and decrease in rainfall have had an equally adverse impact.

Keywords : *Biodiversity, Uncontrolled fishing practices, Gulf of Mannar, Habitat loss.*

Introduction

Biodiversity encompasses the variety of all life on earth. India has rich biological diversity and one of the 12 diverse counters of the world, with only 2.5 per cent of land area, India accounts for 7.8 per cent of the recorded species at global level. India is also rich in traditional and indigenous knowledge both coded as well as informal. Biological diversity means the variability among living organisms from all sources including interalia, terrestrial, marine and other aquatic ecosystems and the ecological complexes of which they are part, this includes diversity with in species, between species and of ecosystems (The Biological Diversity Act, 2002).

The need for fisheries management arises as the surplus production from fish stocks is overtaken by the catching capacity of fishing fleets. Catching capacity is the product of the fishing effort and the combined efficiency of the fishing gear and the fishing vessel (*e.g.* loading capacity, engine power, range capacity, fish finding and navigational equipment) as well as the skills of the crew.

Fisheries management includes different management measures. Among these are technical regulations on fishing gears in order to obtain the overall goal of high sustainable yield in the fisheries. These are regulations *e.g.* on mesh size to improve the selective properties of a fishing gear so that by catches of juvenile fish are reduced – in order to safeguard recruitment to the larger size groups of a fish stock including the spawning stock.

In recent years there has been a growing focus on "ecosystem effects of fisheries", addressing the impact of fishing operations not only on the target species, but also on by catch of or other effects on non-commercial species or habitats. Energy efficiency, reduced pollution and improved quality of the catch are also important aspects related to fishing gears and fishing operations (Code of Conduct for Responsible Fisheries, Article 7.2). From a situation where the development of fishing gears and methods only focused on the highest possible catching efficiency for the target species, now fisheries research, fisheries management and the fishing industry are challenged to develop gear, methods and regulations that meet the different considerations mentioned above. This is part of an emerging ecosystem approach to fisheries management.

Biological Resources

Bioresources are important for progress and economic activities of a nation. But bioresources management and utilization for human welfare is very important and critical component. The optimum utilization of the bioresources awareness of the importance and implications of bioresources among common people as well as elite educated citizens for safeguarding and protecting the optimum and balanced way of using the bioresources well for the benefit of not only the present generation of our people but also to our future generations for their better, healthy and peaceful living on the earth. Biological resources include genetic resources, organisms or parts thereof, populations, or any other biotic components of ecosystems with actual or potential use for humanity.

Marine Biodiversity

India is among 12 mega biodiversity countries and 25 hotspots of the richest and highly endangered eco-regions of the world. Among 5 the Asian countries, India is perhaps the only country that has a long record of inventories of coastal and marine biodiversity dating back to at least two centuries. In terms of marine environment, India has a coast line of about 8000 km and exclusive economic zone of 2.02 m sq km adjoining the continental regions and offshore islands and a wide range of coastal ecosystems such as estuaries, lagoons, mangroves, backwaters, salt marshes, rocky coasts, sandy stretches, coral reefs which are characterized by unique biotic and abiotic properties and processes. In India the accelerated loss of coastal and marine biodiversity components over the last few decades has been of great concern. Environmental changes, over exploitation and habitat loss are among the major causes of species loss that, according certain estimates is of the order of species per day.

Marine Ecosystems

The oceans cover over 70 per cent of the planet surface area and account for 99 per cent of the volume that is known to sustain life. Coastal ecosystems such as estuaries, coral reefs and mangrove forests, also contain significant diversity and are highly valuable for coastal communities. Much of the world's wealth of biodiversity is found in highly diverse marine and coastal habitats. Coastal zone represents 18 per cent of the earth surface, providing space for 60 per cent of the human population since about 70 per cent of the world's cities with population more than 1.6 million are located in coastal zone. 90 per cent of the world fish catch is obtained from this zone. Interestingly the hydrosphere of the coastal zone is only about 8 per cent of that of the world ocean but represents about 18-33 per cent of total primary production. This zone bio-geochemically more important as it buries and mineralizes 80 to 90 per cent of organic matter and serves as a sink for an estimated 50 per cent of the global carbonate deposition. The coastal water also receives discharges of suspended matter associated with elevated levels of pollutants from major rivers and this accounts for 75-90 per cent. This zone has high biological potential as it serves as feeding, nursery and spawning grounds with rich biodiversity and as in intermediary biotope between marine and freshwater environments.

Importance of Marine and Coastal Ecosystems

Marine and coastal ecosystems provide a wide range of important products and services. Fish, crustaceans and molluscs are major food. Marine fish provided about 84 million tones of human food and live stock supplements in 1993. Fish accounts for about 16 per cent of the average individual intake of protein at global level and the proportion is much higher in many developing countries.

Biodiversity

The marine environment has high biodiversity which is less known than that of terrestrial systems. GOM is endowed with three distinct Marine ecosystems namely corals, seagrass and mangroves. It has six of the World's twelve sea grass genera and eleven of the world's 50 species. About 3600 species of fauna and flora have been identified in this region. The GOM has the highest concentration of sea grass species along India's 7500 km coastline. The sea grass beds are some of the largest remaining feeding grounds for the globally endangered sea cow (Dugong dugon). All five species of Marine Turtles - Green - Loggerhead, Olive ridley, Hawksbill and Leather back have been nesting on the Islands. Moreover, it harbours a total of 117 species of Coral belonging to 37 genera. Coral reefs serve as the spawning grounds for fisheries and Sea grass beds as nursery grounds and Mangroves as shelters from a unique component of life, support systems of coastal biodiversity that relates global benefits and Local needs.

Marine habitat posses unique mangrove vegetation which consists of numerous species like *Rhizophora, Avicennia, Ceriops,* etc. Among the 160 species of sea weeds known so far, about 40 species of algae are found growing in abundance in this region. The mean density of algae for the entire Gulf of Mannar region is $0.11 kg/m^2$ wet weight. Nearly 5000 – 7000 tones (dry weight) of sea weeds are harvested annually from this area for the production of agar agar, cellulose and algin which are used in food processing and pharmaceutical industries. A total of 13 species of sea grasses like *Enhalus acorvoides, Halophila ovalis, Cymadocea rotundata,* etc have been recorded. These sea grass bed are some of the largest remaining feeding grounds for the globally endangered dugong (Duging dugon) and many species of Crustaceans, Molluscs, Gastropods and fish larvae have also been observed to inhabitant the sea grass beds.

The Phytoplankton, Zooplankton, Marine algae, Corals, Gorgoonids, Squids and Cuttle fish, Holothurians, Shrimps, Lobsters, Crabs, Stomatopods, Teleostes, Mammals, Baleen whales, Toothed whales, Dolphin, etc., are very much abundance in this area. But, nowadays, they are becoming highly vulnerable species. Marine environment supports the plankton, plants and other organisms. It provides breeding ground and food for many fishes.

Fish Resources

Fishes comprise about half the total number of vertebrates. The number of estimated living fish species might be close to 28,000 in the world. The distribution of marine fishes is rather wide and some genera are common to the Indo-Pacific and the Atlantic regions. 57 per cent of the Indian marine fish genera are common to the Indian Seas and to the Atlantic and Mediterranean. The exact number of species associated with coral reefs of India are still to be found, however, the number of fishes in Indian Ocean is 1367. The Lakshadweep Islands have a total of 603 species of fishes about 750 species are found in the Andaman and Nicobar Islands (Rajan, Zoological Survey of India, Port Blair, Personal communication) and in Gulf of Mannar Biosphere Reserve it is 538 (Unpublished, Zoological Survey of India, Chennai). The category of fishes occurring in coral reef ecosystem includes groups such as the damsel fishes, butterfly fishes, trigger fishes, file fishes, puffers, snappers, hawk fishes, triple fins and most of the wrasses, groupers and gobies. Another 20 per cent are composed of cryptic and nocturnal species that are confined primarily to caverns and reef crevices during daylight periods. This assemblage includes such families as the cusk eels, some groupers and their relatives, most of the moray eels and some scorpion fishes, wrasses and nocturnal families including the squirrel fishes, cardinal fishes and sweet lips. Another 10 per cent of fishes dwell primarily on reefs covered with sand and rubble including snake eels, worm eels, various rays, lizard fishes, grabs fishes, flat fishes, and some wrasses and gobies. A relatively small percentage (about 5 per cent) of the fauna is composed of transient mid water reef species that roam over large areas. This group includes most sharks, jacks, fusiliers, barracudas and a scattering of representatives of other families.

Impact of Fishing Activities Affect the Biodiversity Especially in Gulf of Mannar

Fishing is the major activity in the Gulf of Mannar and nearly 225 fishing village are located along the Gulf of Mannar coast. Several types of net operated in this region cause over exploitation of marine resources especially finfishes and shellfishes. Studied on the impact of fishing on marine diversity is vary limited.

Global studies indicated that direct impact of fishing is that it reduces the abundance of target species. It has often been assumed that this does not impose any direct threat of species extinction as marine fish generally are very fecund and the ocean expanse is wide (Pitcher, 1998). But the past few decades have witnessed a growing awareness that fishes can not only be severely depleted, but also be threatened with extinction through overexploitation (Casey and Myers, 1998). Among commercially important species, those particularly at risk are species that are highly valued, large and slow to mature, have limited geographical range, and/or have sporadic recruitment (Sadovy, 2001). There is actually little support, though, for the general assumption that the most highly fecund marine fish species are less susceptible to overexploitation; rather it seems that this perception is flawed (Hutchings, 2000).

The productivity of the benthic organisms at the base of food webs leading to food fishes is seriously impacted by bottom trawling (Hall, 1998), as is the survival of their juveniles when deprived of the biogenic bottom structure destroyed by that form of fishing (Watling and Norse, 1998). Hence, given the extensive coverage of the world's shelf ecosystems by bottom trawling (Watling and Norse, 1998), it is not surprising that generally longer-lived, demersal (bottom) fishes have tended to decline faster than shorter-lived, pelagic (open water) fishes, a trend also indicated by changes in the ratio of piscivorous (mainly demersal) to zooplanktivorous (mainly pelagic) fishes (Caddy and Garibaldi, 2000).

Gopinatha Menon (1996) studied the impact of bottom trawling on the exploited fishery resources. Overfishing, unscientific exploitation and destruction of habitat caused serious threat in the marine germ plasm resources. Menon (1996) indicated nearly 1.3 lakhs tones of unmarketable benthic organism has caught from Kerala, Tamilnadu and Andhrapradesh. He also stressed the reduction in discard could improve the yield to 1.55 lakhs tones from the yield of 6200t (as on 1980-84 survey). Most of the nets with small mesh size capture pre-reproductive females which cause severe destruction in their population (Venkataraman and Wafar, 2005). The trawls operated along the coast in Indian waters generate higher amount of by catch. Arun Kumar and Vinod Khanna (2000) indicated that exploitation, including hunting, collecting, fisheries and fisheries bycatch, and the impacts of trade in species and species' parts, constitutes a major threat for birds (37 per cent of all), mammals (34 per cent of all). Further they insisted the incidental take, particularly the drowning of aquatic reptiles and mammals in fishing nets.

Most of the marine turtle populations found in the Indian region are in decline. The principal reason for the decrease in numbers is deliberate human predation. In the Gulf of Mannar turtles are still reasonably common near seagrass beds where shrimp trawlers operate, but off the coast of Bengal the growing number of mechanized fishing boats has had the effect of increasing incidental catch rates (Kar and Bhaskar, 1981).

Exploitative fishing, and especially over fishing, threatens many species besides the fishes themselves. The accidental capture of marine reptiles, mammals and other threatened resources are in the rise. The use of purse seine reduced the catch of cat fishes (Menon and Pillai, 1996).

Kumaraguru (2006) indicated that the reduction in capture of chank and almost extinction of pearl oyster is due to overexploitation and introduction of trawl net in Gulf of Mannar. He further stated that use of trawl net and seine nets in and around shallow depths of coastal waters causes severe damage to the reefs.

Conservative Measures Should be Implemented to Save the Biodiversity

- ☆ Provide appropriate technical and legal conditions for the implementation of marine protected areas, and create marine reserves;
- ☆ Carry out technical and legal studies to mitigate the impact of trawling;
- ☆ Intensify environmental education efforts for oceanic and coastal ecosystems, especially reefs and islands with the greatest tourism potential;
- ☆ Identify new fishery resources and still-underexploited stocks and introduce appropriate technologies to reduce by catch;
- ☆ Focus exploitation and sustainable use of living marine resources exclusively on the production of food in the form of fisheries resources, but also consider the resource in terms of its biodiversity (genetic patrimony and biotechnology);

☆ Intensify oceanographic studies, faunal and floral surveys, studies of population and community dynamics, and stock assessments;

☆ Intensify studies of artificial habitats and their effects on the marine environment;

☆ Reducing by catch through regulating mesh size, closuring of trawling in shallow region and peak breeding season.

Biodiversity Convention

The biodiversity convention is the most recent development which has occurred in response to increase of environmental crisis. The need to protect marine biodiversity was considered after the conservation of terrestrial species became widespread. The agenda of biodiversity conservation of the Earth Summit in Rio de Janeiro has been given International recognition. Some plans have been done specifically on controlling plans marine pollution and the protection of the marine environment from land based activities.

The biodiversity convention is directed specifically on protecting biological diversity and a new decision on marine and coastal bio-diversity was negotiated which is known as Jakarta Mandate advocated the conservation of bio-diversity and the sustainable use of its components. It includes establishing a system for regulating biological resources, rehabilitating degraded ecosystems, legislating the protection of threatened species and promoting their recovery and preventing the introduction of non-native species and its adverse effect on biodiversity. An Integrated Coastal Zone Management (ICZM) in 1989 was developed for the management and planning. Effective ICZM involves the establishment of a system for regulating land use, the protection of area of greatest ecological landscape and amenity value and undertaking of environmental impact assessment on both individual and integrated development and guidelines in order to prevent environmental impacts (CAMPN, 1989). This is the right time to initiate steps to conserve the Gulf of Mannar Bioreserve Area and preventing them from pollution.

References

Arun Kumar and Khanna, Vinod, 2000. *Globally Threatened Indian Fauna: Status, Issues and Prospects*.

Caddy, J.F. and Garibaldi, L., 2000. Apparent changes in the trophic composition of world marine harvests: the perspective from the FAO capture database. *Ocean Coastal Mgmt.*, 43: 615–655.

Casey, J.M. and Myers, R.A., 1998. Near extinction of a large, widely distributed fish. *Science*, 281: 690–692.

Das, P., Mahanta, P.C. and Kapoor, D., 1988. *Marine Fish Genetic Resource Conservation*. CMFRI Special Publication, 40: 24.

Food and Agricultural Organizations (FAO) of the United Nations, 1995. *The State of World Fishery and Aquaculture*, FAO, Rome, Italy.

Gopinatha Menon, N., 1996. Impact of bottom trawling on exploited resources. In: *Marine Biodiversity: Conservation and Management*, (Eds.) N.G. Menon and C.S.G. Pillai, p. 97–102.

Hall, S.J., 1998. *The Effects of Fisheries on Ecosystems and Communities*, Blackwell, Oxford.

Hutchings, J.A., 2000. Collapse and recovery of marine fishes. *Nature*, 406: 882–885.

Kar, C.S. and Bhaskar, S., 1981. Status of sea turtles in the Eastern Indian Ocean. In: *Biology and Conservation of Sea Turtles*, (Ed.) K. Bjorndal. Proc. World Conf. Sea Turtle Cons., Smithsonian Institute Press, Washington, pp. 373–383.

Kumaraguru, A.K., Joseph, V. Edwin, Marimuthu, N. and Wilson, J. Jerald, 2006. *Scientific Information on Gulf of Mannar: A Bibliography*, p. 672.

Menon, N.G. and Pillai, N.G.K., 1996. The destruction of young fish and its impact on inshore fisheries. In: *Marine Biodiversity: Conservation and Management*, (Eds.) N.G. Menon and C.S.G. Pillai, p. 89–96.

Sadovy, Y., 2001. The threat of fishing to highly fecund fishes. *J. Fish Biol.*, 59: 90–108.

The Biological Diversity Act, 2002. National Biodiversity Authority, Chennai, Tamil Nadu, pp. 57.

The Convention on Biological Diversity, 1992. An Conference on Environment and Development held at Rio De Janeiro in June.

Turner, S.J., Thrush, S.F., Hewitt, J.E., Cummings, V.J. and Funnell, G., 1999. Fishing impacts and the degradation or loss of habitat structure. *Fish. Mgmt Ecol.*, 6: 401–420.

Venkataraman, K., 2003. *Natural Aquatic Ecosystems of India*. Thematic Working Group, The National Biodiversity Strategy Action Plan, India, p. 1–275.

Venkataraman, K. and Wafar, M., 2005. Coastal and marine biodiversity of India. *Indian Journal of Marine Science*, 34(1): 57–75.

Watling, L. and Norse, E.A., 1998. Disturbance of the seabed by mobile fishing gear: A comparison to forest clearcutting. *Conserv. Biol.*, 12: 1180–1197.

Chapter 4

First Record of Ribbon Fish (Crested Hairtail) *Tentoriceps cristatus* (Klunzinger, 1884) from Gujarat, West Coast of India

☆ *J.B. Solanki, B.S. Peje, P.V. Parmar, A.R. Dodia and H.V. Parmar*

ABSTRACT

Tentoriceps cristatus was recorded for the first time from Gujarat, India. A single specimen of *T. cristatus* (78.8 cm total length and weighing 186 g) was collected from Veraval landing centre, on the north-west coast of India during January 2012. The distinguishing characters of the species from other species of the family are discussed. Morphometric and meristic character of *T. cristatus* are presented in this chapter.

Keywords: First record, Ribbon fish, Tentoriceps cristatus, Gujarat.

Introduction

In Gujarat, Saurashtra coastal belt is replete with fisheries activity related to marine fish capture (Solanki *et al.*, 2011). Ribbon fishes are represented in Indian waters by nine species (Rizvi *et al.*, 2010), out of which three species namely *Trichiurus lepturus* (Linnaeus), *Lepturacanthus savala* (Cuvier), *Eupleurogrammus muticus* (Gray) forms the major component of the ribbon fish catches along the Saurashtra coast (Ghosh *et al.*, 2009). Ribbon fish was landed as bycatch of shrimp trawler earlier but on account of emerging export demand it is now increasing targeted (Khan, 2006). Ribbonfish, one of the important resources off Veraval (Sudhakara and Kasim, 1985) was represented in the gillnet landing.

This report is about the first record of capture of ribbon fish (*Tentoriceps cristatus*) from Veraval, Gujarat. The fish caught measured 78.8 cm in total length and weight 186 g. The fish of family Trichiuridae are characterized by extremely elongate body and strongly compressed, ribbon-like, tapering to a point.

Tentoriceps cristatuts (Klunzinger) has long been a phantasmal fish. Nobody has collected the fish since Klunzinger (1884) reported his *Tentoriceps cristatus* from the Red sea ninety years ago. Whitley (1948) established a new genus *Tentoriceps* for the fish and Tucker (1956) gave a diagnosis of the genus in his revision of the trichiurid fishers, however, nothing new was added to Klunzinger's description on the fish and many taxonomically important characters are left unknown (Senta, 1975).

Material and Methods

A single specimen (78.8 cm total length and weight 186 g) of *Tentoriceps cristatuts* (Klunzinger, 1884) was collected (Figure 4.2) from Veraval landing centre (20° 54' N 70° 22' E) on the west coast of India (Figure 4.1) on January 30, 2012 and identified based on the available literature (FAO, 1984).

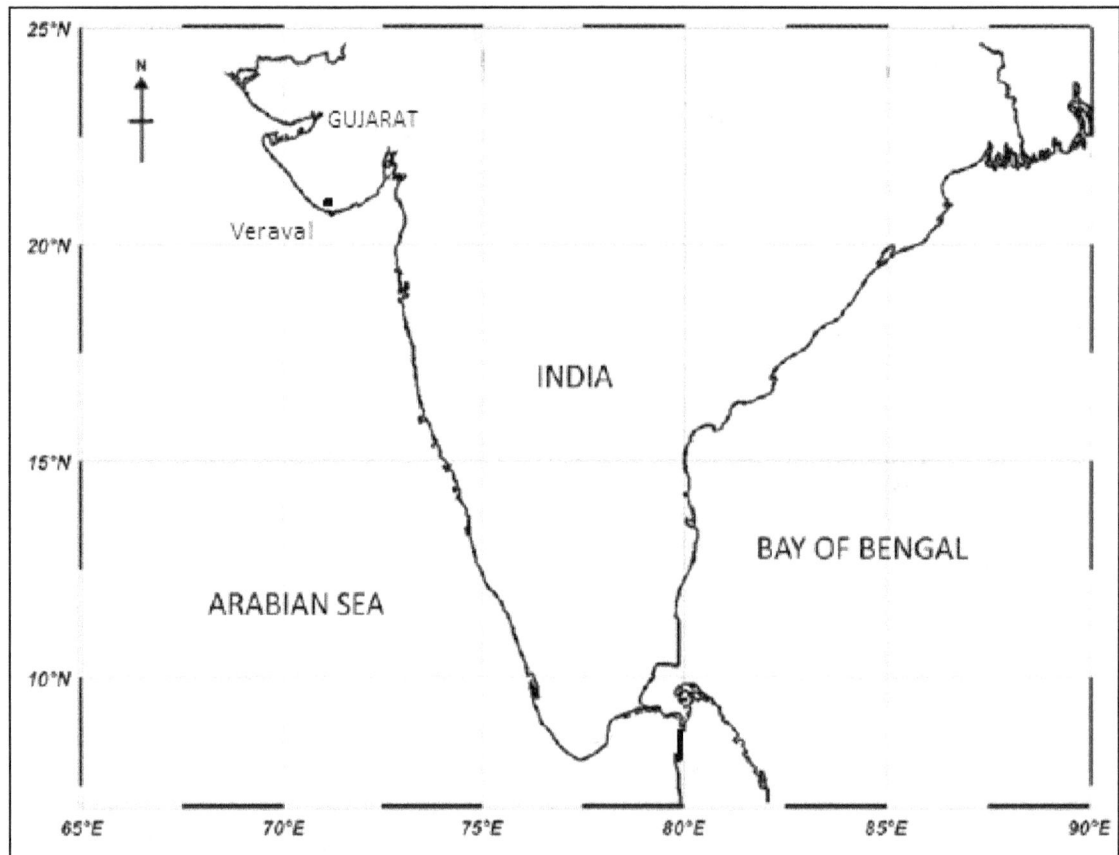

**Figure 4.1: Map of India showing the Location of Veraval,
the Place of Collection of the Present Specimen.**

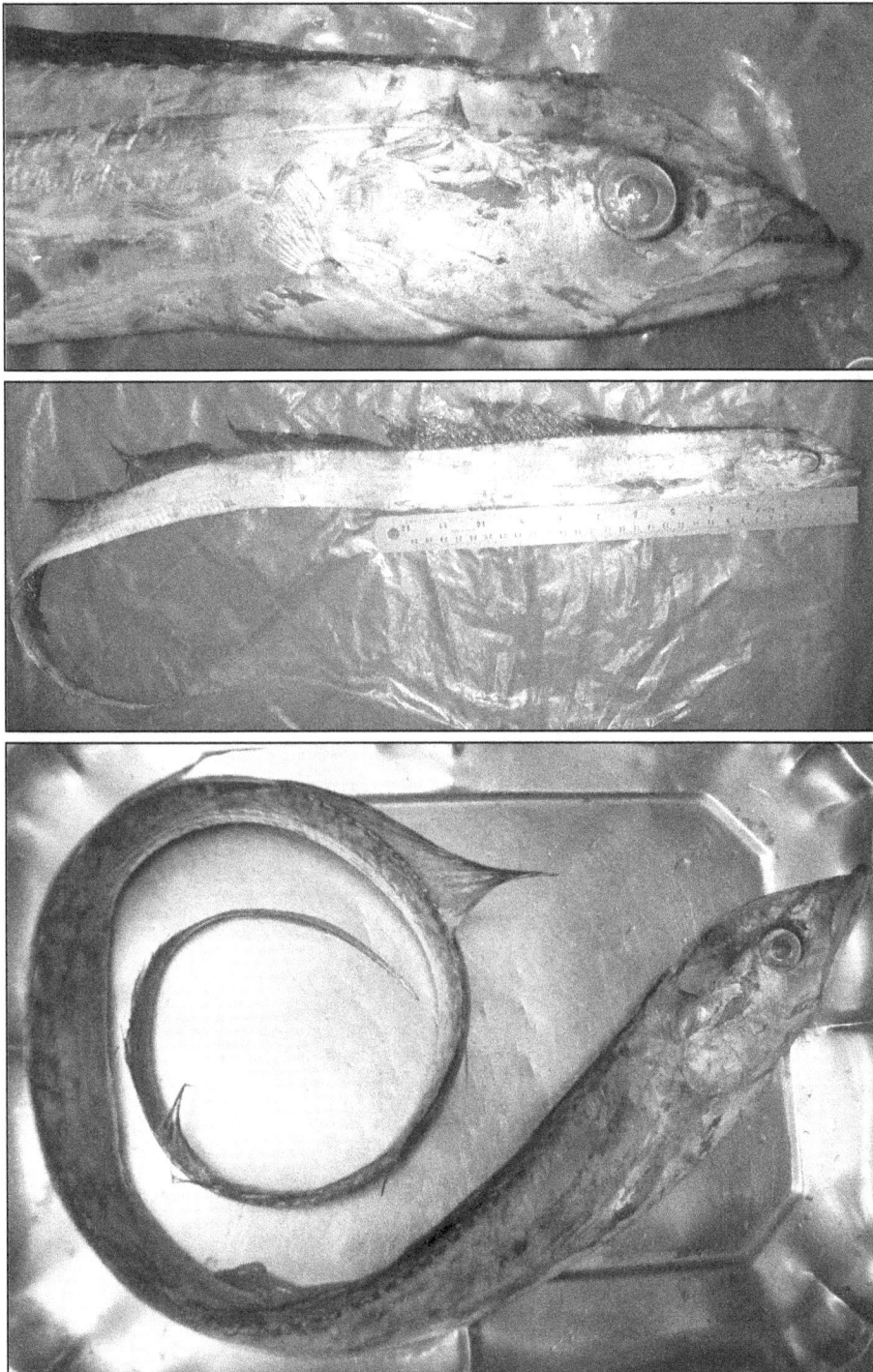

Figure 4.2: *Tentoriceps cristatuts,* 78.8 cm Total Length, Caught Off Veraval,
North-West Coast of India.

The fish was caught in mechanized trawler using trawl net, at 100-120 m while conducting fishing for 8-10 days. The morphometric and meristic characters of the present specimen are given in Table 4.1.

Results and Discussion

Specimen Sex

Female

Description of the Specimen

The species (*Tentoriceps cristatuts*) was first reported from the Red Sea (Klunzinger, 1884). Available of this species is given in Table 4.2.

The present species was identified as *Tentoriceps cristatuts* based on its morphological and meristic characteristics like elongate, ribbon-like body which is tapering to a point; large mouth with a dermal process at tip of each jaw; convex head; gill cover with convex lower hind margin; 2 fangs in each lower and upper jaw; laterally situated large eye; lateral line running slightly nearer ventral than dorsal contour; pectoral fins short, not reaching lateral line; dorsal fin with 136 soft rays; pelvic fins presents and caudal fin absent. The specimen was identified as female by presence of ovary in the body. The distinguishing characters of the species from other species of the family Trichiurinae are dorsal profile of head not evenly convex, pectoral fin tip reaching to above lateral line in other species (FAO, 1984).

Table 4.1: Morphometric and Meristics Measurements of Present Specimen of *Tentoriceps cristatuts*.

Characters	
Morphometric Counts	*Measurements (cm)*
Total length (TL)	78.8
Head length (HL)	8.9
Body depth	4.0
Pre-dorsal length	5.9
Pre-pectoral length	8.7
Eye diameter	1.4
Pre orbital length	4.5
Length of pectoral fin	1.6
Metristic Counts	*(Numbers)*
Pectoral fin rays	16
Dorsal fin rays	136

Table 4.2: Countries/Area where *Tentoriceps cristatus* is Found.

Andaman Island	*Hong Kong*	*Philippines*
Australia	Indonesia	Ryukyu Island
British Indian Ocean Territory	Japan	Saudi Arabia
Chagos Island	Korea Democratic Peoples Rp	Seychelles
China Main	Korea Republic	Singapore
Comoros	Madagascar	Sudan
East Timor	Malaysia	Taiwan
Eritrea	Mozambique	Thailand

Source: www.fishbase.org.in.

The records of ribbonfish from Indian waters are important for their Ichthyological studies. The present report is significant in the sense that even though *Tentoriceps cristatuts* is considered as circumtropical, a huge gap existed in the distribution of this species. The distribution range of this species until this record was made extended from the Andaman Sea, the Philippines and South China Sea. With the present report, the distribution range of this species now extends to the West coast of Indian waters but may be considered as rare occurrence only.

Acknowledgements

The authors wish to express their gratitude to Mr. Manoj Solanki (Active fisherman, Veraval) and Mr. Narasi Solanki (Active fisherman, Veraval) for collection of ribbon fish sample.

References

FAO, 1984. *Species Identification Sheets for Fishery Purposes*. Western Indian Ocean (Fishing Area 51), Volume 4, FAO, Rome.

Ghosh, S., Pillai, N.G.K. and Dhokia, H.K., 2009. Fishery and population dynamics of *Trichiurus lepturus* (Linnaeus) off Veraval, north-west coast of India. *Indian J. Fish.*, 56(4): 241–247.

Khan, M.Z., 2006. Fishery resource characteristics and stock assessment of ribbonfish, *Trichiurus lepturus* (Linnaeus). *Indian J. Fish.*, 53(1): 1–12.

Klunzinger, C.B., 1884. Die fische des Rothen Meeres. E. Schweizerbartische Verlagshandlung (E. Koch), *Stuttgart*, 9(133): 1–13.

Rizvi, A.F., Deshmukh, V.D. and Chakraborty, S.K., 2010. Stock assessment of *Lepturacanthus savala* (Cuvier, 1829) along north-west sector of Mumbai coast in Arabian sea. *Indian J. Fish.*, 57(2): 1–6.

Senta, 1975. Redescription of Trichiurid fish *Tentoriceps cristatus* and its occurrence in the south china sea and the Strais of Malacca. *Japanese Journal of Ichthyology*, 21(4): 175–182.

Solanki, J.B., Kotiya, A.S., Jetani, K.L., Dodia, A.R. and Parmar, H.L., 2011. Traditional storage method of dried fishes by sea sand for consumption along Saurashtra coast, Gujarat. *Fishing Chimes*, 31(2): 50–52.

Sudhakara, R.G. and Kasim, H.M., 1985. Commercial trawl fishery off veraval coast during 1979–82. *Indian J. Fish.*, 32(3): 269–308.

Tucker, D.W., 1956. Studies on the Trichiuroid fishes–3: A preliminary revision of the family Trichiuridae. *Bull. Brit. Mus. (N. H.), Zool. Ser.*, 4: 73–130.

Whitley, G.P., 1948. Studies in Ichthyology, No. 13. *Rec. Aust. Mus.*, 22: 70–94.

Chapter 5

Length-Weight Relationship of the Worm Eel, *Myrophis platyrhynchus* Breder, 1927 from Parangipettai, South East Coast of India

☆ *T. Veeramani, V. Ravi, R. Bharathi,*
G. Mahadevan and S. Balachandar

ABSTARCT

The length-weight relationship of the worm eel, *Myrophis platyrhynchus* were analyzed (1185 specimen: 25.1 cm – 85.6 cm TL) during January to December 2010. Linear regression equation obtained in the present study was in male Log W= –2.280+2.626 log L (r² 0.903), in female Log W= –2.236+2.614 log L (r² 0.933). The exponent value for worm eel was around the hypothetical value (3) 'b' value was 2.626 in male and 2.614 in female. This suggested isometric growth (b=3) in male and a negatively allometric pattern (b<3) in female which indicate that the rate of increasing in body length is not proportional to the rate of increase in the body weight with differed by the months and seasons, thus it is clear that these fishes maintains its shape throughout its life.

Keywords: Worm eel, Myrophis platyrhynchus, Length-weight.

Introduction

Determination of the exact nature of the relationship that exists between length-weight relationships of fishes has been recognized as an important part of fishery biological studies. Length-weight relationships (LWR) are very important in fisheries biology because they allow the estimation of the average weight of fish of a given length group by establishing a mathematical relation between the two variables. They are also useful for the conversion of growth-in-length equations to growth-in-weight for use in stock assessment models and to estimate stock biomass from limited sample sizes (Binohlan and Pauly, 1998; Koutrakis and Tsikliras, 2003; Valle *et al.*, 2003; Ecoutin *et al.*, 2005). Many studies are carried out in freshwater species in streams, rivers, canals and ponds (Mirza, 1982 and Rahman, 1989).

The worm eel *Myrophis platyrhynchus* (Anguilliformes: Ophichthyidae) inhabits tidal creeks or protected/semi-protected bays and can be identified through its broad nose along with posterior nostril labial; pectoral fin large and tail tip flexible. Frequent landings of the worm eel occur regularly in Parangipettai region, southeast coast of India and there will be no comprehensive account on the length-weight relationship of *M. platyrhynchus* available from this region and hence the present study is attempted to determine the length-weight relationship of the worm eel.

Materials and Methods

Monthly sampling of the worm eels (*M. platyrhynchus*) were collected from Mudasalodai landing centre of Parangipettai region. Total length and weight of the fishes were noted to the nearest 1.0 mm and 1.0 g respectively. Sex was determined by examining the gonads. The study was based on the length and weight data of 1185 specimen, 513 males of the length range 25.1 cm – 85.6 cm (weight 12.58 gm – 554.48 gm), 672 females of the length range 25.1 cm – 85.1 cm (weight 13.62 gm – 558.28 gm) collected during the study period from January 2010 to December 2010. The method suggested by Le Cren (1951) was followed to compute the length and weight relationship. Accordingly, the length-weight relationship can be expressed as:

$$W = aL^b$$

where,

W and L are weight (g) and length (cm) of the fish respectively and 'a' and 'b' are two constants (initial growth index and regression constants respectively). When expressed logarithmically be above equation becomes a straight line of the formula:

$$\log W = \log a + b \log L$$

where,

a: Intercept; $y = \log W$; $x = \log L$ and b = slope.

Results and Discussion

The parameters of the length-weight relationships of *M. platyrhynchus* were noticed for 1185 specimen. The length and weight of the worm eel were ranged from 25.1cm to 85.6cm TL (12.58g to 554.8g) and 25.1cm to 85.1cm TL (13.62g to 558.28g) in male and female fishes respectively. The regression line derived from the data for the worm-eel showed a linear relationship between the two variables of length and weight (Table 5.1).

The mathematical relationship between total length and weight of male, female and intermediate stage of worm eel *M. platyrhynchus* obtained by logarithmic regression equations are follows:

$$W = \log a + b \log L$$

The r^2 values were differed based on the length – weight and sex, in male 0.903 and female 0.933 were recorded respectively during the study period. The regression parameters are 'n' = number, 'a, =

weight of specimen and 'b' = total length and r^2 = regression, among the linear equation the 'a' value in male –2.280 and female –2.236 noticed respectively, the 'b' value in male 2.626 and female 2.614 were found respectively during the study period (Table 5.1 and Figure 5.1).

Table 5.1: Regression Parameters for Length-Weight Relationship of Male and Female Worm Eel *M. platyrhynchus*

Sex/Linear Regression	Regression Equation	a	b	R^2
Male	Log W = –2.280+2.626 log L	–2.280	2.626	0.903
Female	Log W = –2.236+2.614 log L	–2.236	2.614	0.933

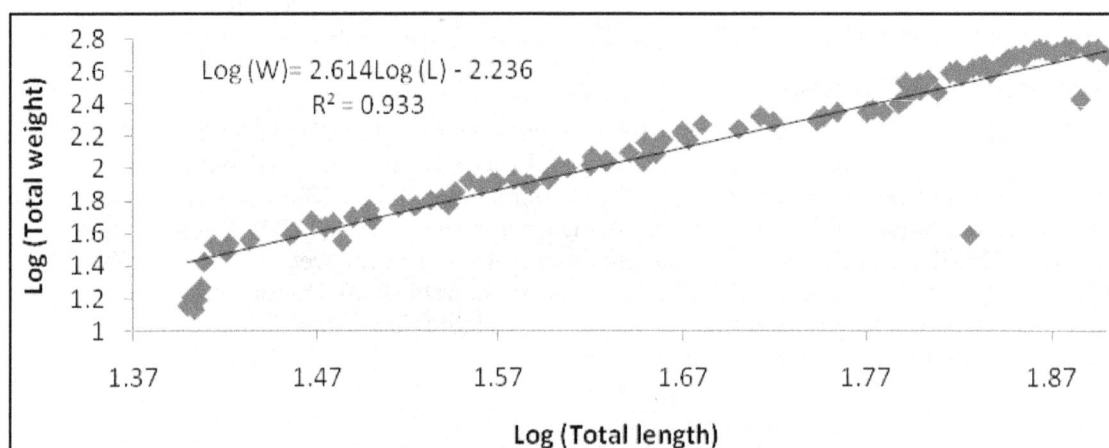

Figure 5.1: Length-Weight Relationship in Females of *M. platyrhynchus*.

Length has an important function for the weight of fish (Weatherley and Gill, 1987). The specific gravity of the flesh is known to undergo changes, while LeCren, (1951) stated that the density of fish might be maintained in the surrounding water by means of swim bladder. Hence change in weight is due to change in form but not in specific gravity.

In the present study it was observed that the length and weight of *M. platyrhynchus* increases at the rate of cube. In male and female specimen of the 'b' value varied, male 2.626 and female 2.614 noticed respectively. According to Wootton (1990), the 'b' values were almost equal to the expected value '3' for ideal fish both the sexes, these values were differed based on the such factors are season, food and feeding of the species, in the male specimens indicate lesser weight than female the difference shows maturation and spawning. Such differences in value 'b' can be ascribed to one or a combination of most of the factors including differences in the number of specimens examined, area/season effects and distinctions in the observed length range of the specimens caught, to which duration of sample collection can be added. According to Jhingran, (1968) and Frosta *et al.* (2004) the slope value 'b' indicates the rate of weight gain relative to growth in length and varies among different populations of the same species or within the same species. Similar observations on LWR are reported by Gulnaz Ozcan (2008) in the freshwater fish species indicating the 'b' value range 2.767 in Eel fish (*Anguilla anguilla*) and 2.967 in catfish (*Clarias gariepinus*).

Acknowledgement

Authors are thankful to the Director, CAS in Marine Biology and authorities of Annamalai University for the facilities provided.

References

Binohlan, C. and Pauly, D., 1998. The length-weight table. In: *Fishbase 1998: Concepts, Design and Data Sources*, (Eds.) R. Froese and D. Pauly. ICLARM, Manila, Philippines, pp. 121–123.

Ecoutin, J.M., Albaret, J.J. and Trape, S., 2005. Length-weight relationships for fish populations of a relatively undisturbed tropical estuary: The Gambia. *Fish. Res.*, 72: 347–351.

Frosta, I.O., Costa, P.A.S. and Braga, A.C., 2004. Length-weight relationships of marine fishes from the Central Brazilian Coast. *Naga, World Fish Centre Quarterly*, 27(182): 20–26.

Gulnaz Ozcan., 2008. Length-weight relationships of five freshwater fish species from the hatay province, Turkey. *J. Fish. Sci. Com.*, 2(1): 51–53.

Jhingran, V.G., 1968. Synopsis of biological data on Catla, *Catla catla* (Hamilton, 1822). FAO Fish Synopsis, 32: Rev. 1.

Koutrakis, E.T. and Tsikliras, A.C., 2003. Length-weight relationships of fishes from three northern Aegean estuarine systems (Greece). *J. Appl. Ichthyol.*, 19: 258–260.

Le Cren, C.D., 1951. The length-weight relationship and seasonal cycle in gonad weight and condition in the perch (*Perca fluviatilis*). *J. Anim. Ecol.*, 20: 201–219.

Mirza, M.R., 1982. *A Contribution to the Fishes of Lahore*. Polymer Publications, Urdu Bazar, Lahore, pp. 48.

Rahman, A.K.A., 1989. *Freshwater Fishes of Bangladesh*. Zool. Soc. Bangladesh University, Dhaka, pp. 346.

Valle, C., Bayle, J.T. and Ramos, A.A., 2003. Weight-length relationships for selected fish species of the western Mediterranean sea. *J. Appl. Ichthyol.*, 19: 261–262.

Weatherley, A.H. and Gill, H.S., 1987. *The Biology of Fish Growth*. Academic Press, London, pp. 443.

Wootton, R.J., 1990. *Ecology of Teleost Fishes*. Chapman and Hall, London, pp. 404.

Chapter 6

Finfish Diversity in Two Landing Centers of Karwar, West Coast, Karnataka

☆ *B. Vasanthkumar*

ABSTRACT

A study was undertaken to access the status of available finfishes in two important landing centers of Karwar. During the present observation 33 varieties of fishes belonging to 16 families were noticed at station–I. Whereas, in station–II there are 21 types of fishes, representing the 15 families were recorded. The present study was carried out for a period of 13 months from January 2006 to January 2007. The important fishes available are mackerel, seer fish. Prawns, flat fish, sciaenids etc., there is no river opening into the sea in station–II.

Keywords: Finfishes, Diversity, Baithkol, Majali, Karwar.

Introduction

In the present investigation two stations were selected namely station–I Karwar and station–II Majali representing two fish landing centers in the Uttara Kannada District to study the qualitative and quantitative study of fin and shell fishes and hydrological parameters. The present study was carried out for a period of 13 months from January 2006 to January 2007. From two different study stations *i.e.* station–I (Karwar) and station–II (Majali). During the present investigation an attempt was made to study the biotic *i.e.* with reference to species diversity and catch statistics was observed from two study locales from two study locales. During the present observation 33 varieties of fishes belonging to 16 families were noticed at station–I. Whereas, in station–II there are 21 types of fishes, representing the 15 families were recorded.

Material and Methods

Monthly collection was carried out for a period of 13 months from January 2006-January 2007 for both landing centers (Figure 6.1). Karwar is one of the major fishing centers situated (14° 50′ N latitude and 74° 03′ E longitude) in the district of Uttara Kannada of Karnataka State. The Kali river is an important river opening into the sea in Karwar. The landing centre is situated approximately 2 km away from Karwar town on the Karwar Mangalore highway. The fish landing Majali is located at 14° 52′ N latitude and 74° 04′ E longitude, this beach is located to the extreme north of the Karwar. The collection was made by random sampling method *i.e.* total boats engaged in fishing were calculated first. On that basis, random sampling was done.

For identification of fishes standard keys such as Day (1978), Talwar and Jhingran (1991) and F.A.O sheets were used. Identification of pelagic, mid-water and benthic fishes was carried out simultaneously for their systemic study.

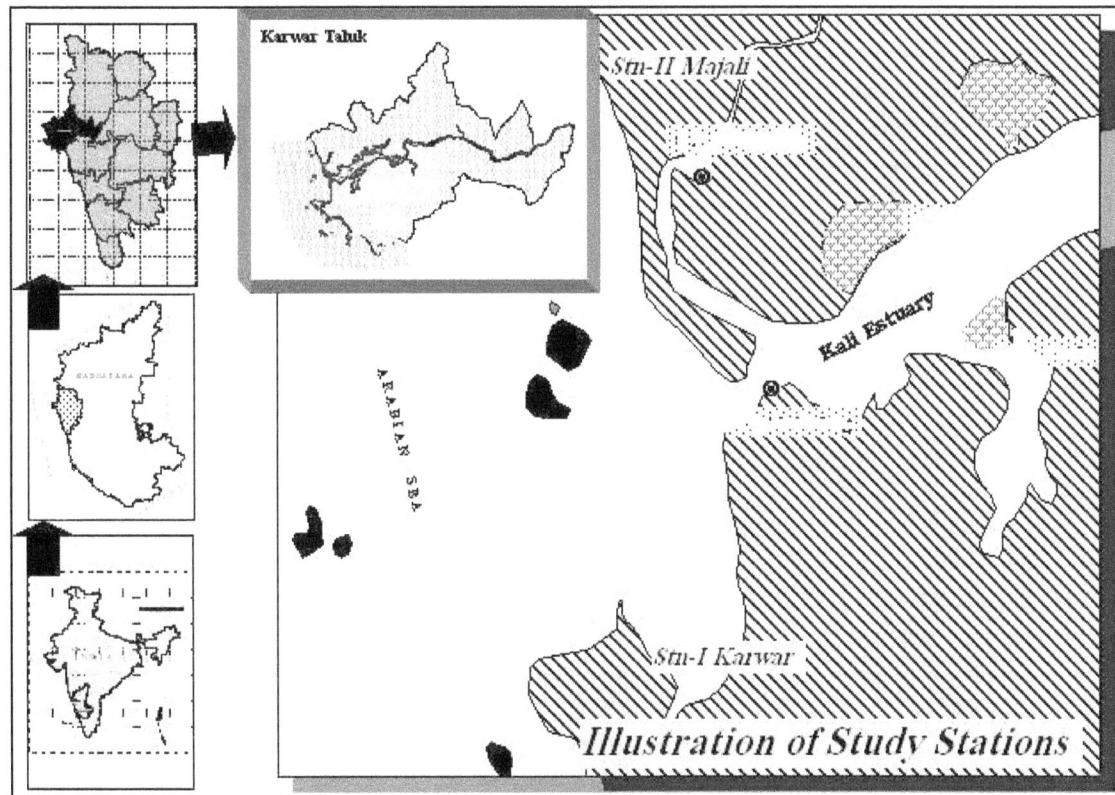

Figure 6.1: The Study Stations.

Results and Discussion

During the present observation 33 varieties of fishes belonging to 16 families were noticed at station–I (Table 6.1). Whereas, in station–II there are 21 types of (Table 6.2). Blarford (1901) studied occurrence and distribution of fish and supported Day's view. Bhat (1981) fishes, representing the 15 families were recorded studied the biology of some freshwater fishes, Dow (1981) studied the temperature influence on the abundance and the availability of marine and estuarine fishes, Rajalaxmi (1980) studied the fishery resources of Godavary estuary. Similar observation was made by Rajesh (1994) while studying the distribution, abundance and composition of fin fishes along Kali estuary.

Table 6.1: Chek List of Finfishes Observed at Station–I Baithkol

Sl.No.	Scientific Name	Family	Sl.No.	Scientific Name	Family
1.	*Priacqanthus hamrur*	Priacanthide	2.	*Decapterus russelli*	Carrangidae
3.	*Alectis indicus*	Carrangidae	4.	*Alepes djedaba*	Carrangidae
5.	*Megalaspis corydyala*	Carrangidae	6.	*Caranx rotteleri*	Carrangidae
7.	*Therapon jarbua*	Teraponidae	8.	*Johnius voqleri*	Sciaenidae
9.	*J. aneus*	Sciaenidae	10.	*Pampus argenteus*	Stromateidae
11.	*Lactarius lactarius*	Lactaritidae	12.	*Leioqnthus dussumieri*	Leiognathidae
13.	*Secutor insidiator*	Leiognathidae	14.	*S. ruconius*	Leiognathidae
15.	*Rastrelliger kanagurta*	Scombridae	16.	*Scomberomours linoclats*	Scombriade
17.	*Ambasis commersioni*	Ambassidae	18.	*Mugil cephalus*	Mugillidae
19.	*Pempheris moluca*	Pempheridae	20.	*Sardinella longiceps*	Clupeidae
21.	*Esualosa thoracata*	Clupeidae	22.	*Dussumieria acuts*	Clupeidae
23.	*Opisthopterus* sps.	Clupeidae	24.	*Sardinella fimbriata*	Clupeidae
25.	*Thryassa malabarica*	Engraulidae	26.	*T. Setirostris*	
27.	*Ancohoviella* sps.		28.	*Stolephorus devisi*	
29.	*Hemiramphus devisi*	Hemiramphide	30.	*Arius* sps.	Ariidae
31.	*Cynoglossus macrostomus*	Cyonglossidae	32.	*C. dubius*	Cyonglossidae
33.	*C. puncticeps*	Cyonglossidae			

Table 6.2: Chek List of Finfishes Observed at Station–II Majali, Karwar

Sl.No.	Scientific Name	Family Name	Common Name
1.	*Scoliodon laticaudus*	Rhinodontidae	Sharks
2.	*Himantura bleekeri*	Dasyatidae	Rays and Skates
3.	*Sardinella longiceps*	Clupidae	Oil sardine
4.	*Sardinella fimbriata*	Clupidae	Lesser sardine
5.	*Clupidae*	Clupidae	Clupidae
6.	*Silverbar*	Silverbar	Silverbar
7.	*Rostrelliger kangurta*	Silverbar	Mackerel
8.	*Scomberomorus commerson*	Scombridae	Seer fish
9.	*Thannus albacares*	Scombridae	Tuna
10.	*Lacterius lacterius*	Lactaritidae	Lactarius
11.	*Megalapsis cordila*	Carrangidae	Carrangids
12.	*Pampus argentius*	Stromatidae	Pomfrets
13.	*Anchoviella commersoni*	Engraulidae	Silver bellies

Contd...

Table 6.2–*Contd...*

Sl.No.	Scientific Name	Family Name	Common Name
14.	*Johnious* sps.	Sciaenidae	Sciaenids
15.	*Trichiurus lepturus*	Trichiuridae	Ribbonfish
16.	*Sphyraena jello*	Soleidae	Soles
17.	*Mugil cephalus*	Mugilidae	Mullets
18.	*Metapenaeus monoceros*	Penaeidae	Brown prawn
19.	*Penaeus semisulcatus*	Penaeidae	Red prawn
20.	*Portunus pelagicus*	Nephropidae	Crabs
21.	*Shell fish*	Penaeidae	Shell fish

Table 6.3: Total Fish Catch in Stations-I and II (Karwar) from January 2006-January 07 (Quantity in Metric tonnes)

Sl.No.	Fishes	Station–I	Station–II	Sl.No.	Fishes	Station–I	Station–II
1.	Sharks	5	7.3	2.	Rays and Skates	6	4.2
3.	Oil sardine	160	47.5	4.	Lesser sardine	66	24.2
5.	White sardine	14	0	6.	Clupidae	84	42
7.	Anchovies	11	0	8.	Silverbar	23	10
9.	Mackerel	82	29.5	10.	Seer fish	47	20.3
11.	Tuna	15	11	12.	Lactarius	30	20.2
13.	Carrangids	73	34.5				
14.	POMFRETS	16	0				
	a. Black pomfret	11	3		b. Silver pomfret	7	4
15.	Silver bellies	97	13.9	16.	*Gerrus* sps.	2	3
17.	Sciaenids	58.5	43.5	18.	Ribbonfish	43	36
19.	Flatfish	15	20	20.	Soles	129	15
21.	Catfish	17	14.2	22.	Eels	4	3
23.	Jawfish	0	0	24.	Pinkperch	0	0
25.	Bigeye	0	0	26.	Lizardfish	0	2
27.	Rockcod	14	10	28.	Threadfin breams	0	0
29.	Ladyfish	17	1.3	30.	Mullets	16	9.7
31.	Pearlspot	0	0	32.	Lobsters	4	1
33.	Crabs	65.5	27.7	34.	Squilla	221	0
35.	MOLLUSCS	0	0				
	a. Squids	20	5		b. Cuttlefish	0	0
	c. Octopus	0	0		d. Shellfish	49	44.3
	e. Miscellaneous	180.6	40.4				

To understand the trends in fish landings for a period of 13 months was given in Table 6.3. The oil Sardine spp. catch statistics was highest (160mt, 47.5 respectively) in station–I and II and lowest catch was recorded *Gerrus* spp. (2mt) and Lizard fish (2mt) in station–I and II respectively. It can be concluded that station–I is more productive in terms of fish landing than that of station–II. More numbers of active fishermen are engaged in the station–I (Karwar). In station–I almost 319 mechanized boats and 1304 non-mechanized boats are actively involved in fishing activity, which includes, the mechanized gears like Purse seines, Trawlers, out board motor nets and gill nets. The station–II is comparative less productive because 251 mechanized boats and 1575 non-mechanized boats are engaged in fishing activity. Gill net fishing is the major gear operated in the station–II.

References

Bhat, U.G., 1981. Studies on some biological aspects and fishery of the Fringe Scale Sardine, *Sardinella fimbriata* from Karwar waters. *M.Sc. Dissertation Thesis.*

Day, F., 1978. *The Fishes of India: Being a Natural History of the Fishes known to Inhabit the Seas and Freshwaters of India, Burma and Ceylon.* Today and Tomorrow's Book Agency, Vol. 1, New Delhi.

Dow, R.L., 1981. Influence of sea temperature cycles on the abundance and availability of marine and estuarine species of commerce. *Marine Dept. Mar. Resour.,* 2: 775–779.

FAO Fish Identification Keybook, 1980.

Rajesh, Naik, 1994. Studies on the distribution, abundance and composition of fin-fishes along along Kali River.

Rajyalakshmi, T., 1980. Biology of culturable species of prawn *Penaeus monodon, P. indicus* and *Metapenaeus dodsoni.* Proc. Summ. CIFRI Barrackpore.

Chapter 7

GIS Based Assessment of Benthic Faunal Resources in Thondi Coastal Area, Palk Bay, South-east Coast of India

☆ *J. Sesh Serebiah, M. Rajkumar, M.M. Rahman,*
M.K. Abu Hena, A. Saravanakumar and C. Stella

ABSTRACT

The investigation of benthic faunal resources in Thondi coastal area was carried out using GIS. Twelve sampling points were fixed by global positioning system (GPS). A total of 54 macro benthic species were recorded in five major goups: gastropods, 24 species; bivalves, 15; amphipods, 5; decapods, 6 and echinoderms, 4. Fifty three nematodes species were recorded, of this 8 species belonged to desmodoridae, 7 species to choromodoridae, 5 species to tripyloididae, 5 species to oncholaimidae, 6 species to xylaidae, 8 species to leptolaimidae, 6 species to oxystominidae and 8 species to diplopeltidae. Thirty polychaetes species belonged to 13 families were identified. The diversity, seasonal variations and dominance of benthic faunal resources in relation to environmental parameters were also studied using GIS software.

Keywords: Benthic macro fauna, Nematode, Polychaete, GIS assessment, Thondi, Palk Bay.

Introduction

Macro benthos are generally greater than 0.5 mm size, resides in sea bottom sediment, performing varieties of ecological function. They act as a link between the biotopes of substratum and water column in the aquatic systems. They take part in breakdown of particulate organic material and export energy to higher trophic level and can potentially support off–shore and pelagic communities (Schrijvers *et al.*, 1996 and Lee, 1997). The developmental stages of many macro benthic organisms are pelagic, forming important components of plankton community, which in turn is consumed by fish and thus having high influence on pelagic fisheries. Thus, the estimation of benthic production is useful to assess the fishery production of a particular area (Sultan Ali *et al.*, 1983). Its distribution highly depends on the nature of the substratum, nutritive content, degree of stability, oxygen content, and level of hydrogen sulphide. The small changes in the environment will have considerable response on the benthic community, which is frequently used to measure the degree of pollution (Coull, 1973). However, it is regarded as indicator organisms hinting the condition, nature and characteristics of the ecosystem. Therefore, assessment of ecosystem health can be achieved through a careful analysis of benthic fauna.

Nematodes are small vermiform and elongated animals, which pass through 0.5 mm sieve. They occupy the interstitial places provided by minute particles by wriggling around and between. They consume largely bacteria, microalgae and detritus and in turn act as potential live feed for growth and survival (Coull, 1990; Nilsson *et al.*, 1993 and Toepfer and Fleeger, 1995) of macro fauna, pelagic predators, crustaceans and their larvae (Ingole and Parulekar, 1998). Thus, they play a crucial role in trophic dynamics and rapid turnover of elements and nutrient recycling (Platt and Warwick, 1980). They have diverse feeding groups such as herbivores, detritus feeder and suspension feeders in several tropic level that greatly involved in biodegradation process by which they breakdown the materials and present before to microorganisms for mineralization. It aids in decomposition process greater than macrobenthos in view of their small size, higher metabolic activity, higher turnover and productivity per unit biomass and have ability to occupy favourable microhabitats. Exposure time, desiccation, availability of food, sediment granulometry, tidal zonation, nutrients and interstitial water quality are the physico-chemical parameters regulate the abundance of nematodes. However, they are key groups sensitive to environmental changes and they act as bioindicators of the ecosystem (Geetanjali *et al.*, 2001 and Chinnadurai and Fernando, 2006). An important feature of nematode populations is the large number of species present in any one habitat, and their numbers are often in order of magnitude greater than for any other taxon (Platt and Warwick, 1980).

The estuarine and coastal water benthic communities on west and east coast of India were well documented (Jagadeesan and Ayyakkannu, 1992; Chakraborty *et al.*, 1992; Venkatesh Prabhu *et al.*, 1993; Jose and Rajagopalan, 1993; Balasubramanian, 1994; Chakraborty and Choudhury, 1994; Ansari *et al.*, 1994; Santhakumaran and Sawant, 1994; Sunilkumar and Antony, 1994; Sunil Kumar, 1995, 1997, 2001; Prabha Devi *et al.*, 1996; Ray *et al.*, 2000; Kathiresan *et al.*, 2000; Sesh Serebiah, 2003; GUIDE, 2000; Zhang *et al.*, 2004; Sajan and Damodaran, 2007; Ganesh and Raman, 2007; Chinnadurai and Fernando, 2006; Saravanakumar *et al.*, 2007 and Anbuchezhian *et al.*, 2012). Many authors have studied density, abundance, seasonal fluctuation and its composition of nematodes in east and west coast of India (Rao and Satpathy, 1996; Goldin *et al.*, 1996; Ingole and Parulekar, 1998 and Ansari and Parulekar, 1998).

Polychaetes are vermiform structural organisms with many legs, dwelling on or in bottom habitats and can be retained in 500 µm sieve. Polychaetes have traditionally been considered either as the dominant group, in terms of abundance or biomass or as an important contributor to the structure and functioning of the macrobenthic communities (Lee, 2009 and Mutlu *et al.*, 2010). In the recent past, it evoked considerable scientific attention to explore it's potentially for monitoring environmental quality in various habitats (Grall and Glemarec, 1997; Hutchings, 1998; Solis-Weiss *et al.*, 2004 and Arvanitidis *et al.*, 2005) as well as in zoogeographical studies (Musco and Giangrande, 2005 and Dogan *et al.*, 2005). They exhibit variety of body shapes, feeding styles, reproductive modes and comprise a critical link in the marine food web. They can penetrate into brackish water habitats such as estuaries, backwaters due to euryhaline in nature. They consume decomposed organic matter (bacteria, planktonic and benthic organisms, detritus etc.,) and in turn act as a food for fishes, birds and other marine invertebrates. They have been used as biological indicators of water quality for assessing the effects of industrialization and urbanization (Pocklington and Wells, 1992; Dernie *et al.*, 2003; Ravichandran and Rameshkumar, 2008 and Anbuchezhian *et al.*, 2012). Polychaetes are worldwidely used in environmental impact assessment studies (Clarke, 1993; Eaton, 2001; Samuelson, 2001; Dauvin *et al.*, 2004; Harkantra and Rodriguez, 2004 and Kress *et al.*, 2004).

Only one paper has been published on the assessment of soft bottom polychaete diversity in Thondi coastal waters (Anbuchezhian *et al.*, 2012). As, there is no detailed study on benthic faunal resources in Thondi coastal waters. The present study has been undertaken to deal with the occurrence, spatial and temporal distribution and abundance of benthic faunal resources in relation to environmental parameters. It would also help us to gain a holistic view of the coastal waters.

Materials and Methods

The study was carried out for a period of one year from January to December 2006. In this tropical region, based on the meteorological events, the year is divisible into four climatic seasons: monsoon (October–December); post monsoon (January–March); summer (April–June) and premonsoon (July–September). A total of 12 stations were made by GPS (Global Positioning System) in Thondi sub tidal region covering 0.7 sq. km area (Figure 7.1). The water and sediment samples were collected monthly at all stations. For benthic faunal resources after retrieval of sediment from Peterson's grab (0.08 m^2),

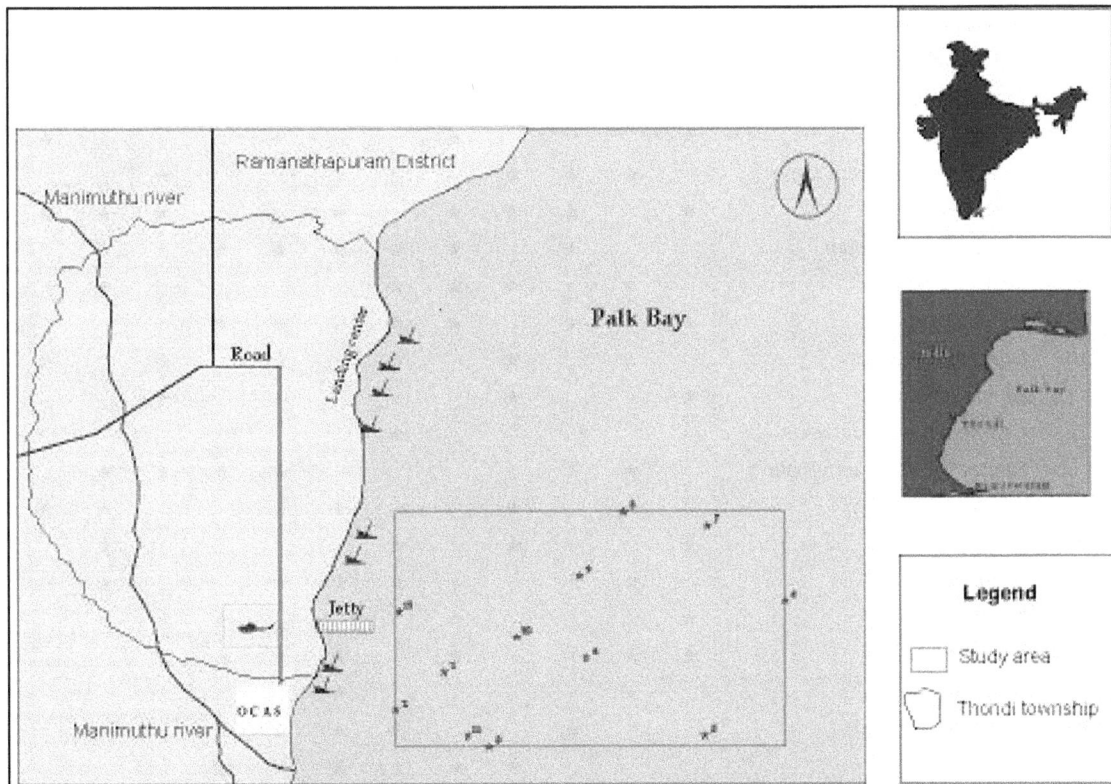

Figure 7.1: Study Area Location Showing the Sampling Stations in Thondi Area, Palk bay, South East Coast of India.

sieved and retained through 0.5, 0.064 mm screen and preserved in 5 per cent and 2 per cent formalin. The animals were separated, counted, identified up to species level using standard references and expressed for macro benthos No./m^2, for polychaetes No./m^2 and for nematodes No./10 cm^2. The temperature, salinity, pH, dissolved oxygen, nitrate, nitrite, phosphate and silicate in water were estimated following standard method of Strickland and Parsons (1972). The total organic carbon (TOC) in sediment was estimated according to El–Wakeel and Riley (1956). The contour maps were drawn for all hydrobiological parameters and faunal density.

Results and Discussion

The discernible temporal and spatial variations were observed in benthic faunal resources communities of Thondi sub tidal regions due to physico-chemical characteristics of the habitat. A total of 54 macrofauna species were identified in 12 random stations. Of this, gastropods, 24 species; bivalves, 15; amphipods, 5; decapods, 6 echinoderms, 4 (Table 7.1).

Table 7.1: List of Macrobenthic Fauna Recorded in Twelve Stations during January 2006 to December 2006.

Sl.No.	Species	Stations											
		1	2	3	4	5	6	7	8	9	10	11	12
	Gastropods												
1.	Cerithidea cingulata	★	★	★	★	★		★	★	★	★	★	★
2.	Potamides cingulatus	★		★		★	★	★		★		★	
3.	Cerithium rubus		★	★	★		★	★	★		★		
4.	C. obeliscus	★				★			★				★
5.	Telescopium telescopium		★	★	★			★			★		
6.	C. fluviatilis	★		★	★	★		★			★	★	★
7.	Umbonium vestiarium			★		★	★	★	★	★	★		
8.	Mauritia arabica		★		★	★	★					★	
9.	Marginella sp.	★		★		★	★		★	★			
10.	Drupa tuberculata		★		★		★		★		★		★
11.	D. margariticola	★			★						★		
12.	Cantharius erythrostomus						★						
13.	Nassarius arcularius plicata		★	★					★	★	★	★	
14.	Strombus canarium		★		★		★	★	★			★	★
15.	S. urceus	★			★	★	★					★	
16.	Natiga tigerina		★	★			★	★	★			★	
17.	Nassa hepatica	★				★				★	★	★	
18.	N. jacksoniana	★	★			★		★	★	★			
19.	Murex tribulus		★					★					
20.	Chicoreus ramosus			★				★	★	★			
21.	Bursa sp.				★	★	★	★	★	★			
22.	Trochus stellatus					★	★	★					

Contd...

Table 7.1–*Contd...*

Sl.No.	Species	1	2	3	4	5	6	7	8	9	10	11	12
							Stations						
23.	*M. hustellam*		★					★	★				
24.	*C. brunneus*						★			★			
	Bivales												
25.	*Lucina ovum*							★					
26.	*Meretrix casta*	★			★			★					★
27.	*Vepricardium asiaticum*						★	★					
28.	*Anadara granosa*	★	★	★								★	★
29.	*Gafrarium* sp.		★	★	★		★						
30.	*Lunulicardia retusa*		★			★	★	★	★				
31.	*Gafrarium tumidum*	★	★										★
32.	*Mactra cuneata*			★	★	★							
33.	*Katelysia opima*	★	★		★	★						★	★
34.	*Paphia textile*		★	★	★	★	★						
35.	*Mesodesma trigonum*		★					★	★	★			
36.	*Tellina bruguieri*	★			★		★		★	★			
37.	*Modiolus tulipa*	★						★	★	★			
38.	*Crassostrea madrasensis*		★			★	★					★	★
39.	*Donax cuneatus*				★			★		★			★
	Amphipods												
40.	*Grandidierella gilesi*	★								★			★
41.	*Corophium triaenonyx*						★	★					
42.	*Urothoe platydactyla*										★		★
43.	*Ampelisca scabripes*		★		★			★					
44.	*Ampithoe raimondi*		★								★		
	Decapods												
45.	*Portunus sanguinolentus*					★	★		★		★		
46.	*P. pelagicus*				★	★		★	★	★	★		
47.	*Scylla serrata*		★			★			★				
48.	*Metapenaeus dobsoni*			★	★	★			★	★			★
49.	*Penaeus monodon*	★	★		★		★			★			
50.	*P. indicus*	★	★	★	★							★	★
	Echinoderms												
51.	*Pentaceraster regulus*	★			★					★		★	
52.	*Coniodiscaster scaber*			★									
53.	*Pentaceraster* sp.							★			★		
54.	*Coniodiscaster* sp.	★	★		★					★		★	

The density of gastropods, bivalves and decapods were higher than the density of amphipods and echinoderms. In the present study, the dominance of *Cerithidea fluviatilis* and *C. cingulata* in the shoreward regions with high temperature, salinity and pH (Plate 7.1) were also observed in Vellar estuary and Tuticorin Bay (Kathiresan *et al.*, 2000 and Kailasam and Sivakami, 2004). Kailasam and Sivakami (2004) and Kathiresan *et al.* (2000) observed extreme adaptability of these species with frequent fluctuation of environmental characteristics. In the present study, *Lucina ovum* (a bivalve) was observed once, while *Vepricardium asiaticum* (Bivalve), *Murex tribulus* (Gastropods), *Cantharius erythrostomus* (Gastropods) and *Corophium triaenonyx* (Amphipod) were observed twice at station 7. However, these were rare species at seaward side stations (Stations 6 and 7) showing the preference of sandy mud bottom.

Plate 7.1: GIS Mapping of the Physico-chemical Parameters in Thondi Area, Palk Bay, South East Coast of India, (a) Temperature (°C), (b) Salinity (ppt), (c) pH, (d) Dissolved oxygen (mg/L), (e) Organic carbon (per cent), (f) Nitrate (mg/L), (g) Nitrite (mg/L), (h) Phosphate (mg/L), (i) Silicate (mg/L).

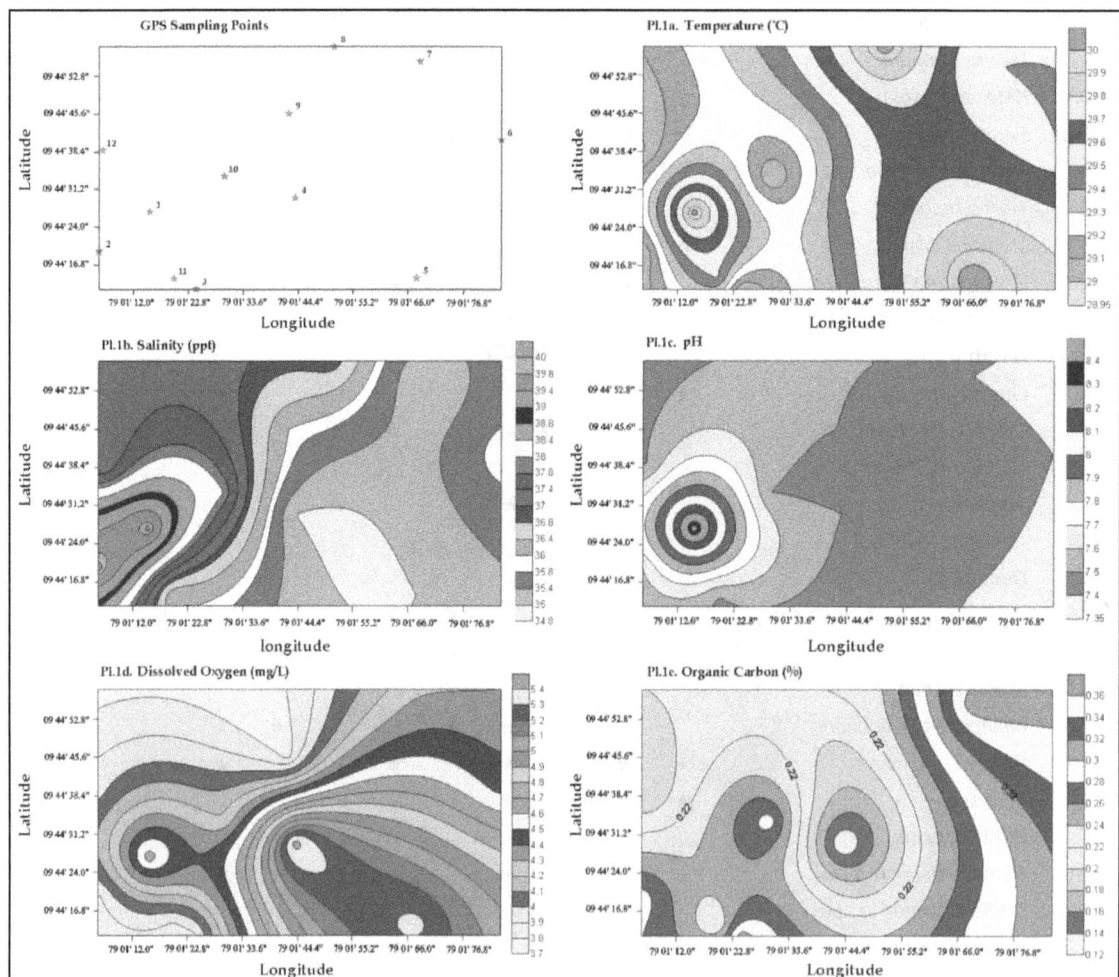

Contd...

Plate 7.1–*Contd...*

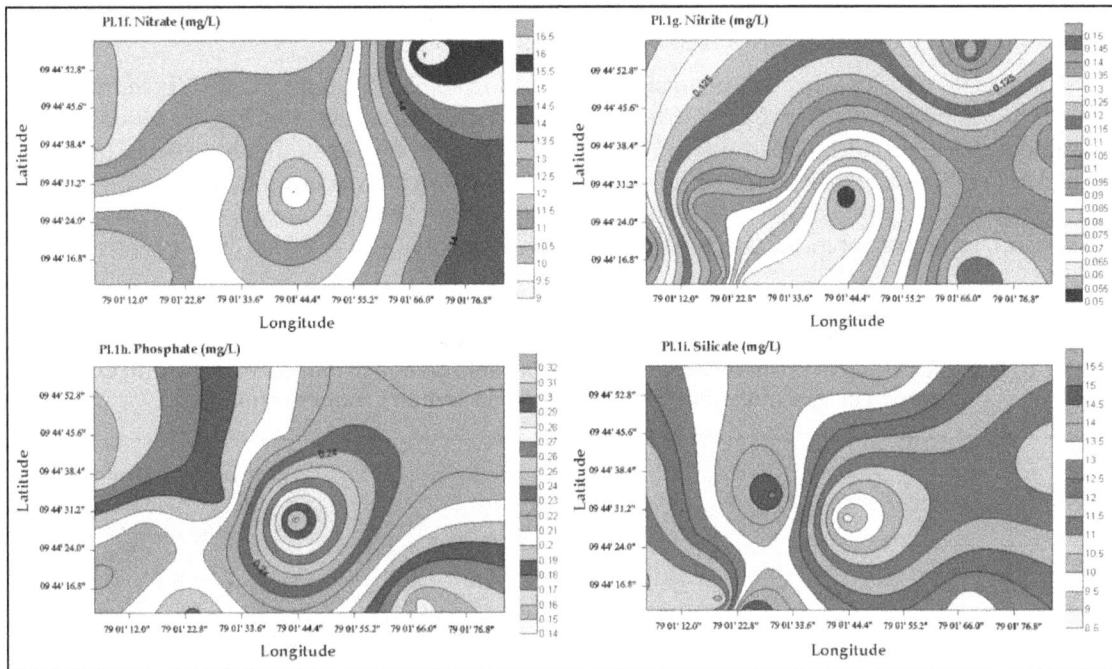

Benthic macrofaunal community is characterized by temporal and spatial changes in the population. The benthic invertebrates were found to have a linkage with fishes through food web (Bell, 1979). In present study, macrofaunal distribution pattern seems to be fully governed by the physico-chemical characteristics of the environment (Sunilkumar, 1995). Intertidal fauna at the studying area have to cope with harsh environmental condition marked by high salinity, increased evaporation, wide seasonal variation of temperature and tidal amplitude. These unique physico-chemical factors exert a strong influence on faunal assemblages, which is withstanding such a situation. Based on substratum, the percentage assemblages of benthic faunal resources in different stations were shown in Figure 7.2. Higher percentage of benthic faunal resources were found in station 7 (10.4 per cent) followed by station 6 (10.12 per cent), station 8 (9.54 per cent) and station 1 (9.54 per cent). It probably be due to abundance of

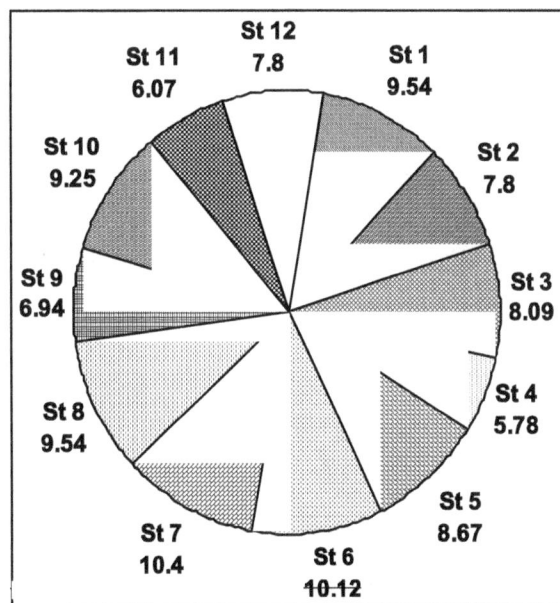

Figure 7.2: Percentage Assemblages of Macro Benthic Macro Faunal Resources in different Stations.

organic carbon content at station 7 and its nearby stations (Plate 7.2). The station 1 seems to have high numbers of *C. fluviatilis* and *C. cingulata* alone to contribute high percentage which are observed to be actively take part the breakdown of stranded seaweeds and sea grass.

The species density (No./m²) in all stations varied from 532 (Summer) to 847 (Post monsoon) showed the dominance of gastropods (468–327) superseded by bivalves (188–111), decapods (98–37), amphipods (50–27) and echinoderms (43–30) (Figure 7.3). Almost similar studies have been conducted by Sankar (1998) in Muthupet lagoon, Sunilkumar (1995) in Cochin backwaters, Prabha Devi (1994) in Coleroon estuary and Ansari *et al.* (1986) in Mandovi estuary, Saravanakumar *et al.* (2007) in Gulf of Kachchh, west coast of India. Our results concur with Athalye and Gokhale (1998), who reported

Plate 7.2: Organic Carbon during January 2006 to December 2006, (a) Organic matter (per cent), (b) Turbidity (NTU), (c) Electrical conductivity, (d) Total suspended solid (mg/L), (e) Total dissolved solid, (f) Organic carbon in pre monsoon (mg/L), (g) Organic carbon in monsoon (mg/L), (h) Organic carbon in post monsoon (mg/L), (i) Organic carbon in summer (mg/L).

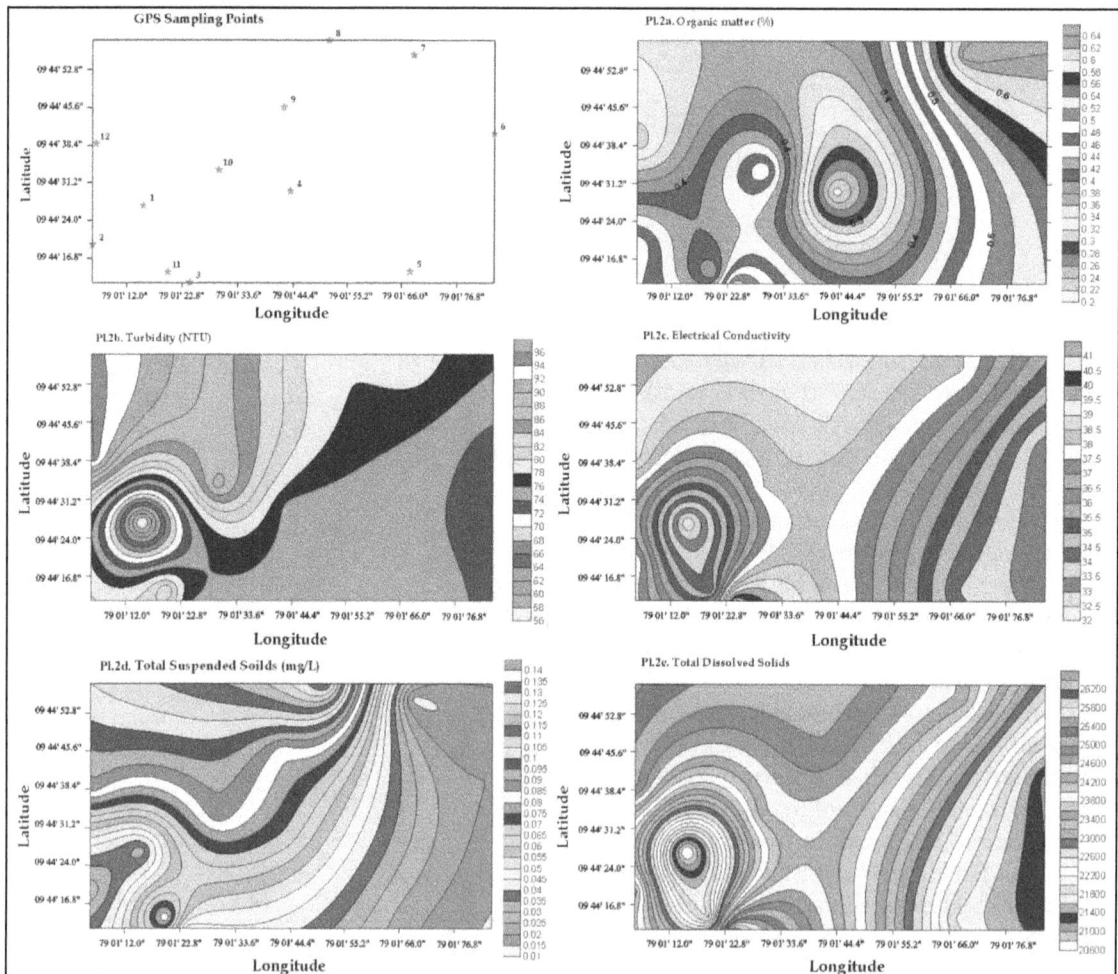

Contd...

Plate 7.2–*Contd...*

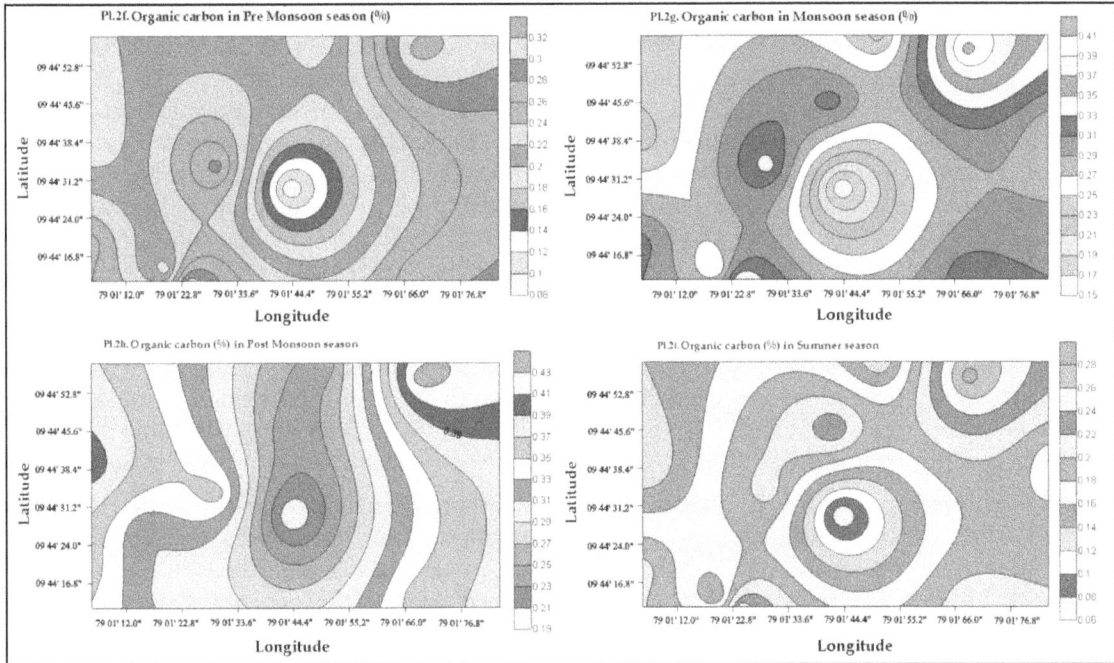

the dominance of polychaetes, gastropods, bivalves and other macro faunal species in post monsoon season and less density in summer in Thane creek, Bombay. This could be due to organic loading in the sea in monsoon season and subsequent proliferation of benthos in post monsoon season and

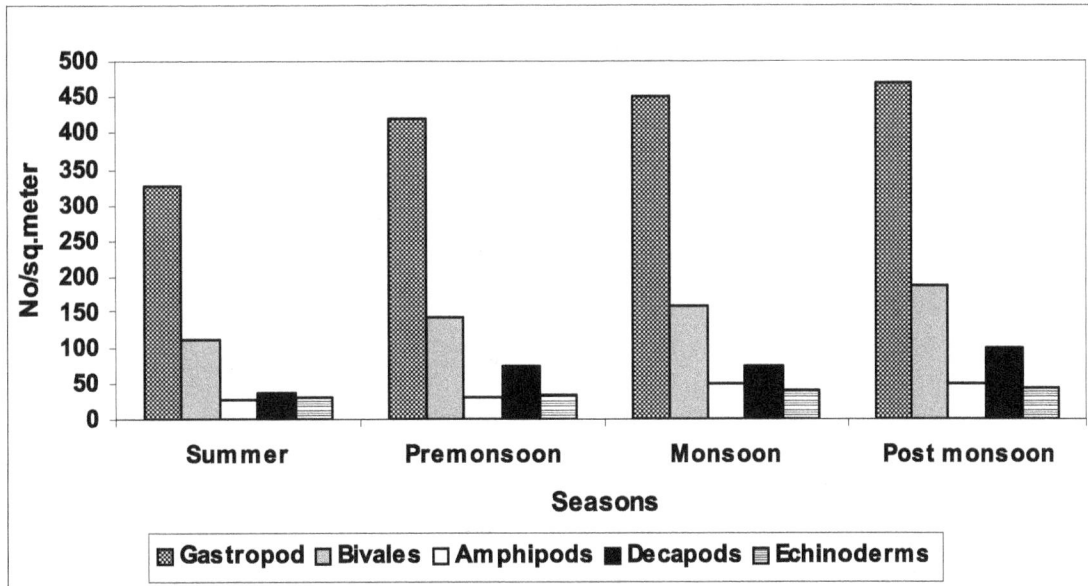

Figure 7.3: Abundance of Macro Benthic Fauna in Thondi Sub Tidal Area.

diminishing in summer for lack of nutrients (Sesh Serebiah, 2003). Ansari *et al.* (1986), Sunilkumar and Antony (1994), GUIDE (2000) and Sunilkumar (1995 and 2001) observed high macrofaunal density during post monsoon seasons (November–February) in the west coast. Low population density recorded in monsoon could be due to the effect of heavy rainfall (Sarvanakumar *et al.*, 2007). Similarly, Seshappa (1953) reported a 'severe decline' in the shallow water macro–benthos during the southwest monsoon and the decrease was attributed to lowered salinity. This observation coincides with the previous findings of Sunilkumar and Antony (1994) and Sunilkumar (2001). But the spatial observation of benthos in four season at different stations varied from 10 No./m² (Summer) to 115 No./m² (Post monsoon) (Plate 7.3). This density was higher than the macrobenthic faunal densities reported by Parulekar and Waugh (1975) in Zuari eatuary (50 to 1437 ind./m²) and by Parulekar and Ansari (1981) in Andaman seas (80 to 998 ind./m²). The macro benthic faunal density reported by Saravanakumar *et al.* (2007) in Gulf of Kachchh was 424 ind./m² to 2393 ind./m². It is comparable also with Harikantra *et al.* (1980), who reported 50 to 3715 ind./m² in the shelf region along the west coast of India. However, the density observed in the present study is lower than the density observed by Parulekar and Waugh (1975) and Salzwedal *et al.* (1985) (1253 to 5723 ind./m²) in northwestern Arabian Sea shelf in northern sea.

The contour map was drawn between the organic carbon and macro benthos revealed that high correlation between them in all seasons. In all season except summer, the benthos with density of 70–75 No./m² was distributed in larger area in the study site. In particular, the benthos with density of 75 No./m² in monsoon season (Plate 7.3) shown the initiation of proliferation of benthos. But in summer, the density with 40 No./m² (Pink colour) had larger area distribution shown the declination of faunal density.

A total of 53 nematode species were identified in twelve random stations. Of this 8 species belonged to desmodoridae, 7 species to choromodoridae, 5 species to tripyloididae, 5 species to oncholaimidae, 6 species to xylaidae, 8 species to leptolaimidae, 6 species to oxystominidae and 8 species to diplopeltidae (Table 7.2). Among the nematodes families, desmodoridae (19.84 per cent) choromodoridae (17.41 per cent) and diplopeltidae (12.15 per cent) were dominant families (Figures 7.4 and 7.5). *Acontiolaimus zostericola* belongs to choromodoridae family found to be avilable in 8 stations. *Metachromadora macroutera*, *Desmodora* sp., *Spirina* sp., *Eubostrichus* sp. belongs to desmodoridae and *Tripyloides marinus*, belongs to tripyloididae were reported in station 7 indicated that all above species were common and evenly distributed in Thondi subtidal regions.

Similarly, Sesh Serebiah (2003) and Sultan Ali *et al.* (1983) observed *Tripyloides* sp., and *Metachromadora* sp. are the dominant species in Kachchh coast and Pichavaram mangroves. However, they were rarely sighted species in stations 6 and 7 indicating their sandy mud bottom preference.

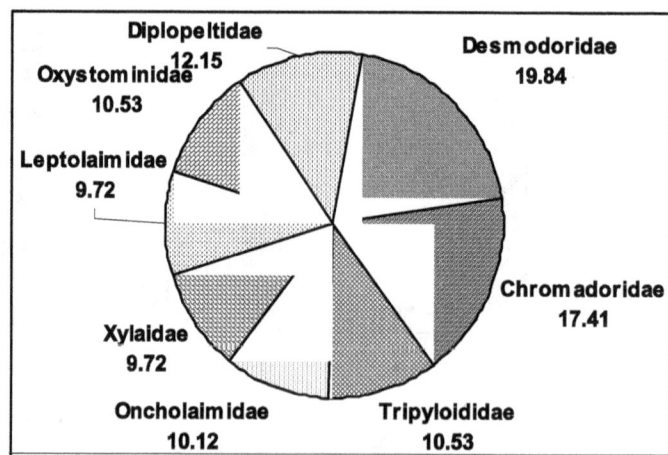

Figure 7.4: Percentage Occurrence of Nematode Fauna in different Families.

**Plate 7.3: Macro Benthic Faunal Density and Organic Carbon in different Seasons,
(a) Monsoon, (b) Post monsoon, (c) Pre monsoon, (d) Summer.**

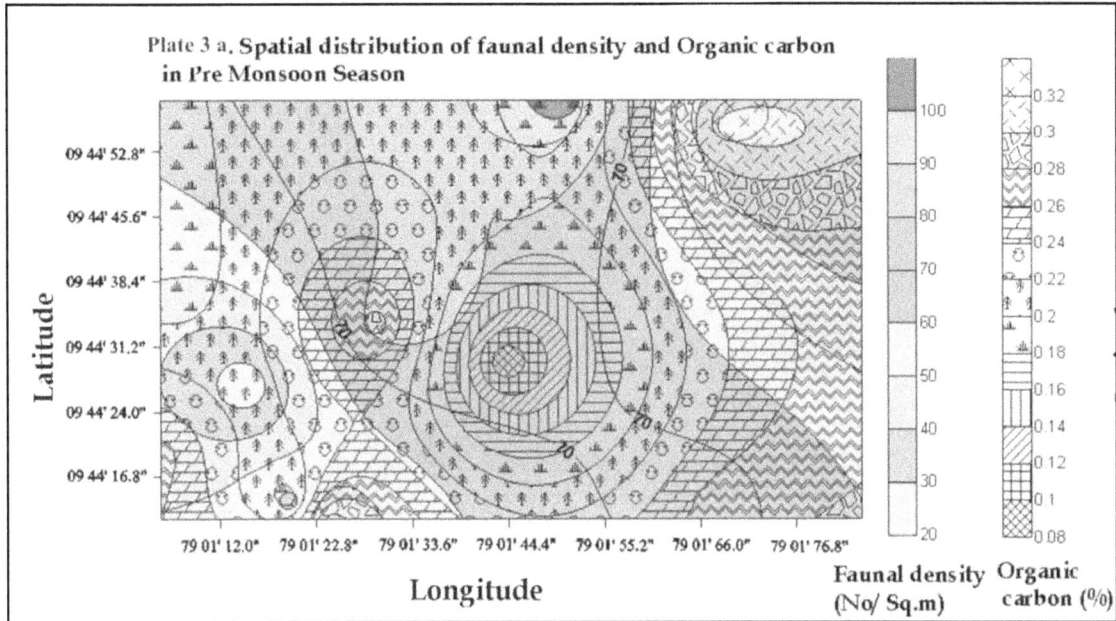

Plate 3 a. Spatial distribution of faunal density and Organic carbon in Pre Monsoon Season

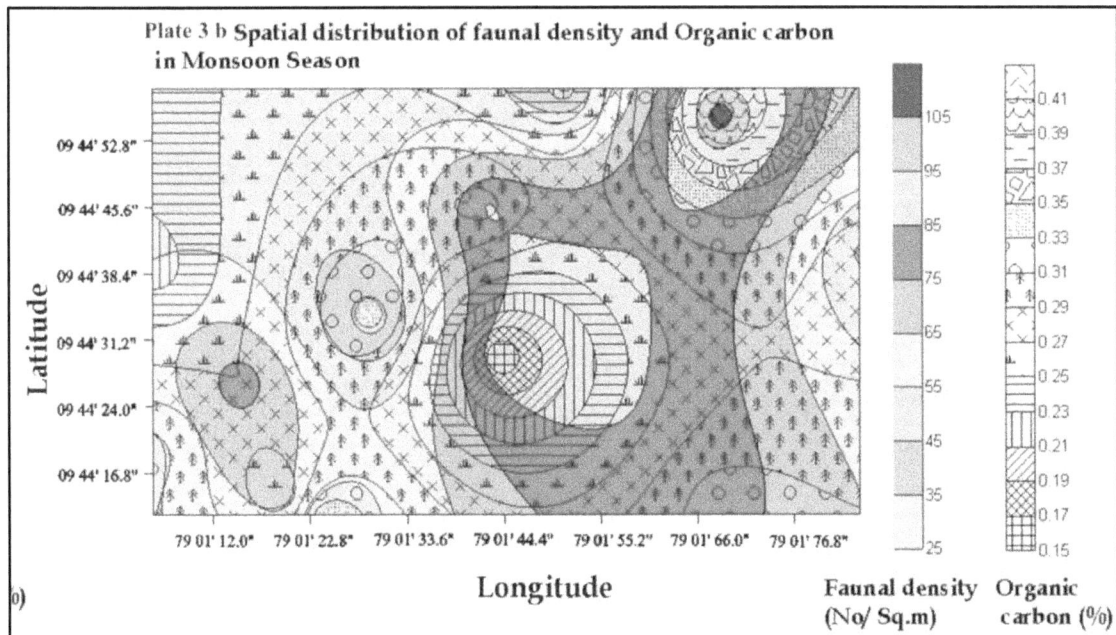

Plate 3 b Spatial distribution of faunal density and Organic carbon in Monsoon Season

Contd...

Plate 7.3–*Contd...*

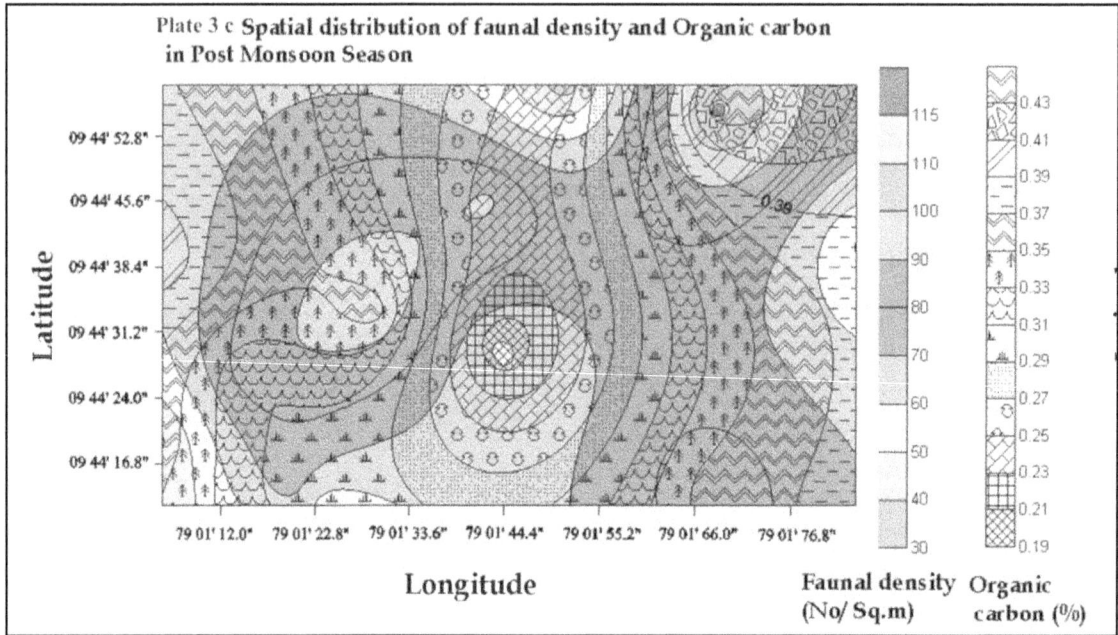

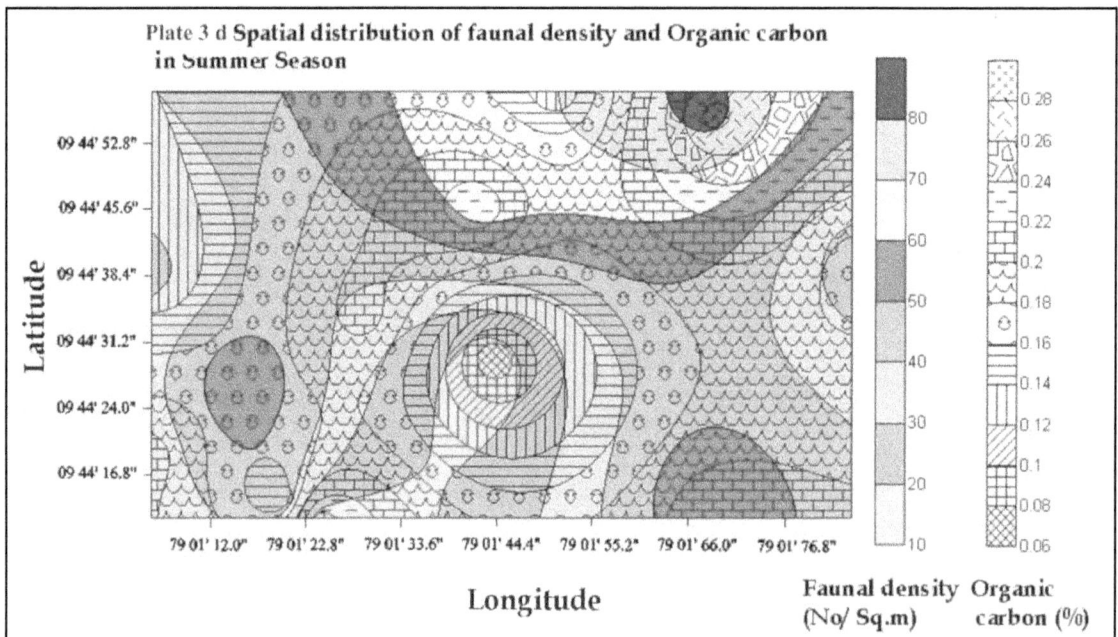

Plate 3 c **Spatial distribution of faunal density and Organic carbon in Post Monsoon Season**

Plate 3 d **Spatial distribution of faunal density and Organic carbon in Summer Season**

Table 7.2: List of Nematodes Recorded in Twelve Stations during January 2006–December 2006

Sl.No.	Species	1	2	3	4	5	6	7	8	9	10	11	12
	Desmodoridae												
1.	*Metachromadora macroutera*	★			★	★		★	★	★			★
2.	*Desmodora* sp.	★		★		★	★	★		★		★	
3.	*Spirina* sp.		★	★	★		★	★	★		★		
4.	*D. luticola*	★				★			★				★
5.	*Chromaspirina* sp.		★	★	★			★			★		
6.	*Pseudonchus* sp.		★	★			★	★	★			★	
7.	*Polysigma* sp.	★				★				★	★	★	★
8.	*Eubostrichus* sp.	★	★			★		★	★	★			★
	Chromadoridae												
9.	*Acontiolaimus zostericola*			★		★	★	★	★	★	★		★
10.	*Prochromadora* sp.		★	★			★	★	★			★	
11.	*Trochus* sp.	★				★				★	★	★	★
12.	*Spillophorella* sp.	★	★			★		★	★	★			
13.	*Neochromadora Izhora*		★		★	★	★					★	
14.	*Spillophorella paradoxa*	★		★		★	★		★	★			
15.	*Prochromadorella brachyura*		★		★		★		★		★		★
	Tripyloididae												
16.	*Tripyloides marinus*	★				★		★	★	★		★	★
17.	*Bathlaimus* sp.	★	★			★		★		★			
18.	*T. gracilis*		★		★		★		★			★	★
19.	*Bathlaimus inermis*		★	★				★	★	★	★		
20.	*Gairleanema* sp.							★					
	Oncholaimidae												
21.	*Oncholaimus oxyuris*					★	★	★	★				
22.	*Viscosia* sp.			★			★				★		★
23.	*O. fuscus*		★	★			★	★	★			★	
24.	*Oncholaimus* sp.	★				★				★	★	★	
25.	*Viscosia viscosa*	★	★			★		★	★	★			
	Xylaidae												
26.	*Xyala* sp.	★			★		★	★	★				
27.	*Linhystera* sp.						★						
28.	*Cobbia* sp.	★	★					★	★	★	★		
29.	*Daptonema conicum*				★	★	★	★	★	★			
30.	*D. oxycera*					★	★	★					
31.	*Theristus pertenuis*		★					★	★				

Contd...

Table 7.2–*Contd...*

SI.No.	Species	Stations											
		1	2	3	4	5	6	7	8	9	10	11	12
	Leptolaimidae												
32.	*Leptolaimus* sp.			★			★		★	★	★		
33.	*Leptolaimoides* sp.							★					
34.	*Antomicroon* sp.			★	★					★			★
35.	*Dagda* sp.						★			★			
36.	*Procamacolaimus* sp.							★					
37.	*Stephanolaimus* sp.	★			★			★					★
38.	*Onchium* sp.	★	★	★								★	★
39.	*Leptolaimoidus* sp.		★				★						
	Oxystominidae												
40.	*Nemanema* sp.		★			★	★	★	★				
41.	*Halalaimus longicollis*			★			★		★	★	★		
42.	*Halalaimus* sp.							★					
43.	*Paroxytomina* sp.			★	★					★	★	★	★
44.	*Wiesera* sp.	★	★			★						★	★
45.	*Halalaimus gracillis*		★		★	★	★						
	Diplopeltidae												
46.	*Areolaimus tongicauda*	★			★		★		★	★			
47.	*Diplopetis* sp.	★						★	★	★			
48.	*Camplaimus* sp.		★			★	★					★	★
49.	*Sabptieria* sp.				★			★		★			★
50.	*Synochus fasciculates*	★								★			★
51.	*Southernill* sp.				★			★					
52.	*Diplopeltula* sp.									★		★	★
53.	*Diplopetis* sp.		★		★	★		★					

Station 7 (11.74 per cent) shown higher abundance than station 8 (10.53 per cent) followed by station 9 (9.72 per cent) and station 6 (9.72 per cent). It probably be due to high organic carbon content at station 7 and nearby stations. The species density (No./10 cm^2) in all stations varied from 243 (Summer) to 840 (Monsoon) showed the dominance of Desmodoridae (54–132), followed by choromodoridae (39–114), oxystominidae (30–108), diplopeltidae (33–120), leptolaimidae (27–96), tripyloididae (21–96), xylaidae (21–84) and oncholaimidae (18–90). Similar studies have been carried out in Zuari estuary, Goa (Ansari and Parulekar, 1998) and Jakhau, Kachchh coast; Gujarat (Sesh Serebiah, 2003) reported that nematodes were the most dominant group in Goa than Kachchh coast of Gujarat, west coast of India. The spatial density of nematodes in four seasons at different stations varied from 10 No./cm^2 (Pre–summer) to 104 No./cm^2 (Monsoon). The gradual increase of faunal density and organic carbon in seaward side has been highly predicted spatially.

Temperature, sediment granulometry, and tidal inundation are the main environmental factors that influence the distribution of meiofaunal communities in tropical mangroves (Alongi, 1987a, b). The temperature in the study site observed to be least variation from 29 to 30°C (Plate 7.1). The higher temperature was observed in shore regions owing to shallowness. The present study is in agreement with the statements of Kurian (1972) and Prabha Devi and Ayyakkannu (1989), who reported that the temperature is not an important factor that affects the distribution of benthos in Cochin and Coleroon estuary. However, salinity was highly positively correlated with temperature and pH. The elevated salinity in shore regions could be due to shallowness and high temperature (Stations 1 and 2) (Plate 1) caused moderate faunal assemblage. The moderate salinity and pH were noticed in seaward side could be contributed the increased faunal density. Lower pH (Plate

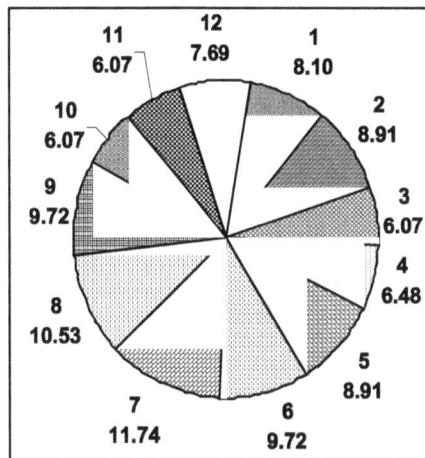

Figure 7.5: Percentage Occurrence of Nematode Fauna in different Stations.

7.1) in seaward side could be due to low temperature and decomposition of organic matter (Shriadah *et al.*, 1999). The dissolved oxygen varied from 3.7 to 5.4 mg/L (Plate 7.1). The increasing trend of dissolved oxygen was observed from shore to seaward side (also reported by Jacob *et al.*, 1982 and

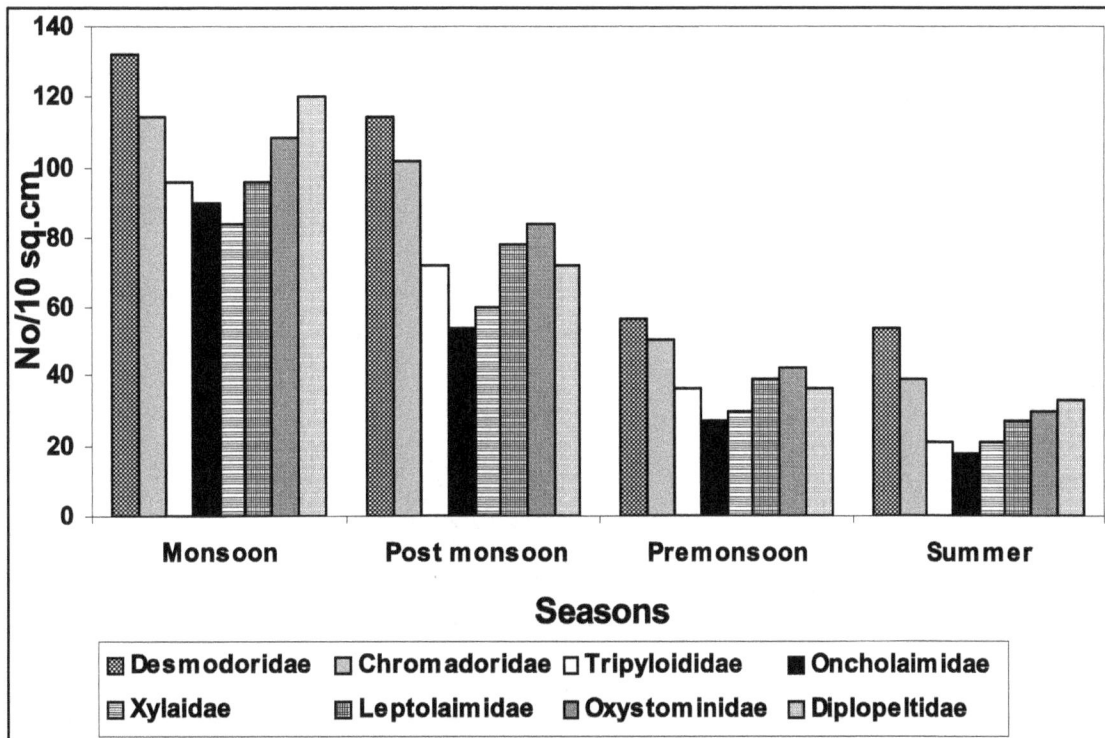

Figure 7.6: Abundance of Nematodes in Thondi Sub Tidal Area.

**Plate 7.4: Nematode Density and Organic Carbon in different Seasons,
(a) Monsoon, (b) Post monsoon, (c) Pre monsoon, (d) Summer.**

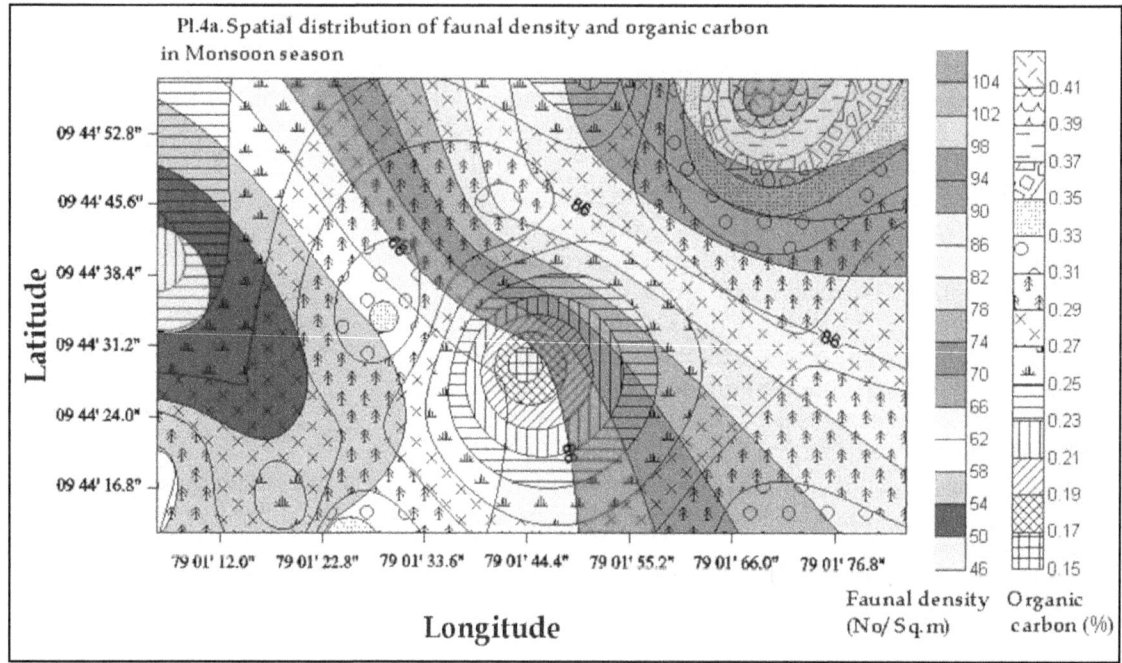

Pl.4a. Spatial distribution of faunal density and organic carbon in Monsoon season

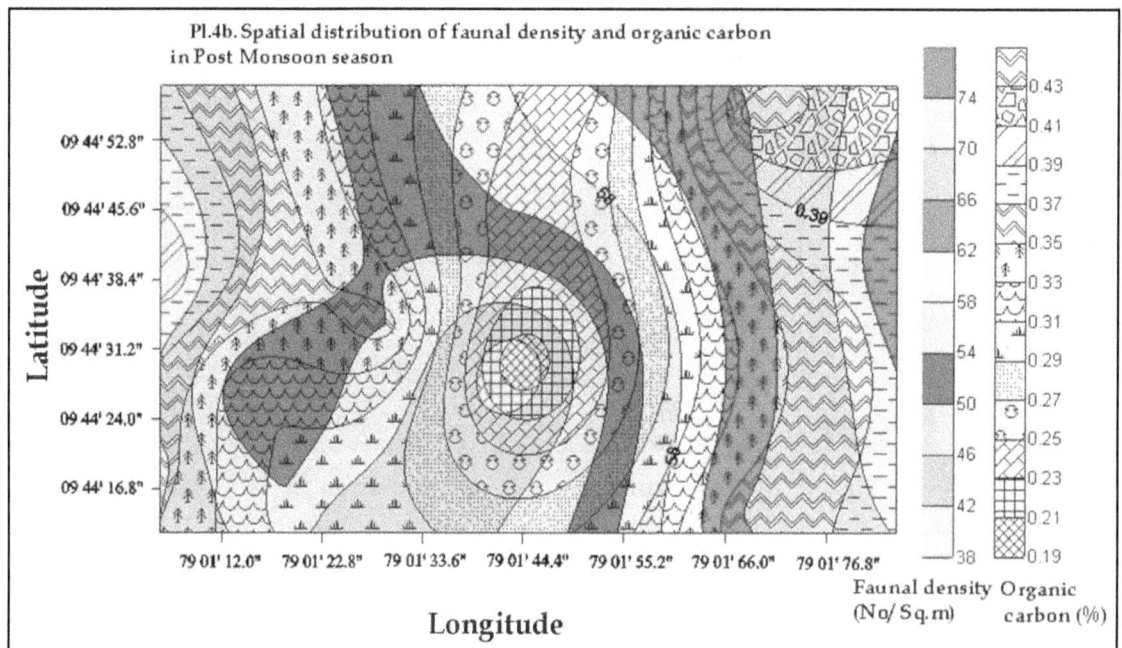

Pl.4b. Spatial distribution of faunal density and organic carbon in Post Monsoon season

Contd...

Plate 7.4–*Contd...*

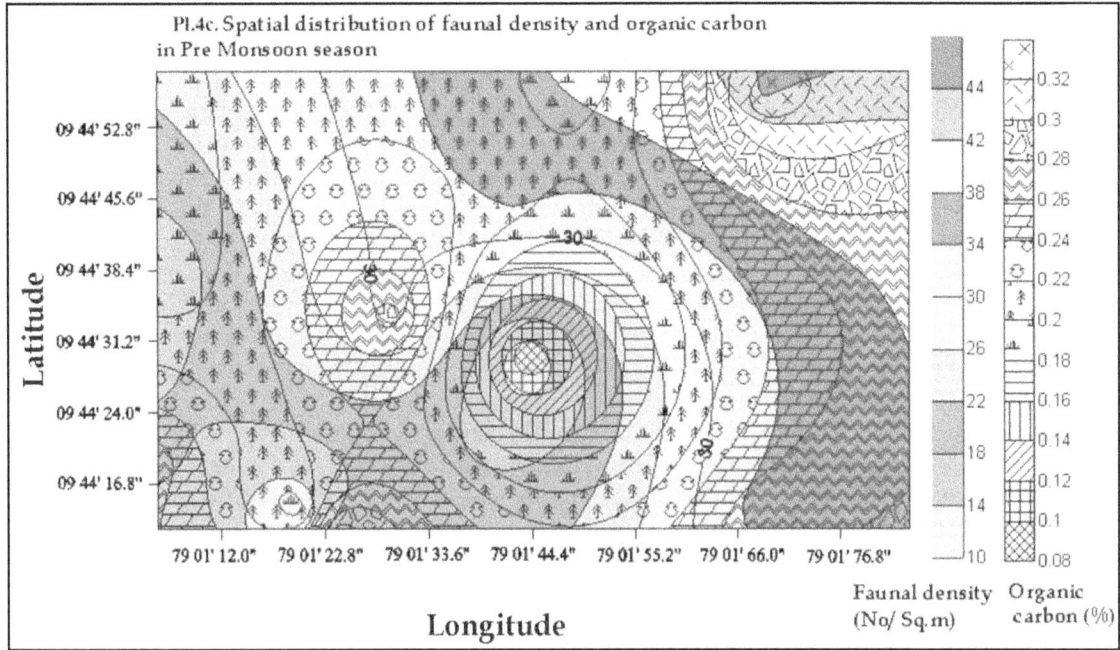

Pl.4c. Spatial distribution of faunal density and organic carbon in Pre Monsoon season

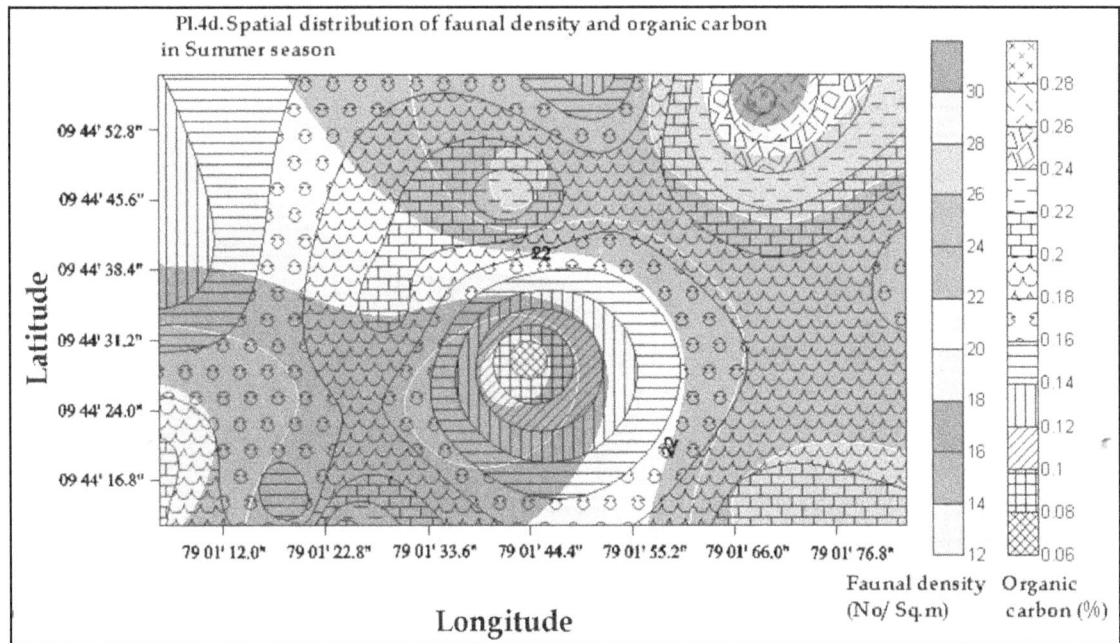

Pl.4d. Spatial distribution of faunal density and organic carbon in Summer season

Shriadah *et al.*, 1999). However, the high temperature could be the reason of low dissolved oxygen in shore regions of present investigation.

The organic carbon, nitrate and nitrite were observed with maximum levels in station 7 and near by areas are due to high flourished growth of seaweeds and sea grass, both of these caused higher nutrients level in water. This could be the reason of high density faunal assemblage (Plate 7.4). Evidently lower pH in station 7 and near by areas was mainly due to decomposition of organic material come from detritus and derivatives of seaweeds and sea grass (Badarudeen *et al.*, 1996 and Shriadah, 2000). The high values of nitrate and nitrite were correlated with organic carbon, which was high in seaward side in the present investigation as also clearly documented by Balakrishanan Nair *et al.* (1983, 1984) in the Ashtamudi estuary and Tam *et al.* (1995) in Hong Kong mangroves. Dye (1983) reported that the plant density is a determining factor in meiofaunal distribution. But lower levels of organic carbon, nitrate, nitrite and silicate were observed in station 4 is owing to absence of seaweeds and sea grass (Plates 7.1, 7.2 and 7.4). The phosphate was observed with controversial to other nutrients had high level in station 4 could be due to the availability of more darters (snake head bird) and sea birds. So the present study concluded that the nematodes distributed in high density at station 7 and near by stations. Food acts as a limiting factor in the distribution and abundance of benthic faunal resources (Ingole *et al.*, 1987). Organic carbon serves as a food sources for many meiobenthic organisms including nematodes (Coull, 1973). High density of nematodes was observed with high content of organic carbon and organic matter in Mahanadi system, east coast of India (Sarma and Wilsanand, 1994). The large amount of organic matter (high in monsoon) being converted into organic carbon by fungal and bacterial population, furnishes a rich source of food to polychaetes (Sunilkumar and Antony, 1994 and Chakraborty and Choudhury, 1997).

Table 7.3: List of Polychaetes Recorded in Twelve Stations during January 2006–December 2006

Sl.No.	Species	Stations											
		1	2	3	4	5	6	7	8	9	10	11	12
	Pisionidae												
1.	*Pisionicdens indica*		★				★		★		★	★	★
	Pilargidae												
2.	*Ancistrosyllis* sp.				★				★	★			
3.	*Thalasapia annandalai*	★	★				★						
	Syllidae												
4.	*Syllides fongocirrats*				★			★		★		★	★
5.	*Irmula spissipes*		★		★	★			★				
6.	*Autolycus prolifer*			★	★	★			★		★	★	
7.	*Syllis cornuta*								★				
	Nereidae												
8.	*Dendronereis aestuarina*	★	★			★	★	★	★	★	★	★	★
9.	*Nereis glandicincta*	★	★			★	★	★	★	★	★	★	★
10.	*Leonnates decipiens*	★			★			★					★

Contd...

Table 7.3–*Contd...*

Sl.No.	Species	Stations											
		1	2	3	4	5	6	7	8	9	10	11	12
	Glyceridae												
11.	*Glycera alba*			★		★	★	★	★	★	★	★	★
12.	*G. prashachi*								★	★		★	
13.	*G. tessellata*			★			★			★	★	★	
	Eunicidae												
14.	*Lumbriconereis laterlli*	★				★	★		★		★		
15.	*Eunice tubifex*		★				★	★	★				
16.	*Diopatra neapolitana*	★	★	★		★		★	★	★	★	★	
17.	*Marphysa gravelyi*	★	★	★				★	★	★	★	★	★
	Spiondae												
18.	*Polydora kempi*	★	★	★								★	★
19.	*Malacorceres indicus*	★	★	★	★		★						★
20.	*Leonates* sp.		★			★	★	★	★				
	Cossuridae												
21.	*Cossura coasta*	★	★	★		★		★					
22.	*C. delta*	★						★		★	★	★	
	Owenidae												
23.	*Owenia fusiformis*				★	★	★	★	★	★			
	Ophelidae												
24.	*Polyophthalmus pictus*	★	★	★		★		★	★				★
25.	*Armandia longicaudatus*	★						★	★	★	★	★	
	Maldanidae												
26.	*Maldane sarsi*		★				★	★	★				
27.	*Euclymene annadalei*	★	★	★		★		★	★				
	Cirratulidae												
28.	*Raphidrilus nemasoma*		★				★	★	★	★	★	★	★
29.	*Cirratulus cirratus*		★		★	★	★	★	★	★			
	Sabellidae												
30.	*Laonome indica*		★	★		★	★	★	★	★		★	★

In the present study, a total of 30 polychaete species belonged to 13 families were identified in twelve random stations. Of this, a species belongs to Pisionidae, Owenidae, and Sabellidae, 3 Nereidae, 4 Syllidae, 3 Glyceridae, 4 Eunicidae, 3 Spiondae, 2 Cossuridae, Ophelidae, Pilargidae, Maldanidae and Cirratulidae (Table 7.3). Among all the polychaetes families, Eunicidae (15.43 per cent), Nereidae (13.71 per cent), Glyceridae (9.71 per cent) Spiondae (9.14 per cent), Syllidae (9.14 per cent) were dominant families. The families like Pisionidae, Owenidae were less number of species and density with the least percentage of 3.43 per cent (Figure 7.7). Dominant polychaetes species were *Nereis glandicincta, Marphysa gravelyi, Dendronereis aestuarina, Glycera alba, Diopatra neapolitana* and *Laonome indica* throughout the year. These species are found to be euryhaline forms except *G. alba* and *D. neapolitana*, which are stenohaline in nature. The influence of environmental parameters especially salinity on the distribution of benthic organisms has been reported for Cochin backwaters (Devi and Venugopal, 1989 and Sunilkumar and Antony, 1994). Among these, *N. glandicincta, D. aestuarina*, were observed in 10 stations. *M. gravelyi, G. alba, D. neapolitana* and *L. indica* were available in 9 stations shows the dominancy and preference of substratum.

The species in the present study *M. gravelyi, N. glandicincta* and *D. aestuarina* were prominent species recorded in Cochin backwaters (Sunilkumar, 1995). *Glycera alba* and *D. neapolitana*

Figure 7.7: Percentage Occurrence of Polychaete Fauna in different Families.

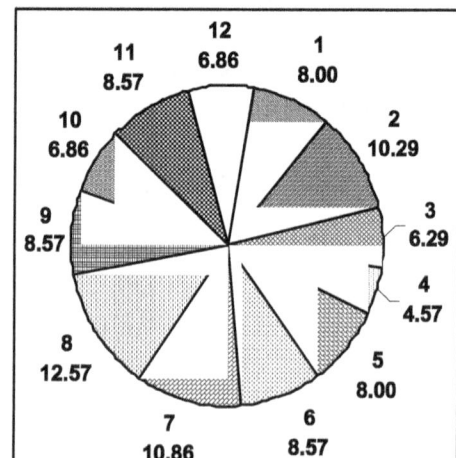

were dominant species recorded in Mandovi estuary (Ansari *et al.*, 1986) and *G. alba* was the dominant species in Coleroon estuary (Prabha Devi, 1994). The reason of dominancy could be the protection against desiccation due to firm substrate provided by roots of seaweeds and seagrass (Mishra and Choudhury, 1985). Untawale and Parulekar (1976) reported the dominance of polychaetes in the mangrove fauna and the less number of molluscs in the silty clay substratum.

Station 8 (12.57 per cent) shown higher percentage of polychaetes than station 7 (10.86 per cent) followed by station 2 (10.29 per cent). Higher abundance of polychaetes in stations 7 and 8 could be due to high content organic carbon (Figure 7.8). But station 2, even observed with unfavourable and fluctuating environmental characteristics, the dominancy leads to potential ability to colonize stressed environments (Raveenthiranaath Nehru, 1990) by mucus-secreting devices to protect themselves (Sadhana, 1993).

The species density (No./m^2) in all stations varied from 173 (Summer) to 497 (Monsoon) showed the dominance of Glyceridae (32–111) superseded by Syllidae (26–88), Pilargidae (15–45), Eunicidae (15–41) and Pisionidae (12–33) (Figure 7.9). Similar results were reported by Sankar (1998) in Muthupet lagoon, Sunilkumar (1995) in Cochin backwaters, Prabha Devi (1994) in Coleroon estuary and Ansari *et al.* (1986) in Mandovi estuary, Athalye and Gokhale (1998) in Thane creek, Bombay. This could be due to organic loading in the sea in monsoon season and subsequent proliferation of polychaetes and diminishing in summer due lack of nutrients. The spatial observation polychaetes in four seasons at different stations varied from 6 No./m^2 (summer) to 73 No./m^2 (monsoon) (Plate 7.5).

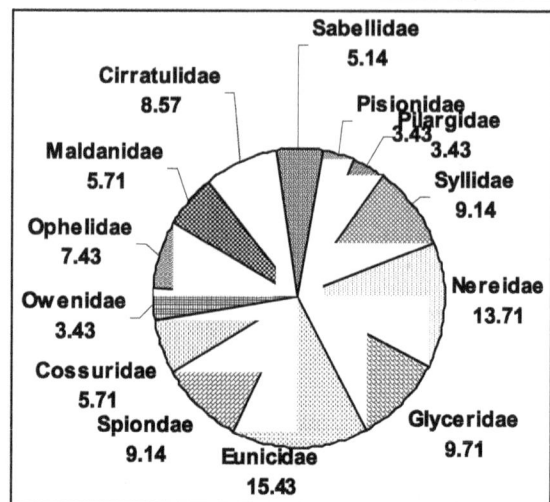

Figure 7.8: Percentage Occurrence of Polychaete Fauna in different Stations.

**Plate 7.5: Polychaete Density and Organic Carbon in different Seasons,
(a) Monsoon, (b) Post monsoon, (c) Pre monsoon, (d) Summer.**

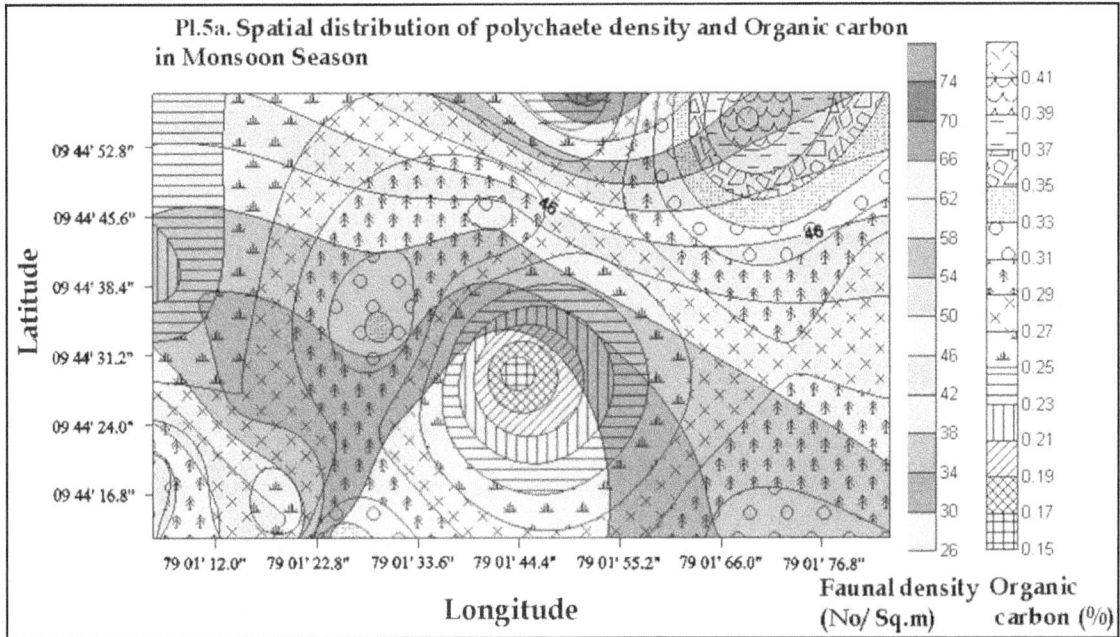

Pl.5a. Spatial distribution of polychaete density and Organic carbon in Monsoon Season

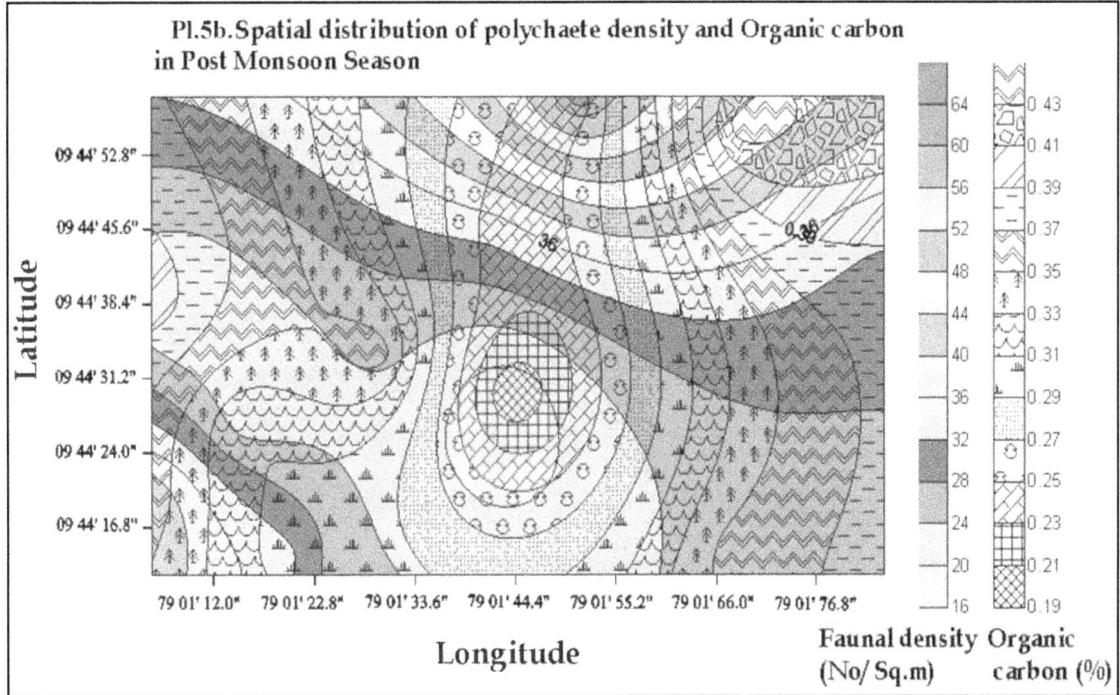

Pl.5b. Spatial distribution of polychaete density and Organic carbon in Post Monsoon Season

Contd...

Plate 7.5–*Contd...*

Pl.5c. Spatial distribution of polychaete density and Organic carbon in Pre monsoon Season

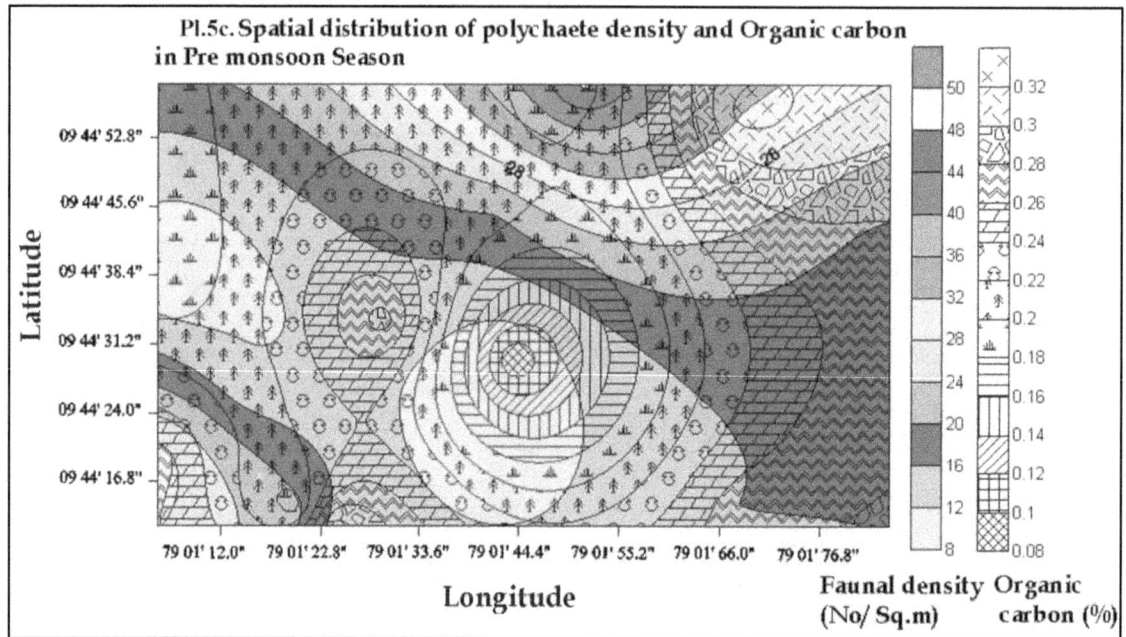

Faunal density (No/ Sq.m) Organic carbon (%)

Pl.5d. Spatial distribution of polychaete density and Organic carbon in Summer Season

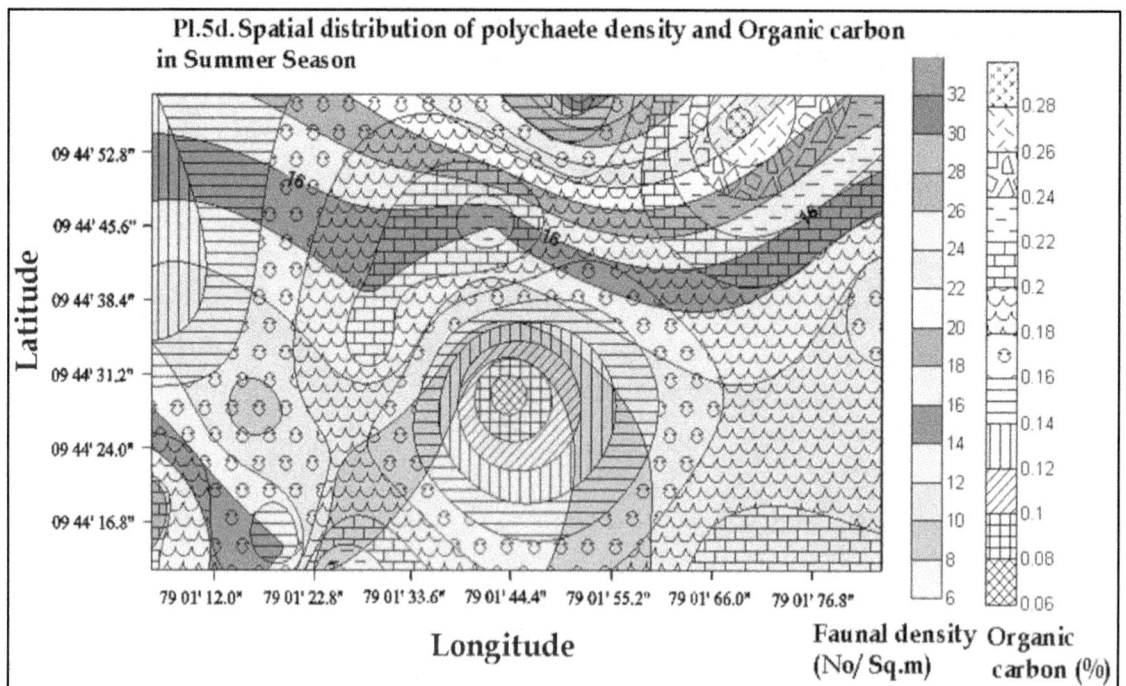

Faunal density (No/ Sq.m) Organic carbon (%)

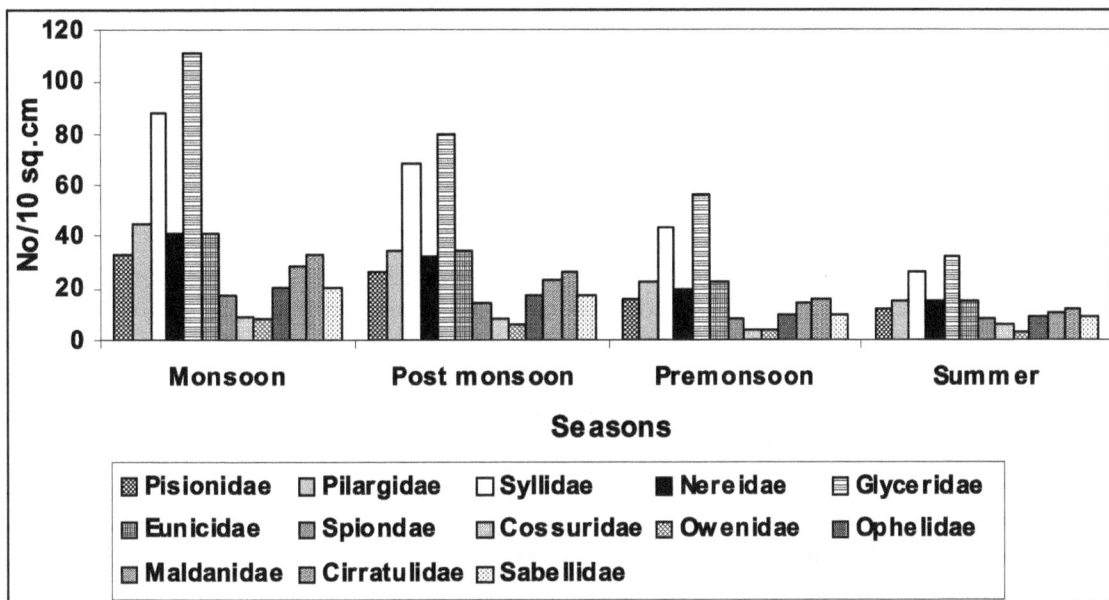

Figure 7.9: Abundance of Polychaetes in Thondi Sub Tidal Area.

It probably be due to large growth of seaweeds and sea grass in the substratum contributed high organic carbon, nitrate and nitrite and silicate with moderate phosphate level. But in shore region, the high temperature, salinity, pH and less dissolved oxygen are limiting factors for benthic faunal resources. They also favourable in seaward side are the other one reason of faunal flourishment.

Acknowledgments

The authors are grateful to authorities of Alagappa University for providing facilities to carry out the work.

References

Alongi, D.M., 1987a. Inter-estuary variation and intertidal zonation of free–living nematode communities in tropical mangrove systems. *Mar. Ecol. Prog. Ser.*, 40: 103–114.

Alongi, D. M., 1987b. Intertidal zonation and seasonality of meiobenthos in tropical mangrove estuaries. *Mar. Biol.*, 95: 447–458.

Anbuchezhian, R. M., S. Ravichandran, P. Murugesan and J. Sesh Serebiah, 2012. Assessment of soft bottom polychaete diversity in Thondi, Palk bay, India. *J. Environ. Biol.*, 33: 917–921.

Ansari, Z. A. and Parulekar, A. H., 1998. Community structure of meiobenthos from a tropical estuary. *Indian J. Mar. Sci.*, 27: 362–366.

Ansari, Z. A., Ingole, B. S., Banerjee, G. and Parulekar, A. H., 1986. Spatial and temporal changes in benthic macrofauna from Mandovi and Zuari estuaries of Goa, West coast of India. *Indian J. Mar. Sci.*, 15: 223–229.

Ansari, Z. A., R. A. Sreepada and A. Kanti, 1994. Macrobenthic assemblage in the soft sediment of Marmugao harbour, Goa (Central Westcoast of India). *Indian J. Mar. Sci.*, 23: 225–231.

Arvanitidis, C., G. Hatzigeorgiou, D. Koutsoubas, C. Dounas, A. Eleftheriou and A. P. Koulouri, 2005. Mediterranean lagoons revisited: Weakness and efficiency of the rapid biodiversity assessment technique in severely fluctuating environment. *Biodiversity Conserv.*, 14: 2347–2359.

Athalye, R. P. and K. S. Gokhale, 1998. Macrobenthos from the mudflats of Thane creek, Maharashtra, India. *J. Bombay Nat. Hist. Soc.*, 95: 259–266.

Badarudeen, A., K. T. Damodaran, K. Sajan and D. Padmalal, 1996. Texture and geochemistry of the sediments of a tropical mangrove ecosystem, southwest coast of India. *Environ. Geol.*, 27: 164–169.

Balakrishnan Nair, N., K. Dharmaraj, P. K. Abdul Azis, M. Arunachalam, K. Krishnakumar and N. K. Balasubramanian, 1984. Ecology of Indian estuaries: VIII. Inorganic nutrients in the Ashtamudi estuary. *Mahasagar–Bull. Natn. Inst. Oceanogr.*, 17: 19–32.

Balakrishnan Nair, N., P. K. Abdul Azis, K. Dharmaraj, M. Arunachalam, K. Krishnakumar and N. K. Balasubramanian, 1983. Ecology of Indian estuaries: Part I. Physico–chemical features of water and sediment nutrients of Astamudi estuary. *Indian J. Mar. Sci.*, 12: 143–150.

Balasubramanyan, K., 1994. Microinvertebrate benthic fauna of Pitchavaram mangroves. In: A Training Manual (Eds). Sanjay V. Deshmukh and V. Balaji, Conservation of Mangroves Forest Genetic Resources: M. S. Swaminathan Research Foundation, Chennai, India, 257–259.

Bell, S. S., 1979. Short and long term variation in a high marsh meiofauna community. *Estuarine Coastal Mar. Sci.*, 9: 331–350.

Chakraborthy, S. K., T. K. Poddar and A. Choudhury, 1992. Species diversity of macrozoobenthos of Sagar Island, Sunderbans, India. *Proc. Zool. Soc., Calcutta*, 45: 435–444.

Chakraborty, S. K. and A. Choudhury, 1994. Community structure of macrobenthic polychaetes of intertidal region of Sagar Island, Hoogly estuary, Sunderbans, India. *Tropical Ecology*, 35: 97–104.

Chakraborty, S. K. and A. Choudhury, 1997. Occurrence and abundance of benthic polychaetes in Hooghly estuary, Sagar Island, Sunderbans, India. *J. Mar. Biol. Ass. India*, 39: 140–147.

Chinnadurai, G. and O. J. Fernando, 2006. Meiobenthos of Cochin mangroves (Southwest coast of India) with emphasis on free–living marine nematode assemblages. *Russian J. Nematology*, 14: 127–137.

Clarke, K. R., 1993. Non parametric multivariate analyses of changes in community structure. *Aust. J. Ecol.*, 18: 117–143.

Coull, B. C., 1973. Estuarine meiofauna a review, tropic relationship and microbial ecology. L. H. Stevenson and Colwell (Eds.). University of South Carolina Press, Columbia, 449–511.

Coull, B. C., 1990. A members of the meiofauna food for higher trophic levels? *Trans. Am. Microsc. Soc.*, 109: 233–246.

Dauvin, J. C. E., J. L. Thiebaut, K. G. Gesteira, F. Ghertsos, M. Gentil, and B. S. Ropert, 2004. Spatial structure of a subtidal macrobenthic community in the Bay of Veys (western Bay of Seine, English Channel. *J. Exp. Mar. Biol. Ecol.*, 307: 217–235.

Dernie, K. M., M. J. Kaiser, E. A. Richardson and R. M. Warwick, 2003. Recovery of soft sediment communities and habitats fol lowing physical disturbance. *J. Exp. Mar. Biol. Ecol.*, 286: 415–434.

Devi, K. S. and P. Venugopal, 1989. Benthos of Cochin backwaters receiving industrial effluents. *Indian J. Mar. Sci.*, 18: 165–169.

Dogan, A., M. E. Cinar, M. Onen, Z. Ergen and T. Katagan, 2005. Seasonal dynamics of soft bottom zoobenthic communities in polluted and unpolluted areas of Izmir Bay (Aegean Sea). *Senckenb. Marit.*, 35: 133–145.

Dye, A. H., 1983. Composition and seasonal fluctuations of meiofauna in a southern African mangrove estuary. *Mar. Biol.*, 73: 165–170.

Eaton, L., 2001. Development and validation of biocriteria using benthic macro invertebrates for North Carolina estuarine waters. *Mar. Pollut. Bull.*, 42: 23–30.

El–Wakeel, S. K. and J. P. Riley, 1956. The determination of organic carbon in marine muds. *Journal du conseil permament Inti. Pourl. Exploration de de mer.*, 22: 180 –183.

Ganesh, T. and A. V. Raman, 2007. Macrobenthic community structure of the northeast Indian slope, Bay of Bengal. *Mar. Ecol. Prog. Ser.*, 341: 59–73.

Geetanjali, S. K., Malhatra, Anshu Malhotra, Zakir Ansari and Anil Chattergi, 2001. Role of nematodes as bioindicators in marine and freshwater habitats. *Curr. Sci.*, 82: 10.

Golding, Q., V. Mishra, V. Uillal, R. P. Athalye and K. S. Gokhle, 1996. Meiobenthos of mangrove mudflats from shallow region of Thane creek. Central West Coast of India. *Indian J. Mar. Sci.*, 25: 137–141.

Grall, J. and M. Glemarec, 1997. Using biotic indices to estimate macrobenthic community perturbation in the Bay of Brest. *Estuarine Coastal Shelf Sci.*, 44: 43–53.

GUIDE (Gujarat Institute of Desert Ecology), 2000. An Ecological study of Kachchh mangroves and its associated fauna with reference to management and conservation. *Gujarat Institute of Desert Ecology*, 1–75.

Harikantra, S. N., Z. A. Ayyappan, Z. A. Ansari and H. Parulekar, 1980. Benthos of the shelf region along the west coast of India. *Indian J. Mar. Sci.*, 9: 106–110.

Harkantra, S. N. and N. R. Rodriguez, 2004. Numerical analyses of soft bottom macro invertebrates to diagnose the pollution in tropical coastal waters. *Environ. Monit. Assess.*, 93: 251–275.

Hutchings, P., 1998. Biodiversity and functioning of polychaetes in benthic sediments. *Biodiversity Conserv.*, 7: 1133–1145.

Ingole, B. S. and A. H. Parulekar, 1998. Role of salinity in structuring the intertidal meiofauna of a tropical estuarine beach: Field evidence. *Indian J. Mar. Sci.*, 27: 356–361.

Ingole, P. S., Z. A. Ansari and A. H. Parulekar, 1987. Meiobenthos of Saphala salt marsh, west coast of India. *Indian J. Mar. Sci.*, 16: 110–113.

Jacob, R. G., V. C. Anderlini, M. A. Zarba and O. S. Mohammed, 1982. Annual Research Report. Kuwait Institute for Scientific Research, Kuwait, 141 pp.

Jagadeesan, P. and K. Ayyakkannu, 1992. Seasonal variation of benthic fauna in marine zone of Coleroon estuary and inshore waters, south east coast of India. *Indian J. Mar. Sci.*, 21: 67–69.

Jose, J. and M. S. Rajagopalan, 1993. Studies on a mangrove habitat dominated by *Bruguiera* spp. Mariculture Research under the Postgraduate Programme in Mariculture. Part–3, K. A. Rengarajan, Noble Prathibha, V. Kripa, N. Sridhar and M. Zakhriah, (Eds.), Cochin, India, CMFRI, 54: 78–84.

Kailasam, M. and S. Sivakami, 2004. Effect of thermal effluent discharge on benthic fauna off Tuticorin bay, south east coast of India. *Indian J. Mar. Sci.*, 33: 194–201.

Kathiresan, K., N. Rajendran, V. Palaniselvam and T. Ramanathan, 2000. Macrofauna population in a mangrove nursery of *Rhizophora apiculata* Blume. *Environ. Ecol.,* 18: 230–232.

Kress, N., B. Herut and B. S. Galil, 2004. Sewage sludge impact on sediment quality and benthic assemblages off the Mediterranean coast of Israel–a long-term study. *Mar. Environ. Res.,* 57: 213–233.

Kurian, C. V., 1972. Ecology of benthos in a tropical estuary. *Proc. Indian Natn. Sci. Acad.,* 38: 156–163.

Lee, S. Y., 1997. Potential tropic importance of the faecal material of the mangrove sesarmine crab *Sesarma messa. Mar. Ecol. Prog. Ser.,* 159: 275–284.

Lee, S. Y., 2009. Mangrove macrobenthos: Assemblages, services and linkages. *J. Sea Res.,* 9: 16–29.

Mishra, A. and A. Choudhury, 1985. Proceedings of national symposium on biology, utilization and conservation of mangroves, 448 pp.

Musco, L. and A. Giangrande, 2005. Mediterranean Syllidae (Annelida: polychaeta) revisited: Biogeography, diversity and species fidelity to environmental features. *Mar. Ecol. Prog. Ser.,* 304: 143–153.

Mutlu, E., M. E. Cinar and M. B. Ergev, 2010. Distribution of soft-bottom polychaetes of the Levantine coast of Turkey, eastern Mediterranean Sea. *J. Mar. Sys.,* 79: 23–35.

Nilsson, P., K. Sundback and B. Jonsson, 1993. Effect of the brown shrimp *Crangon crangon* L. Onendobenthic macrofauna, meiofauna and meiofaunal grazing rates. *Neth. J. Sea. Res.,* 31: 95–106.

Parulekar, A. H. and A. B. Waugh, 1975. Quantitative studies on the benthic macrofauna of northeastern Arabian sea shelf. *Indian J. Mar. Sci.,* 4: 174–176.

Parulekar, A. H. and Z. A. Anzari, 1981. Benthic macrofauna of the Andaman sea. *Indian J. Mar. Sci.,* 10: 280–287.

Platt, H. M. and R. M. Warwick, 1980. The significance of nematodes to the littoral ecosystem. In: The Shore Environment, J. H. Price, D. E. C. Irvine and W. H. Franham (Eds.), Ecosystems, Academic Press, London, 2: 729–759.

Pocklington, P. and P. G. Wells, 1992. Key taxa for marine environmental quality monitoring. *Mar. Pollut. Bull.,* 24: 593–598.

Prabha Devi, L., 1994. Ecology of Coleroon estuary: Studies on benthic fauna. *J. Mar. Biol. Ass. India,* 36: 260–266.

Prabha Devi, L. and K. Ayyakkannu, 1989. Macrobenthos of the Buckingham canal and beachwaters of Coleroon estuary. *J. Mar. Biol. Ass. India,* 31: 80–85.

Prabha Devi, L., P. Natarajan, G. Saraswathy Ammal and P. K. Abdul Azis, 1996. Water quality and benthic fauna of the Kayamkulam backwaters and Arattupuzha coast along southwest coast of India. *Indian J. Mar. Sci.,* 25: 264–267.

Rao, D. G. and S. Satpathy, 1996. Demoecology of kinorhyncha of Chilka lagoon (Bay of Bengal). *J. Mar. Biol. Ass. India,* 38: 15–24.

Raveenthiranath Nehru, 1990. Ecology of macrobenthos in and around Mahendrapalli region of Coleroon estuary, southeast coast of India. Ph. D., Thesis, Annamalai University, India, 219 pp.

Ravichandran, S. and G. Rameshkumar, 2008. Macrobenthos–indicator of the wellbeing of mangroves. Seshaiyana, 16, 4.

Ray, S., R. E. Ulanowicz, N. C. Majee and A. B. Roy, 2000. Network analysis of a benthic food web model of a partly reclaimed island in the Sundarban mangrove ecosystem, India. *J. Biol. Sys.*, 8: 263–278.

Sadhana, R., 1993. Ecology of macrobenthos in lower reaches of river Kaveri, southeast coast of India. Ph. D., Thesis, Annamalai University, India, 163 pp.

Sajan, S. and R. Damodaran, 2007. Faunal composition of meiobenthos from the continental shelf regions off the west coast of India. *J. Mar. Biol. Ass. India*, 49: 19–26.

Salzwedal, H., E. Racher and D. Gerdes, 1985. Benthic macrofauna community in the German bight. *Veroff. Inst. Meeresforch. Bremerh*, 20: 199–267.

Samuelson, G. M., 2001. Polychaetes as indicators of environmental disturbance on sub-artic tidal flats, Iqaluit, Baffin Island, Nunavut territory. *Mar. Pollut. Bull.*, 42: 741–773.

Sankar, G., 1998. Studies on the hydrobiology, benthic ecology and fisheries of Muthupet lagoon, India. Ph. D., Thesis, Annamalai University, India, 105 pp.

Santhakumaran, L. N. and S. G. Sawant, 1994. Observation on the damage caused by marine fouling organisms to mangrove saplings along Goa coast. *J. Tim. Dev. Ass. India*, 40: 9–19.

Saravanakumar, A., J. Sesh Serebiah, G. A. Thivakaran and M. Rajkumar, 2007. Benthic macrofaunal assemblage in the arid zone mangroves, Gulf of Kachchh–Gujarat. *J. Ocean Univ. China*, 6: 303–309.

Sarma, A. L. N. and V. Wilsanand, 1994. Littoral meiofauna of Bhitarkanika mangroves of River Mahanadi system, east coast of India. *Indian J. Mar. Sci.*, 23: 221–224.

Schrijvers, J., H. Fermon and M. Vincx, 1996. Resource competition between macrobenthic epifauna and infauna in a Kenyan *Avicennia marina* mangrove forest. *Mar. Ecol. Prog. Ser.*, 136: 123–135.

Sesh Serebiah, J., 2003. Studies on Benthic faunal assemblage on Mangrove environment of Jakhau, Gulf of Kachchh – Gujarat. Ph. D., Thesis, Annamalai University, India, 147 pp.

Seshappa, G. 1953. Observation on the physical and biological features of the inshore sea bottom along the Malabar coast. *Proc. Natn. Inst. Sci. India*, 19: 257–279.

Shriadah, M. A., 2000. Chemistry of the mangrove waters and sediments along the Arabian Gulf shoreline of the United Arab Emirates. *Indian J. Mar. Sci.*, 29: 224–229.

Shriadah, M. A., and Saif M. Al–Ghais, 1999. Environmental characteristics of the United Arab Emirates waters along the Arabian Gulf: Hydrographical survey and nutrient salts. *Indian J. Mar. Sci.*, 28: 225–232.

Solis–Weiss, V., F. Aleffi, N. Bettoso, P. Rossin, G., Orel and S. Fonda–Umani, 2004. Effects of industrial and urban pollution on the benthic macrofauna in the Bay of Muggia (industrial port of Trieste, Italy). *Sci. Tot. Environ.*, 328: 247–263.

Strickland, J. D. H. and T. R. Parsons, 1972. A practical handbook of seawater analysis. *Bull. Fish. Res. Bd. Canada*, 310 pp.

Sultan Ali, M., K. Krishnamurthy and M. J. Prince Jeyaseelan, 1983. Energy flow through the benthic ecosystem of the mangroves with special reference to nematodes. *Mahasagar–Bull. Natn. Inst. Oceanogr.*, 16: 317–325.

Sunilkumar, R., 1995. Macro benthos in the mangrove ecosystem of Cochin backwaters, Kerala (Southwest coast of India). *Indian J. Mar. Sci.*, 24: 56–61.

Sunilkumar, R., 1997. Vertical distribution and abundance of sediment dwelling macro invertebrates in an estuarine mangrove biotope, southwest coast of India. *Indian J. Mar. Sci.*, 26: 26–30.

Sunilkumar, R., 2001. Intertidal zonation and seasonality of benthos in a tropical mangrove. *Intl. J. Ecol. Environ. Sci.*, 27: 199–208.

Sunilkumar, R. and A. Antony, 1994. Impact of environmental parameters on polychaetous annelids in the mangrove swamps of Cochin, southwest coast of India. *Indian J. Mar. Sci.*, 23: 137–142.

Tam, M. F. Y., S. H. Li, C. Y. Lan, G. Z. Chen, M. S. Li and Y. S. Wong, 1995. Nutrients and heavy metal contamination of plants and sediments in Futian mangrove forest. *Hydrobiologiai*, 295: 149–158.

Toepfer, C. S. and J. W. Fleeger, 1995. Diet of juvenile fishes *Citharichthys spilopterus, Symphurus plaginsa* and *Cobionellus boleosoma. Bull. Mar. Sci.*, 56: 238–249.

Untawale, A. G. and A. H. Parulekar, 1976. Some observation on the ecology of an estuarine mangrove of Goa. *Mahasagar–Bull. Natn. Inst. Oceanogr.*, 9: 57–62.

Venkatesh Prabhu, H., A. C. Narayana and R. J. Katti, 1993. Macrobenthic fauna in near shore sediments of Gangolli, west coast of India. *Indian J. Mar. Sci.*, 22: 168–171.

Zhang, Z. N., K. X. Lin, H. Zhou and R. Z. Wang, 2004. Abundance and biomass of meiobenthos in autumn and spring in the East China Sea. *Acta Ecol. Sin.*, 24: 997–1005.

Chapter 8

Gears and Crafts of Mangroves of Karwar, West Coast of India

☆ *B. Vasanthkumar*

ABSTRACT

The present study conducted at the mangrove ecosystem of Kali estuary. The fishing methods practiced in Kali estuary are conventional in nature and diversified. The important indigenous gears (nets) used in this estuary are gill nets, cast nets, lines-pole and line, hook and line, Scare lines, scoop nets, bag nets and drag nets. Besides this, crab fishing and clam fishing are also done here.

Keywords: Karwar, Kali river, Fishing gears and Crafts.

Introduction

Fishing is the mechanism of capturing fish from the wild. Fishing started long before man existed on earth. In olden days, man selected places near good hunting or good fishing areas and used arrows or spears or traps to capture fishes well as birds or mammals when man began grow plants and near animals, he continued to hunt and fish especially during seasons when no crops were available. The development of modern fishing practices began with improvements in sailing crafts, navigation and crafts.

Uttara Kannada district of Karnataka State has vast coastline of 144 km, which is blessed with four major riverine systems like Kali, Ganagavali, Aghanashini and Sharavathi. These rivers help in escalating the productivity and fishery resources of the Arabian Sea, west coast of India. This coast is known as "Mackerel coast" on fishing atlas besides this the oil sardine is another major fishery is established in this coast and earning good revenue.

There is no well-organized shellfishery is observed in the River Kali since olden days. But, many fisherman families are solely depended on this fishery for their livelihood. They use different type of gears (nets) for fishing purpose such as pole and line, hook and line, Scare lines, Drag nets, Bag nets (in sluice gates), cast nets, drift gill net, set gill net etc. Generally they use dug out canoe, plank-built boat and FRP boats for this purposes. In case of clam fishery, they use scoop net and rakes whereas collecting oysters, they use knife and chisel to detach them from the submerged rocks. During low tide period more than hundred to two hundred people will be engaged in this clam fishery, each person collects about 25 to 50 kg of clams per day.

The capture fisheries of this area are mainly supported by pelagic and benthic fish which are exploited by different types of gears and crafts. A knowledge on fisheries gear, craft and fishing methods is very much essential for scientific and judicious exploitation and management of any capture fishery. There is hardly any published data about the different fishing method of Kali estuary and its adjoining water except few studies by Neelakantan (1981), Naik and Neelakantan (1988) and Mallikesi (2005). The latter study was mainly focused on the shell fish (clams) of the Kali estuary.

Results and Discussion

More commonly used crafts in this estuary/mangrove areas are dug- out canoe and plank built boat. These days the FRP boats are also made use in fishing activities. The dug-out canoe is smaller type of boat used in the estuary, backwaters and mangrove areas and in river. This is cutout of a single log and is usually called as dug-out canoe and locally as 'dhoni'. The size of the craft varies from 1.5 to 2.15 m in length. The slightly bigger craft will have size of 5.5 × 0.9 × 0.6 m manned by 4-5 persons and is generally used in the inshore waters as a scout boat. The slightly higher version of this craft is the plank-built boat having a size of 13.6 × 3.04 × 0.6 m with a crew of 10-12 men and is used for carrying the bigger net. The craft is propelled with oars and these days it is manoeuvered by outboard or inboard engines. This craft is supported by side balancer called as out-rigger for safety purposes.

The fishing season of Karwar region commences from September and lasts till May. The main type of crafts are used during this period are plank built boat, out rigger boat, dug-out canoe etc. But, in the river Kali, fishing is done throughout the year irrespective of time and tide. Some of the common gears that are used in this biotope are explained hereunder.

Drag Net

This is made of small piece of nylon net (3 × 8′) with varying dimension mainly used to catch fish seeds and smaller fishes. The length of the net is double the width. The gear is operated by two persons at time holding at both ends and dragged it above the bed for about 10-20 minutes interval.

Gill Net

These are passive gears and are generally classified as surface, mid water and bottom gillnets and are generally operated during the night hours. The nets are allowed to drift along with the wind, tide or current or the nets are set at a particular depth by anchoring and are likewise referred as drift gill net, set gill net, bottom gill net respectively. The nets are of different dimensions depending on the type of fishes to be caught. Locally these nets are called as 'patte bale'.

Cast Net

This gear is operated from the shore, elevated platform or from a boat in estuarine waters and are locally known as 'beesu bale'. Two types of net are used, one stringless with peripheral pockets and

Figure 8.1: Operation of Drag Net.

the other stringed without pockets while the latter is more common in this region. Mesh size of the gear varies depending on the type of fish to be caught. This type of gear generally used to catch the mullets, *Sillago* sp., silver bellies etc.

Scoop Net

It is a piece of nylon net attached loosely to a rectangular wooden or metal frame and is used to catch shellfishes in shallow waters during the low tide period. This net is used throughout the year and is a common sight during southwest monsoon season.

Bag Net

This is a specially designed gear to collect the clams apart from the hand picking method. It consists of a semi-circular frame attached by horizontal bar with spikes and behind this is a conical bag net is fixed. This bag net is dragged or scooped over the bottom of the river/estuarine bed to collect the clams. This net is also used during the low tide period and is being operated throughout the year in this river basin.

Figure 8.2: A Fisherman Spreading the Cast Net.

Ghol Net

It is another type of conical shaped bag net exclusively used in the backwater area especially in the sluice gate of a culvert. It measure about 2.5–3.25 m in length and mesh size decreases from the mouth region to the cod end. This net is generally operated during the night hours of new moon or full moon period.

Scare Line

These are not the gears as such but are used as accessories in other fishing methods like cast nets and bag net operations. Here the Palm (coconut) leaves are tied in single line to the coir or nylon rope at regular intervals and are used to scare the fish shoals to move towards the area of operation of cast net and bag nets. Mullets are mainly caught by this method.

Hook and Line

In this method, a different numbered hooks with baits are tied to the line to catch varities of fish. It can be operated by single person either from the shore or from the dug-out canoe. Different types of baits are used depending on the type of fish to be caught. For catching the *Sillago sihama* (ladyfish),

Figure 8.3: Long Line Fishing in Progress.

Gerres fialmentosus fishes, baits such as wheat flour, annelid worms, clams and mussels are used. Algal filaments are used to catch the fishes like *Teuthes vermiculatis* and *E. suratensis* species.

Pole and Line

This is another version of line fishing where hook and line are tied to the piece of bamboo pole with a line measuring the length 1-2.75 m so that fishes can be caught from the distance without using the canoe. It is also used in the small boat/dug-out canoe etc.

References

Banse, K., 1959. On upwelling and bottom trawling off the southwest coast of India. *J. Mar. Biol. Ass. India*, 1: 33–49.

Kusuma, N., 1983. Biology of *Johnius belangerii* (Cuvier) with notes on the sciaenid fishery of the North Kanara coast. *Ph.D. Thesis in Marine Biology*, Karnatak University, Dharwad.

Mallikesi, V.B., 2005. Shell industry in Karwar taluka: Prospects for marketing and management. *UGC Minor Research Project*.

Naik, Rajesh, 1994. Studies on the distribution, abundance and composition of finfishes along Kali river. *M.Sc. Dissertation Thesis*, Karnatak University, Dharwad.

Naik, U.G. and Neelakantan, B., 1988. Gears and crafts of Karwar: An overview. *J. Ind. Fishries Assoc.*, 18: 245–252.

Naik, U.G., Reddy, C.R., Shetty, D.C. and Neelakantan, B., 1990. Plankton of Karwar waters with remarks on hydrographic conditions and fishery. *Fishery Technology*, 27: 98–102.

Nelakantan, B., 1981. Studies on biology of *Lactarius lactarius*, false travelly, with notes on the sciaenid fishery of the North Kanara coast. *Ph.D. Thesis in Marine Biology*, Karnatak University, Dharwad.

Payne, A.I., 1986. *The Ecology of Tropical Lakes and Rivers*. John Wiley and Sons, New York, 301 pp.

Thurman, H.V., 1997. *Introductory Oceanography*. Prentice Hall College, New Jersey, USA.

Chapter 9

Age and Growth of *Chicoreus virgineus ponderosus* and *Siratus virgineus ponderosus* (Gastropoda : Muricidae) from Thondi Coast, Palk Bay in Tamil Nadu

☆ *R. Ravichandran, D. Chellaiyan, C. Lathasumathi,*
A. Priya and C. Stella

ABSTRACT

Age and growth of *Chicoreus virgineus ponderosus* and *Siratus virgineus ponderosus* were studied using FISAT -1 software. In the present study the results showed more or less similar growth for the males and females during the study period. The age and growth estimation of *Chicoreus virgineus ponderosus* and *Siratus virgineus ponderosus* has been done through several methods and the outcome of one method will act as a check and control over the other. The L_α values obtained in the present study using various methods did not much varied. The values of asymptotic length (L_α) obtained for *Chicoreus virgineus ponderosus* are 115.16 mm for males and 106.70 mm for females and the growth rate (K) for males (0.213 yr^{-1}) and females (0.230 yr^{-1}) are given by k-scan routine in ELEFAN. Similarly the values of asymptotic length (L_α) obtained for *Siratus virgineus ponderosus* are 94.28 mm for males and 95.80 mm for females and the growth rate (K) for males (0.150 yr^{-1}) and females (0.780 yr^{-1}) are given by k-scan routine in ELEFAN.

Keywords: Gastropoda, Muricidae, Age, Growth, Chicoreus species, Siratus species.

Introduction

Age and growth of the animals are interrelated phenomena which denote the duration of life spent by the individual (age) and the increase in its volume of mass (growth) during the corresponding period of its life history. The body form and shell of Muricids may be substantially modified by environmental conditions and its substratum. The influence of the habitat on the shape of the shell in marine gastropod has been studied by several workers. Thivakaran (1988) has studied the population, age structure, growth and longevity of the gastropod *Littorinids* sp. The variation, which is characteristic of the shell in different population of *Nucella lapillus* has been studied by Largen, (1971) and Crothers, (1973, 1974). The age and environmental variations have profound influence the shell growth in Molluscs by Wilbur and Owen, (1964). Stella(1995) studied the age and growth of *Chicoreus virgineus* and *Muricanthus virgineus* in Cuddalore coast.

Growth rates of intertidal grazing gastropods may differ with tide level or microhabitat were studied (Creese 1980, McCormack 1982, Fletcher 1984, Underwood 1984b, Jardine 1985). Food availability (microalgal abundance) has been shown to influence growth rate, density, and competition among grazing gastropods and this can vary with season and tide level (Branch and Branch 1980, Underwood 1984b). Estebenet, and Cazzaniga, (1992) have studied the Growth and Demography of *Pomacea canaliculata* (Gastropoda: Ampullariidae) under laboratory conditions. Estebenet (1998) has studied the food and feeding in *Pomacea canaliculata* (Gastropoda: Ampullariidae). Grana-Raffuccr and Appeldoorn (1997) have studied the age determination of larval strombid gastropods by means of growth increment counts in *statoliths*. Estebenet (1998) has studied the Allometric growth and insight on sexual dimorphism in *Pomacea canaliculata* (Gastropoda: Ampullariidae). Estebenet and Cazzaniga, Estebenet (1998) have studied the Sex related differential growth in *Pomacea canaliculata* (Gastropoda: Ampullariidae).

Estebenet, and Marti (2000) have studied the inter and intrapopulation variation in growth patterns of *Pomacea*. Ismail *et al.* (2000) have studied the populaton structure and shell morphometrics of the corallivorous gastropods *Drupella cornus* in the Gulf of Aqaba. Rumi (2004) has studied the population structure in *Drepanotrema kermatoides* and *D. cimex* (Gastropoda: Planorbidae) in natural conditions. Ismail and Elkarmi (2006) have studied the age, growth and shell morphometrics of the Gastropod *Monodonta dama* (*Neritidae: Prosobranchia*) from the Gulf of Aqaba. Alejandra Rumi, *et al.* (2007) have studied the growth rate fitting using the von Bertalanffy model: analysis of natural populations of *Drepanotrema* sp. snails (Gastropoda: Planorbidae). The gastropod shell is an ideal tool for documenting the growth throughout the life of the animal and the snail grows the aperture deposits new calcium carbonate shell material. This shell acts like a bio-recorder of environmental conditions and the individual's development and the entire ontogeny of a Mollusc is represented in its shell by (Goodwin *et al.,* 2001). Knowledge of the age and growth refers to the relative length and weight of the individuals. This investigation has also been undertaken to provide information on the stocking policy, grouping of the year classes, the dominant year groups, the environmental suitability and the fluctuations in the rate of growth. Studies on age and growth also reveal that the age of sexual maturity and the age at which marketable sizes are attained. Hence the present attempt has been made to study the age and growth of *Chicoreus virgineus ponderosus* and *Siratus virgineus ponderosus* was determined by employing different methods.

Description of the Study Area

During the Quaternary period the Palk Strait must have originated introducing a close connection to the Southern Gulf of Mannar and to the northern Bay of Bengal with in the latitude of 90° and 10° N and longitude of 79° and 80° E. Northern boundary of the strait is of Kodiyakkarai Map. The Southern one is restricted to the Adams Bridge and the Eastern limit is to the Sri Lanka and Thalaimannar region. The Palk Bay is influenced mainly by the North east monsoon. The Bay has strong potential of living and non living resources. Thondi is a small village situated in the Palk Bay region of Tamil Nadu. The study area lies in the latitude of 9°44'N and longitude of 79°19' E. The present study was carried out at Thondi coast, South east coast of India. The species of *Chicoreus virgineus ponderosus* and *Siratus virgineus ponderosus* are exclusively marine in

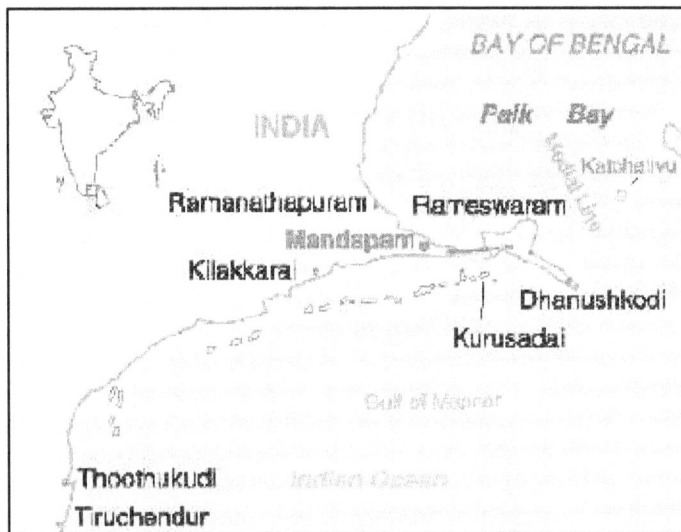

Figure 9.1: Study Area–Palk Bay.

distribution. The collection of these species was made from 10 to 15 fathom lines with muddy bottom. The species live in sandy mud benthic zones. The animals were collected from the fishing trawlers along with other benthic gastropods like *Rapana bulbosa*, *Babylonia spirata*, *Murex trunculus*, *Conus* sps. etc.

Materials and Methods

Random samples of *Chicoreus virgineus ponderosus* and *Siratus virgineus ponderosus,* were collected once in a month from trawlers of Thondi coast for a period of one year (Jan 2006 to December 2006). The shells were measured with the help of Venier caliper to the nearest 0.1 mm. The method of determining growth and longevity in this study was by size frequency analysis of both shell length and weight. Data on shell length were arranged in size groups and plotted against percentage size frequencies. Growth rate was determined by employing Powel Wetherall, ELEFAN, Automatic scan, K-scan and Shepherd's method.

ELEFAN – 1

Growth parameters were estimated using FISAT -1 software by Gayanilo *et al.* Growth was modeled following von Bertalanffy's growth function (VBGF). An initial estimate of L_α was obtained using Powell Whetherall plot. The length frequency data was run on ELEFAN–1 sub package available in FISAT using the automatic search routine, response surface analysis and scan of "K" values, the best fitting curve was estimated.

Shell Measurements

Length: The vertical distance from the apex to umbilical base of the shell is defined as length.
Width: This is taken as the maximum dimension at right angles to the length of the shell in transverse plane. The length and width of the operculum was also measured. To find out weight, the soft parts were removed from the shell, blotted to remove the excess moisture and then weighed to the nearest 0.1 mg using a single pan electric balance.

Results and Discussion

ELEFAN – 1

Estimation of Growth Parameters from Length Frequencies of *Chicoreus virgineus ponderosus*

The optimized growth parameters (L_∞ and K) and the goodness of fit index (Rn) for the obtained growth parameters for males and females of *Chicoreus virgineus ponderosus* by ELEFAN–I method in the FISAT–II package are given in Tables 9.1 and 9.2. The non-seasonalized length frequency histograms with growth curves for males and females of *Chicoreus virgineus ponderosus* are shown in Figures 9.2 and 9.3 the automatic search routine in FISAT package derived the L_α and K values of 96.60 mm and

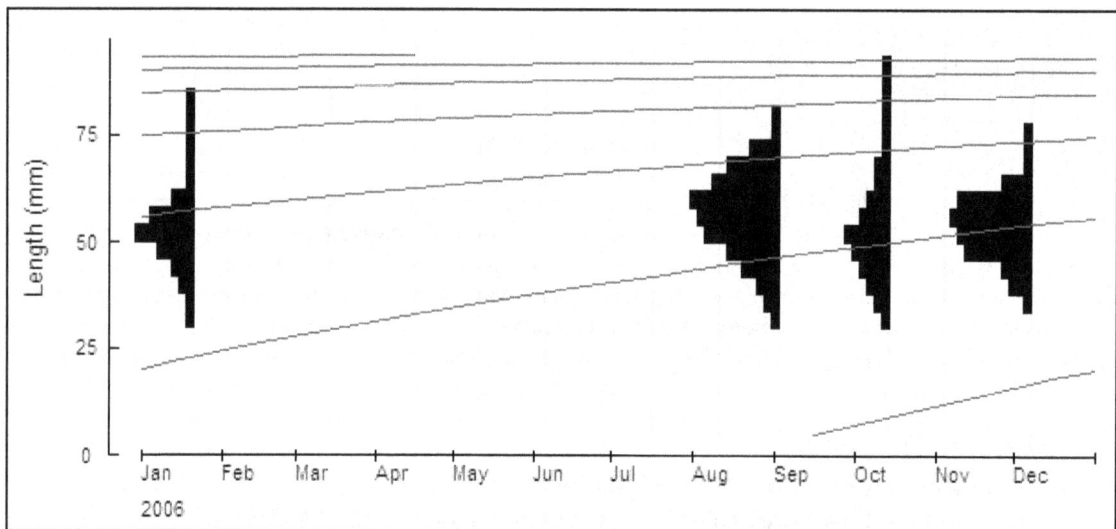

Figure 9.2: ELEFAN Growth Curve of Male *Chicoreus virgineus ponderosus*.

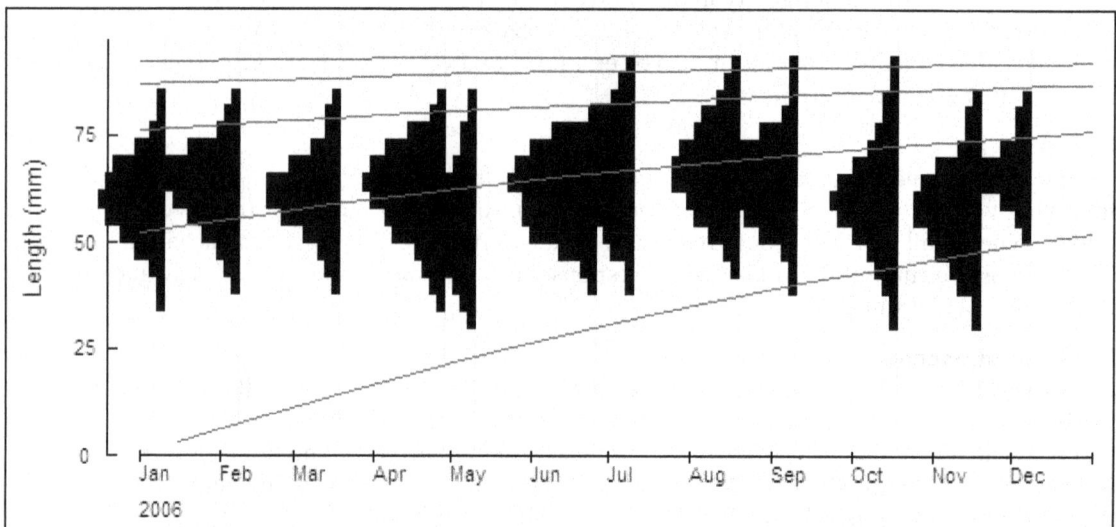

Figure 9.3: ELEFAN Growth Curve of Female *Chicoreus virgineus ponderosus*.

1.40 yr^{-1} for males and 96.6 mm and 1.40 yr^{-1} for females respectively, whereas, the K-scan routines gave the values of 96.60 mm and 3.210 yr^{-1} for males and 96.50 mm and 0.230 yr^{-1} for females respectively.

Estimation of Growth Parameters from Length Frequencies of *Siratus virgineus ponderosus*

The optimized growth parameters (L$_\infty$ and K) and the goodness of fit index (Rn) for the obtained growth parameters for males and females of *Siratus virgineus ponderosus* by ELEFAN–I method in the FISAT–II package are given in Tables 9.1 and 9.2. The non-seasonalized length frequency histograms with growth curves for males and females of *Siratus virgineus ponderosus* are shown in Figures 9.4 and 9.5 the automatic search routine in FISAT package derived the L$_\alpha$ and K values of 96.05 mm and 2.30

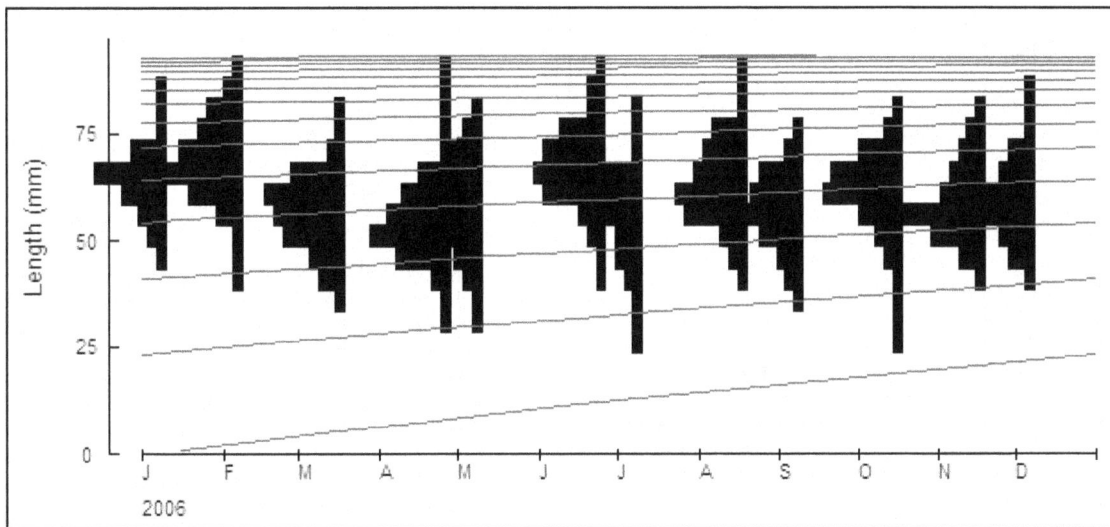

Figure 9.4: ELEFAN Growth Curve of Male *Siratus virgineus ponderosus*.

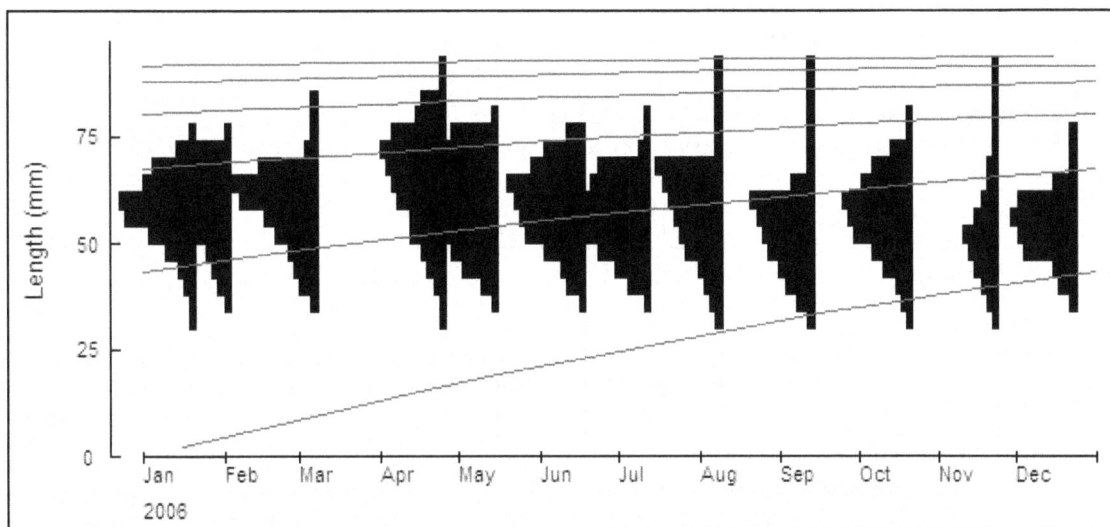

Figure 9.5: ELEFAN Growth Curve of Female *Siratus virgineus ponderosus*.

yr⁻¹ for males and 95.80 mm and 1.00 yr⁻¹ for females respectively, whereas, the K-scan routines gave the values of 96.05 mm and 0.150 yr⁻¹ for males and 95.80 mm and 0.780 yr⁻¹ for females respectively.

Powell-Wetherall Method of *Chicoreus virgineus ponderosus*

Powell-Wetherall plots for the estimation of L_α and Z/K of males and females of *Chicoreus virgineus ponderosus* are given in Figures 9.6 and 9.7. The L_α values obtained for males and females of *Chicoreus virgineus ponderosus* were 115.16 mm [r=-0.959; regression equation, Y= 15.58 + (-0.130)*X] and 106.70 mm [r=-0.960; regression equation, Y= 19.39 + (-0.185) *X] respectively. The alignment of points on the straight line was quite satisfactory with a good coefficient of correlation (0. 973).

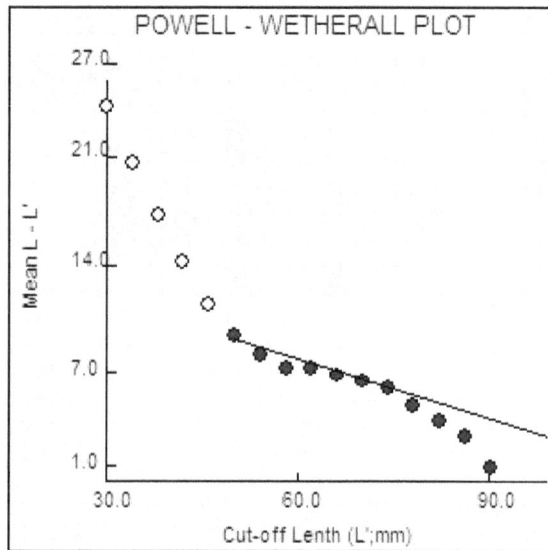

Figure 9.6: Powell-Wetherall Plot of Male *Chicoreus virgineus ponderosus*.

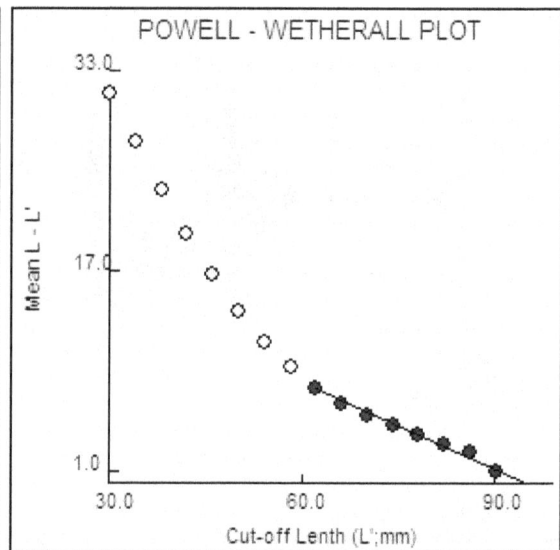

Figure 9.7: Powell-Wetherall Plot of Female *Chicoreus virgineus ponderosus*.

Powell-Wetherall Method of *Siratus virgineus ponderosus*

Powell-Wetherall plots for the estimation of L_α and Z/K of males and females of *Siratus virgineus ponderosus* are given in Figures 9.8 and 9.9. The L_α values obtained for males and females of *Siratus virgineus ponderosus* were 94.28 mm [r=-0.993; regression equation, Y=22.09+ (-0.214)*X] and 106.70 mm [r=-0.993; regression equation, Y=22.73+ (-0.229) *X] respectively. The alignment of points on the straight line was quite satisfactory with a good coefficient of correlation (0.912).

Shepherd's Method

Shepherd's plots for the estimation of L_α and K of males and females of *Chicoreus virgineus ponderosus* are given in Figures 9.10 and 9.11. The L_α values obtained for males and females of *Chicoreus virgineus ponderosus* were 96.60 and 96.50 mm respectively. The K values obtained for males and females were 3.210 and 0.230 respectively.

Shepherd's plots for the estimation of L_α and K of males and females of *Siratus virgineus ponderosus* are given in Figures 9.12 and 9.13. The L_α values obtained for males and females of *Siratus virgineus*

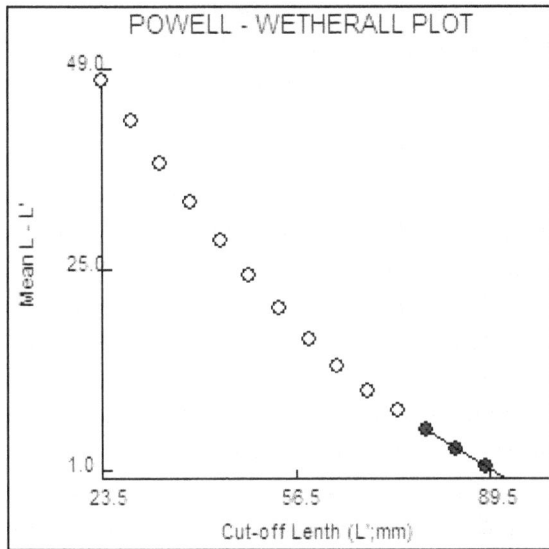

Figure 9.8: Powell-Wetherall Plot of Male *Siratus virgineus ponderosus.*

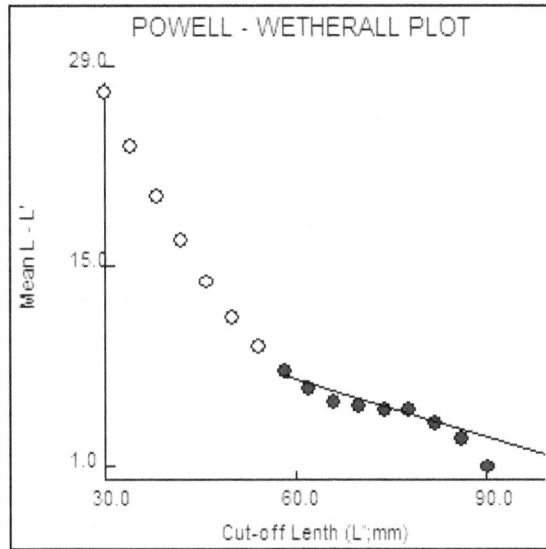

Figure 9.9: Powell-Wetherall Plot of Female *Siratus virgineus ponderosus*

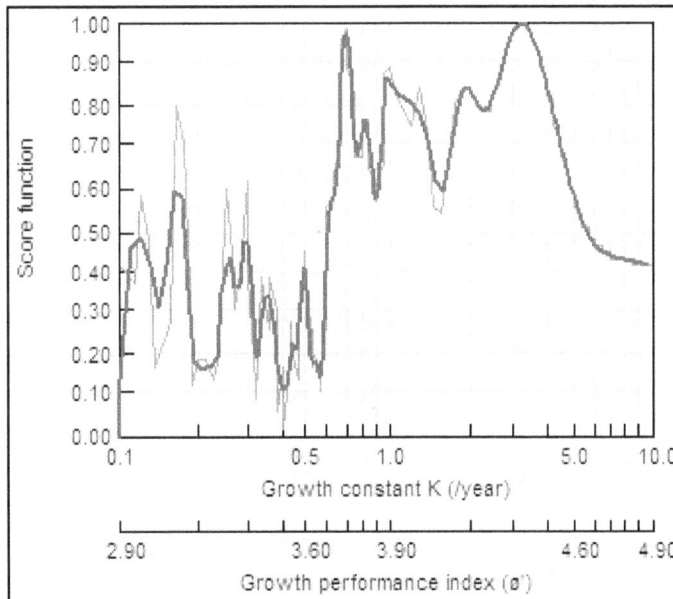

Figure 9.10: K-scan Plot of Male *Chicoreus virgineus ponderosus.*

ponderosus were 96.05 and 95.80 mm respectively. The K values obtained for males and females were 0.150 and 0.780 respectively.

Age and growth of *Chicoreus virgineus ponderosus* and *Siratus virgineus ponderosus* were studied using FISAT -1 software by (Gayanilo *et al.,* 1996). An initial estimate of L_α was obtained using Powell

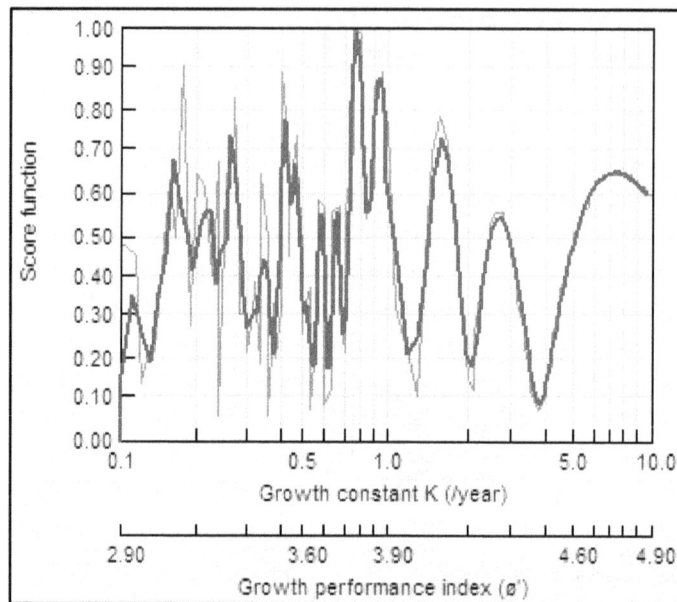

Figure 9.11: K-scan Plot of Female *Chicoreus virgineus ponderosus.*

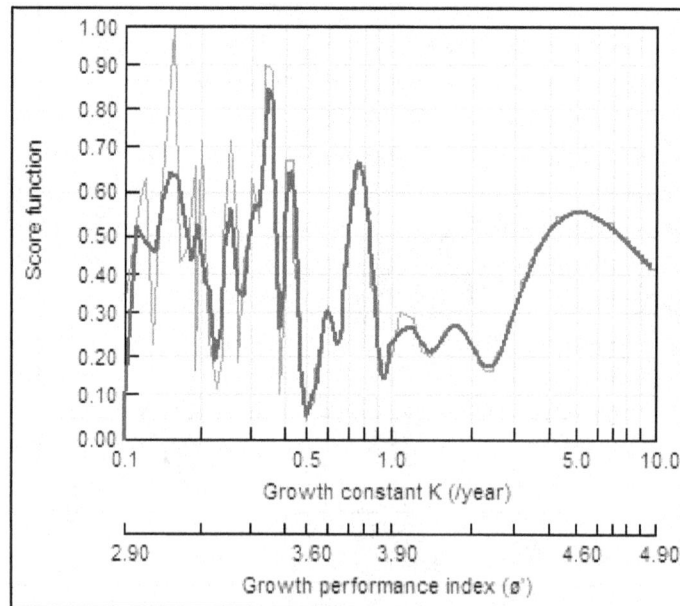

Figure 9.12: K-scan Plot of Male *Siratus virgineus ponderosus.*

Whetherall plot. Then this length frequency data was run on ELEFAN – 1 sub package available in FISAT using the automatic search routine, response surface analysis and scan of "K" values, the best fitting curve was estimated. The results showed more or less similar growth for the males and females

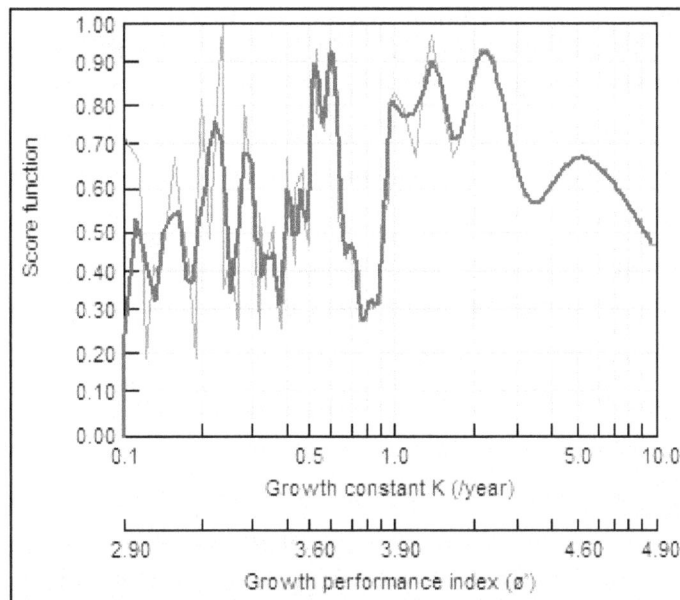

Figure 9.13: K-scan Plot of Female *Siratus virgineus ponderosus*.

Table 9.1: Growth Parameters of Males and Females of *Chicoreus virgineus ponderosus* Obtained from different Methods from Length Frequency Data.

Method	Sex	L_∞ (mm)	Z/K	K(yr⁻¹)	t_0 (yr)	Rn/Score
Powell Wetherall	M	115.16	5.406	–	–	–
	F	106.70	4.803	–	–	–
ELEFAN						
i. Automatic scan	M	96.60	–	1.40	–	0.273
	F	96.60	–	1.40	–	0.227
ii. K-scan	M	96.60	–	0.213	–	1.000
	F	96.50	–	0.230	–	1.000

Table 9.2: Growth Parameters of Males and Females of *Siratus virgineus ponderosus* Obtained from different Methods from Length Frequency Data.

Method	Sex	L_∞ (mm)	Z/K	K(yr⁻¹)	t_0 (yr)	Rn/Score
Powell wetherall	M	94.28	1.269	–	–	–
	F	95.80	1.214	–	–	–
ELEFAN						
i. Automatic scan	M	96.05	–	2.30	–	0.334
	F	95.80	–	1.00	–	0.256
ii. K-scan	M	96.05	–	0.150	–	1.000
	F	95.80	–	0.780	–	1.000

during the study period. The determination of age and growth based on a single method has its own limitations especially when the determination of age and growth is through indirect methods or through statistical analysis. The age and growth estimation of *Chicoreus virgineus ponderosus* and *Siratus vigineus ponderosus* has been done through several methods so that the outcome of one method will act as a check and control over the other. The L_α values obtained in the present study using various methods did not varied much. The values of asymptotic length (L_α) obtained for *Chicoreus virgineus ponderosus* were 115.16mm for males and 106.70 mm for females and the growth rate (K) for males (0.213 yr^{-1}) and females (0.230 yr^{-1}) were given by k-scan routine in ELEFAN. Similarly the values of asymptotic length (L_α) obtained for *Siratus virgineus ponderosus* were 94.28 mm for males and 95.80 mm for females and the growth rate (K) for males (0.150 yr^{-1}) and females (0.780 yr^{-1}) were given by k-scan routine in ELEFAN.

Wilbur and Olsen (1964) have reported that the decrease in the relative growth with an increase in age is known in bivalves. Brown (1957) has stated that specific growth rate decrease as the age of the organism increases. Abraham (1953) have been reported that the growth rate was high in younger clams than in adult one. In the present study the growth of *Chicoreus virgineus ponderosus* and *Siratus vigineus pondersous* calculated using the above methods was presented in Tables 9.1 and 9.2. Both the species have almost similar growth rates and life span. Several investigators, both in tropical and temperate waters, have found that growth pattern was not uniform throughout the year and in the early stage, growth rate is faster than the later part by Kamala, (1983). Various factors are known to influence growth by (Wilbur and Owen, 1964). In some gastropods, such as *Nerita fulguarans, N. peloronta and N. Versicolor,* growth was found to be slow during winter but rapid during summer, parallel with the seasonal changes in *Littorina* spp. (Kolipinski,1964). From tropical and temperate waters, exhibited remarkable variations in growth, size and life span. Moore (1937) has recorded a maximum growth of 27.5 mm with a life span of 4-5 years for in *L. littence* and a life span of 5-6 years and the a maximum height of 18.5 mm in *L. Sukatilis* by Moretean, (1976). Stella, *et al.,* (1992) have recorded that *Chicoreus ramosus* attained a length of 107.7, 163.7, 205.0, and 235.3 in the 1st, 2nd, 3rd, 4th year respectively. This species was found to have a life span of 4-5 years.

In the present investigation, both the species were found to have a life span of more than 4 years. This increased growth rate of these species may be due to increased food availability and the combined effect of hydro biological factors in their habitat. Besides these, it is presumed that the northeast monsoon is said to have considerable influence on the growth rate of this gastropod.

References

Abraham, K.C., 1953. Observation on the biology of *Meretrix casta* (Chemnitz). *J. Zool. Soc., India,* 5(2): 163–190.

Alejandra Rumi, Diego E. Gutiérrez Gregoric and M. Andrea Roche, 2007. Growth rate fitting using the von Bertalanffy model: analysis of natural populations of *Drepanotrema* spp. snails (Gastropoda: Planorbidae). *Rev. Biol. Trop. (Int. J. Trop. Biol.)* 55(2): 559–567.

Branch, G.M. and Branch, M.L., 1980. Competition in *Bembicium auratum* (Gastropoda) and its effect on microalgal standing stock in mangrove muds. *Oecologia,* 46: 106–114.

Brown, A.E., 1957. *The Physiology of Fishes. I. Metabolism.* Academic Press Inc., New York, pp. 371.

Creese, R.G., 1980. An analysis of distribution and abundance of populations of the high-shore Lunpet, *Notoacmea petterdi* (Tenison-Woods). *O. Ecologia,* 45: 252–260.

Crothers, J. H., 1973. On variation *Nucella lapillus* (L.). Shell shape in populations from Pembrokeshire South Wales. *Proc. Mal. Soc. Lond.*, 40: 318–327.

Crothers, J. H., 1974. On variation in *Nucella lapillus* (L.): Shell shape in populations from the Bristol Channel. *Proc. Mala. Soc. Lond.*, 40: 319–327.

Estebenet, A.L. and Martin, P.R., 2000. Inter and intra population variation in growth patterns of *Pomacea canaliculata* (Gastropoda: Ampullariidae). *VI Internet. Conger. Med. Appl. Malacol.* Havana, Cuba.

Estebenet, A.L., 1998. Allometric growth and insight on sexual dimorphism in *Pomacea canaliculata* (Gastropoda: Ampullariidae). *Malacol.*, (39): 207–213.

Estebenet, A.L. and Cazzaniga, N.J., 1992. Growth and demography of *Pomacea canaliculata* (Gastropoda: Ampullariidae) under laboratory conditions. *Malacol. Rev.*, (25): 1–12.

Fletcher, W.J., 1984. Intraspecific variation in the population dynamics and growth of the limpet, Cellana tramoserica. *Oecologia*, 63: 110–121.

Gayanilo, F.C. Jr., Sparre, P. and Pauly, D., 1996. FAO–ICLARM Stock assessment tools, (FISAT) user's manual. *FAO Comp. Infor. Ser. (Fish)*, 7: 126.

Goodwin, David H., K. W. Flessa, B. R. Schone and D. L. Dettman., 2001. Cross-calibration of daily growth increments, stable isotope variation, and temperature in the Gulf of California bivalve Mollusc *Chione cortezi*: Implications for Paleo environmental analysis. *Palaios*, 16: 387–398.

Grana–Raffuccr and Appeldoorn., 1997. Studied the age determination of larval strombid gastropods by means of growth increment counts in statoliths. *Fishery Bulletin*, 95: 857–862.

Ismail, N. S. and Elkarmi, A. Z., 2006. Age, Growth and Shell Morphometrics of the Gastropod *Monodonta dama* (Neritidae: Prosobranchia) from Azraq Oasis, Jordan. *Pak. J. Biol. Sci.*, 9(3): 549–552.

Ismail, N. S., A. Z, Elkarmi and S. M. AL–Moghrabi., 2000. Studies on the Populaton structure and shell morphometrics of the corallivorous gastropods *Drupella cornus* in the Gulf of Aqaba, Red Sea, *Indian. J. Mar. Sci.*, 29: 165–170.

Jardine, I.W., 1985. Height on the shore as a factor influencing growth rate and reproduction of the top-shell *Gibbula cineraria* (L.). In: *The Ecology of Rocky Coast*, (Eds.) R.G. More and R. Seed. Hodder and Stoughton, London, p. 117–135.

Kamala, B., 1983. Studies on some aspects of the biology of the top shell *Euchelus asper* (Gmelin) (Gastropoda: Prosobranchia) of the palm beach shingles of the Visakhapatnam coast. *Ph.D. Thesis*, Andhra University, Waltair, South India.

Kolipinski, M. C., 1964. The life History, growth and ecology of four intertidal gastropods. (Genus: Nerita) of South east Florida. *Ph.D. Thesis*, University of Miami.

Largen, M. J., 1971. Genetic and environmental influences upon the expression of shell sculpture in the Dog whelk (*Nucella lapillus*). *Proc. Malac. Soc. Lond.*, 39: 383.

McCormack, S. M. D., 1982. The maintenance of shore-level size gradients in an intertidal snail (*Littonna sitkana*). *Oecologia*, 54: 177–183.

Moore, H.B., 1937. The biology of *Littorina littorea*. Part I. Growth of shell and tissues, Spawning, Length of life and mortality. *J. Mar. Biol. Assoc.*, 21: 721–742.

Moretean, J., 1976. Study on the growth and life spawn of *Littorina saxatilis* (Olovi) *rudis* (Maton). *Can. Biol. Mar.*, 17(4): 463–484.

Rumi, A., D. Gutiérrez Gregoric, Roche, M. and Tassara, M., 2004. Population structure in *Drepanotrema kermatoides* and *D. cimex* (Gastropoda: Planorbidae) in natural conditions. *Malacologia* 45: 453–458.

Stella, C., Rajkumar, T. and Ayyakkannu, K., 1992. Analysis of size class distributions of *Chicoreus ramosus* collected from the Gulf of Manner area, South East Coast of India. *Phuket Mar. Biol. Cent. Spec. Publ.*, 11: 91–93.

Stella. C., 1995. Studies on the taxonomy and ecobiology of *Chicoreus* species from Parangipettai waters, South East Coast of India. *Ph.D. Thesis*, Annamalai University. pp. 195.

Thivakaran, G.A., 1988. Studies on the littorinids *Littorina quadricentus* (Phillip) and *Nodilittorina pyramidalis* (Quoy and Gaimard, 1833) Gastropoda: prosobranchia littorinidae) from the tranquebar rocky shore (South East Coast of India) *Ph.D. Thesis*, Annamalai University, pp. 179.

Underwood, A. J., 1984b. Microalgal food and the growth on the intertidal gastropods *Nenta atramentosa Reeve* and *Bembiciurn nanum* (Lamarck) at four heights on a shore. *J. Exp. Mar. Biol. Ecol.*, 79: 277–291.

Wilbur, K. M. and Owen, G., 1964. *Growth in Physiology of Molluscs*, (Eds.). K.M. Wilbur and C.M. Yonge. Academic Press, New York, Vol. 2.

Chapter 10

Molluscan-Faunistic Composition in the Mangrove Ecosystem of Kali Estuary, Karwar

ABSTRACT

Mangrove ecosystem plays an important role in escalating bioresources like fin fishes of the estuary in general and shell fishes in particular. As molluscan known as poor man's rich protein food more emphasis has been given to study their community's biology and diversity profile. Keeping this in view, the present study was undertaken for the period of one year from January 2008 to January 2009 from three selected sites. Totally 37 species (19 families) have been recorded during the study period.

Among which family Veneridae has formed the dominant group (1567/m^2) followed by family Littorinidae (1061/m^2). Osteridae (854/m^2) and Buccinidae (20/m^2) families did not contribute much to the total density of this community in this area. It is surmised from the data that the Mavinahole locale has found to be a highly productive in terms of molluscan population (2307/m^2) whereas the least population (943/m^2) has been noticed at Kadwad study area. The variation in the population of this community is mainly governed by the sediment texture and other sedimentological parameters.

Keywords: Kali estuary, Molluscs, Mangrove, Karwar, Karnataka state.

Introduction

The phylum mollusca contains 80,000 species second in dominance in size to the large phylum Arthropoda in the world marine ecosystem. This phylum includes diverse members as clams, oysters, mussels, snails, octopus, and squid. The molluscs play an important role in the trophic ecosystem as it serves as an intermediary link between primary and tertiary rung. As they occupy benthic realm of macro invertebrate community of an aquatic ecosystem, which are considered as bio-indicators of water quality. Besides this, some of the members (bivalves) of this phylum are commercially important and are considered as poor man's rich protein food. The exterior shells of these bivalves are chiefly used for manufacturing the lime. Though these animals thrive in fresh, brackish water and marine ecosystem, very little information is available with respect to the mangrove ecosystem. Many workers studied the population dynamics of macro benthic invertebrates in the Kali River and estuary (Bhat, 1984, Neelakantan *et al.*, 1981; Philipose, 1981; Harakantra, 1975) whereas meio and macro benthos studies were carried out in the Kanasgiri backwaters by Kavita (2010) and Smrutha (2010). But, no work was carried out on molluscs of mangrove ecosystems in the Kali estuary. Keeping in view their importance and significant role played in the water quality, food and fishery, and to fill the lacunae, the present work was undertaken for the period of thirteen months from January 2008 to January 2009 in the mangrove ecosystem of Kali estuary, Karwar (Karnataka).

Materials and Methods

The present study was undertaken for the period of thirteen months from January 2008 to January 2009 to study the composition, abundance and distribution of mollusks in the mangrove ecosystems of Kali estuary, Karwar (Figure 10.1). All mollusc species were collected by hand picking. The clams (family: Veneridae) were collected from an area of 1 × 1m quadrate during the low tide period. The

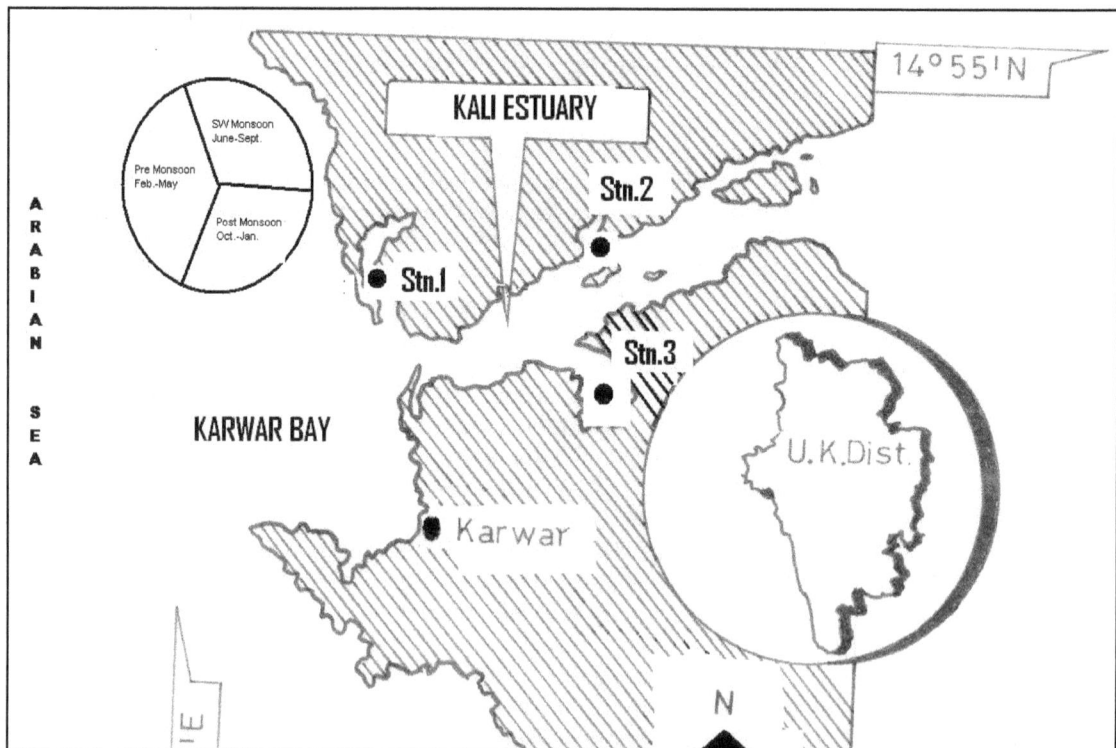

Figure 10.1: Map Showing the Study Stations in the Mangrove Ecosystems of Kali Estuary.

oysters were counted, recorded and represented for 1 × 1m quadrate area of small rocks during low tide period because these rocks are submerged in the water during the high tide period (CMFRI Bulletin, 2000).

The biotic entities reflect the ecological and environmental status and were calculated in terms of number of individuals or specimens (N), number of species (S), total abundance (A), Margalef species richness (d), Pielou's evenness (J'), Shannon index (H') at each study site (Clarke and Gorley, 2001). Bray Curtis similarity for species diversity for all the species belonging to molluscs, was determined analytically by using PRIMER-v5 software.

Results and Discussion

The present study was mainly focused on the composition distribution and abundance of molluscan fauna in the Mangroves ecosystem of Kali Estuary Karwar, Karnataka. Totally 37 species were recorded belongs to 19 families.

The Molluscan density showed wide variations in months and seasons and more or less a uniform pattern of distribution was seen among the stations. Spatially higher density was recorded at station 1 (2299/m²) where as the lowest standing stock of this fauna was observed at station 3 (741/m²) (Figure 10.2). The station 2, has shown an intermediary values of density (1237/m²) over the period of study. Among 19 families, the family Veneridae was found to dominant group (4.58 to 24.8 per cent), but, the lower density of this group was recorded at station 3. The second dominant group

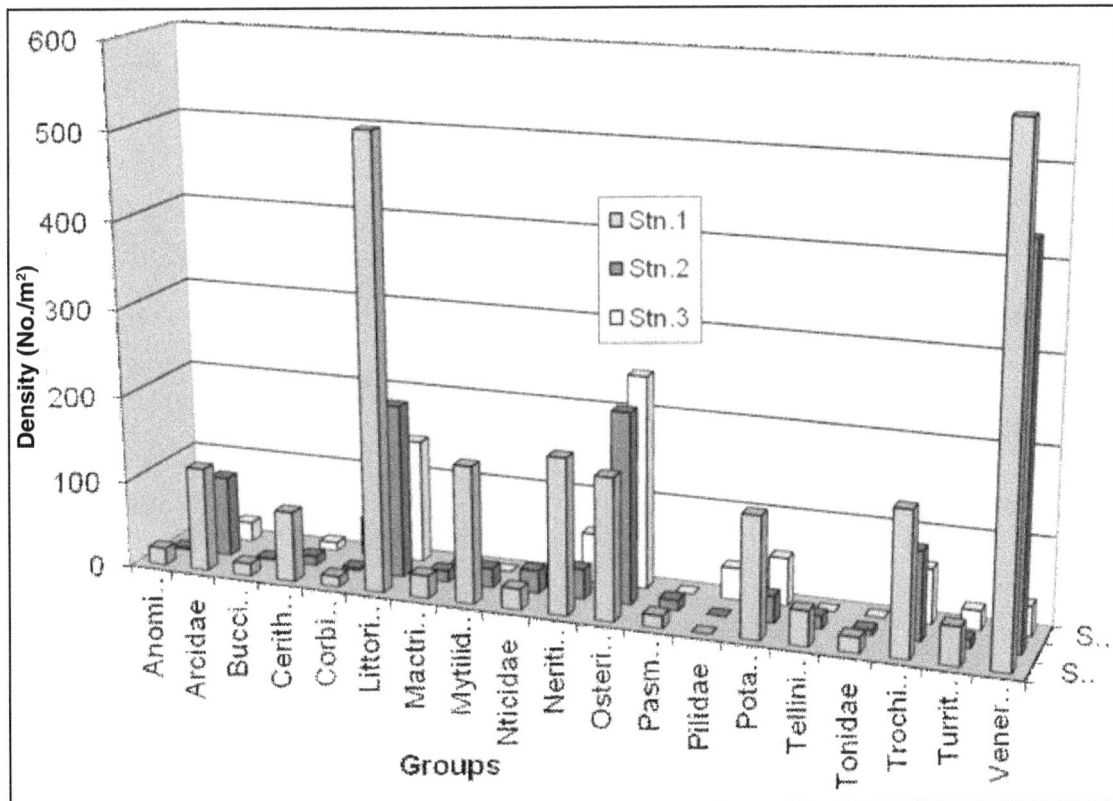

Figure 10.2: Total Density of Molluscan Groups at Study Stations.

was the Littorinidae was found comparatively an uniform pattern of distribution in all the three study sites contributing the percentage share of 22.48, 16.08 and 19.16 per cent at station 1, 2 and 3 respectively. Whereas, the Neritidae, Trochidae, Mytilidae and Osteridae were found higher at station 1, whereas at station 2, the Veneridae, Osteridae, and Littorinidae were found higher in density, whereas at station 3 the Osteridae and Littorinidae were found maximum density during the study period (Figures 10.3–10.5).

The rest of the groups did not contribute much to the total density of molluscs fauna in any of the study sites. It indicates that the some of the members of this faunal community are site specific. It is surmised from the Figures 10.3 and 10.4 that the maximum percentile was exhibited by Veneridae, (24.8 and 35.6 per cent) at Station 1 and 2, where as the next dominant group was Littorinidae (16.08-22.48 per cent) at station 2 and 1 while at station 3 it showed moderately and intermediate percentile (19.16 per cent) during the study tenure (Figure 10.5). As it is said earlier, the molluscan members were site specific *i.e.*, the members of Buccinidae, Niticidae, Pasmodidae, Tellinidae and Tonnidae were

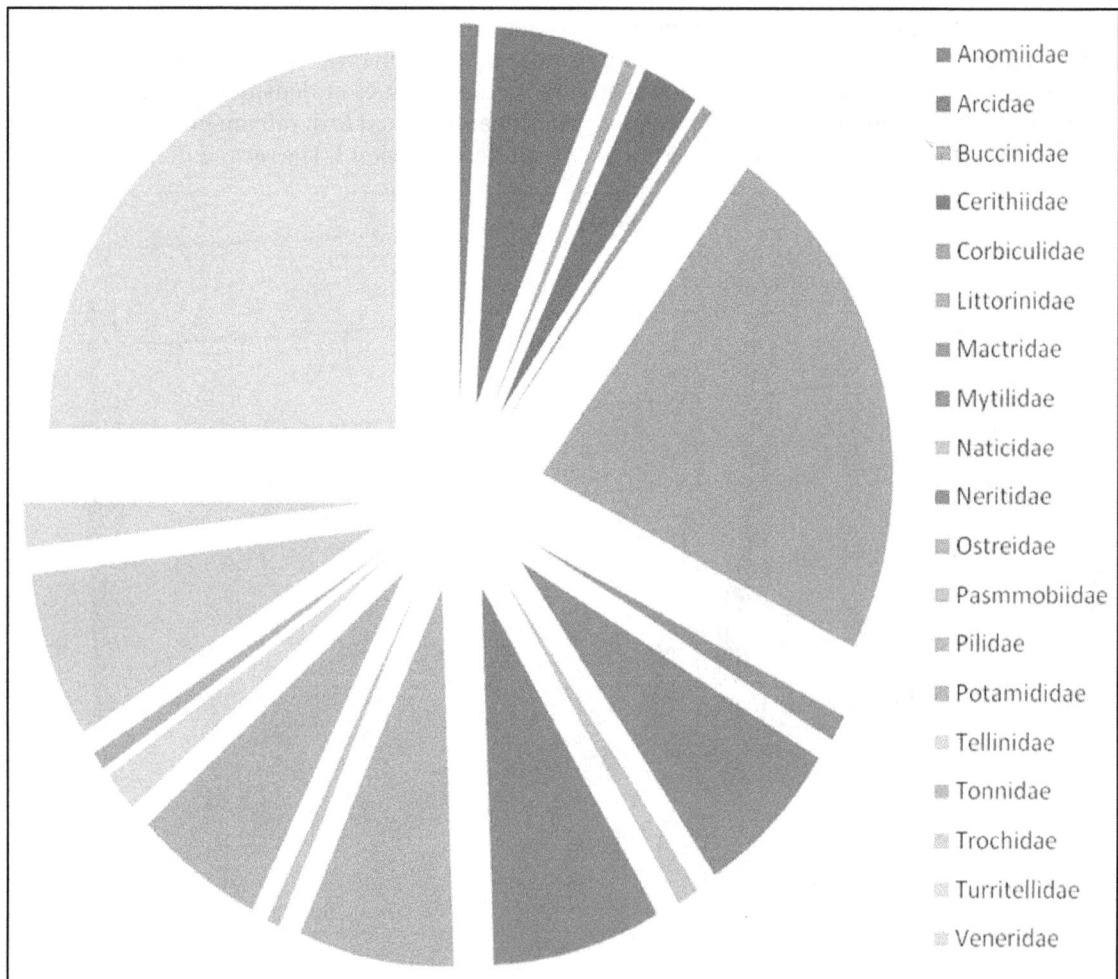

Figure 10.3: Percentage Composition of different Groups of Mollusc at Study Station 1.

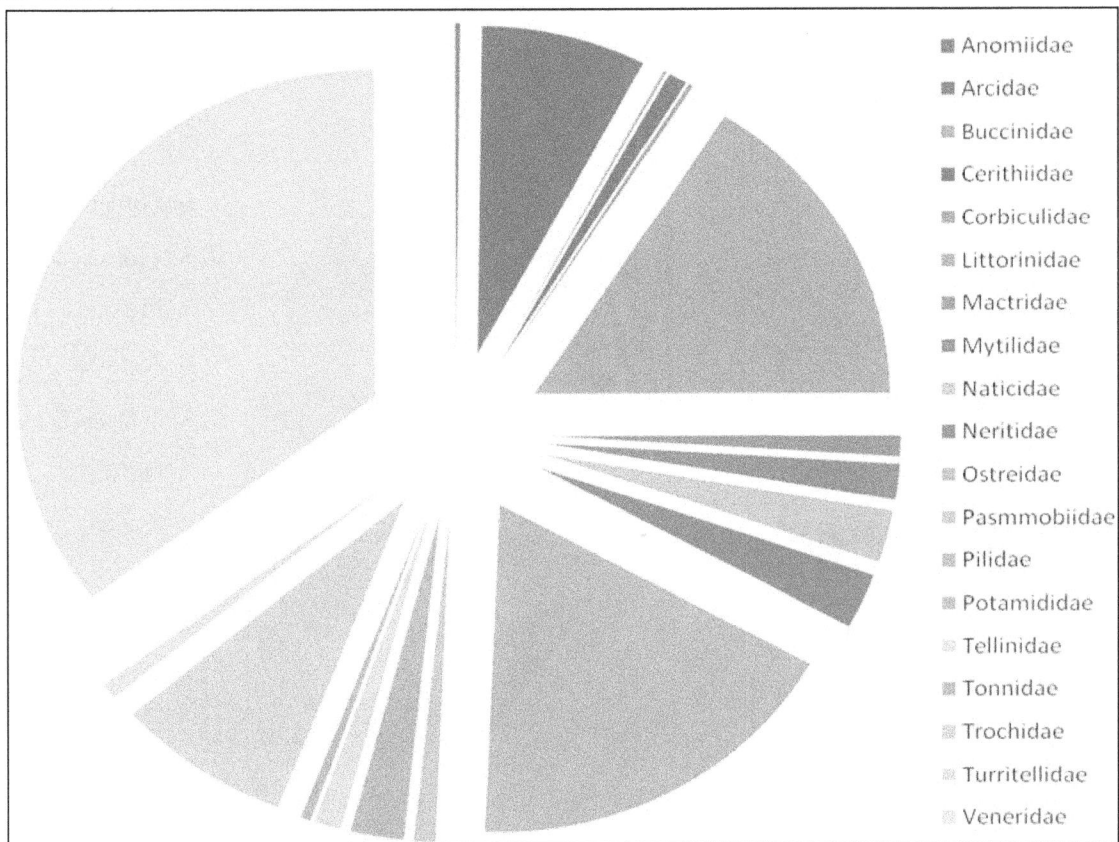

Figure 10.4: Percentage Composition of different Groups of Mollusc at Study Station 2.

totally absent at station 3, but were represented at Station 1 and 2 with contrast this *Pila globosa* (Pilidae) did not represent at station. 1 and 2 but it contributed 4.8 per cent only to the total density of mollusc fauna at station 3 only (Figure 10.5).

Table 10.1 explains the monthly variations in the species diversity, richness and similarity entities. It is inferred from the Table 1 that the Shannon Weiner diversity index varied from 1.2353 (July – 08) to 3.0965 (December'08) the higher values noticed during November (3.0392) and in May (3.0756) where as in the case of Margalef's richness (*d*) it varied between 1.0701 (July) and 6.1392 (November) the higher richness values of 5.3466 (January'08) 5.7975 (February), 5.5973 (March), 5.5227 (April), 5.5377 (May), 5.5102 (October), 5.7689 (December) and 5.9511 (January'09); the Pielou's evenness (*J*), it varied between 0.6308 (August) and 0.9133 (May) at station 1. Similarly at Station 2, the Shannon Weiner index varied between 0.6890 (August) and 2.9698 (December) and the richness values varied between 0.2860 (August) and 6.0360 (December) at Station 2. The species evenness values did not vary much between the months and it ranged from 0.7259 (July) to 0.9940 (August) respectively (Table 10.2).

At Station 3, the Shannon index value did not show much variations in its values during the period of study and the values oscillated from 1.0439 (July) to 2.8219 (April); but the species richness values shown a wide variations between the months ranging from 0.6068 (July) and 4.9199 (April)

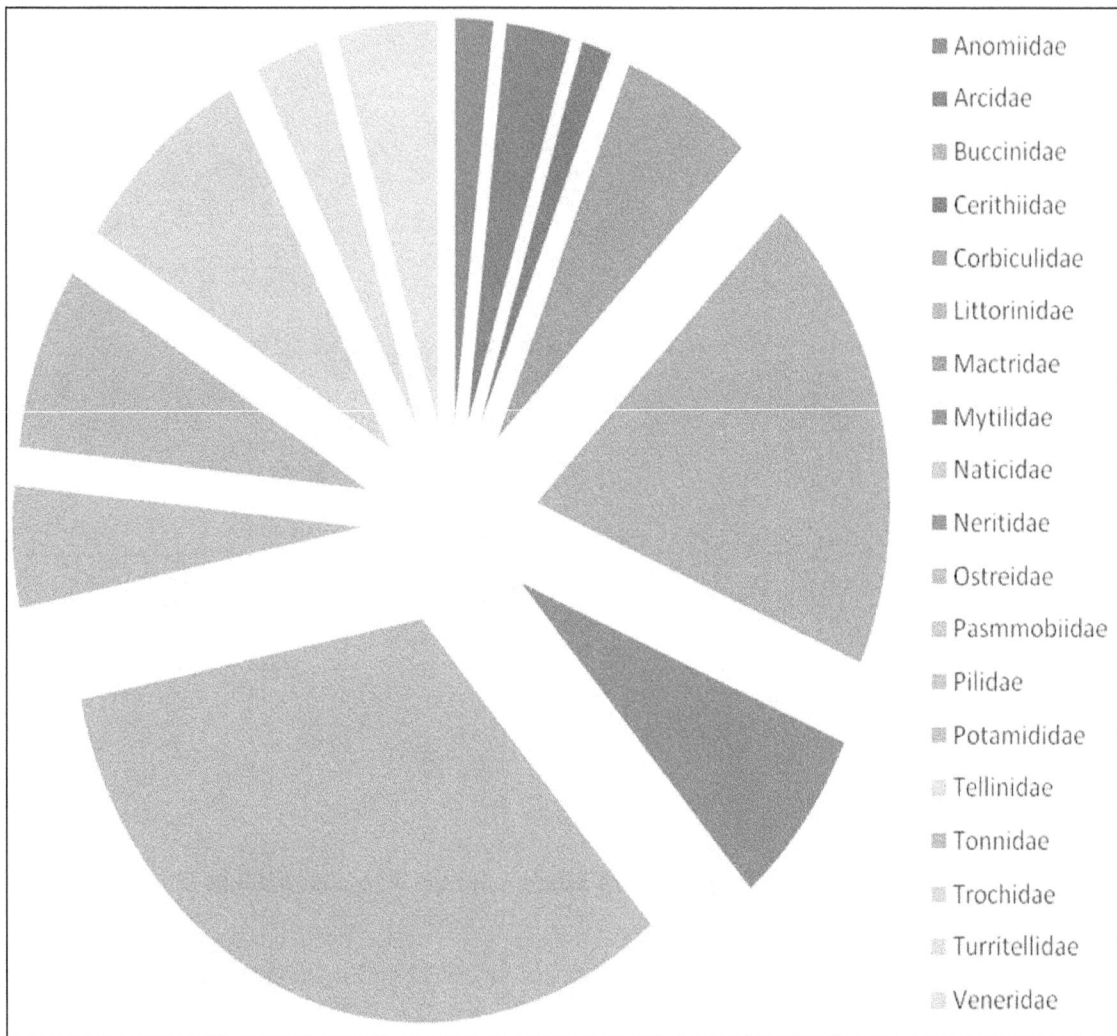

Legend:
- Anomiidae
- Arcidae
- Buccinidae
- Cerithiidae
- Corbiculidae
- Littorinidae
- Mactridae
- Mytilidae
- Naticidae
- Neritidae
- Ostreidae
- Pasmmobiidae
- Pilidae
- Potamididae
- Tellinidae
- Tonnidae
- Trochidae
- Turritellidae
- Veneridae

Figure 10.5: Percentage Composition of different Groups of Mollusc at Study Station 3.

respectively. In contrast to this the species evenness did not show any marked variations in its values and it varied between 0.7859 (Feb.) and 0.9502 (July) respectively (Table 10.3).

Figures 10.6–10.8 explains the faunal similarity indices between the months by means of dendrogram clusters and it is deduced from Figure 10.6 that the three groups of clusters have shown more than 90 per cent similarity during December and March; November, February and May, October and Jan'08. December and March was shown 99.59 per cent and was followed November and February (98.27 per cent). At Station 2, more than 95 per cent was incurred between March and December (97.87 per cent) followed by February and January'09, (96.84 per cent) and January'08 to May (95.69 per cent) respectively (Figure 10.7). At Station 3, the maximum similarity in January'09 and March and May (99.41 and 99.33 per cent) and none of the clusters of the group showed the higher similarity indices among the months.

Table 10.1: Monthly Variation in the Species Diversity and Richness of Molluscan Fauna at Study Station 1

	S	N	d	J'	Brillouin	Fisher	H'(loge)	1-Lambda'
Jan. 08	28	156	5.3466	0.8016	2.4195	9.9497	2.6712	0.8797
Feb.	32	210	5.7975	0.8550	2.7258	10.5155	2.9632	0.9152
Mar.	33	304	5.5973	0.8459	2.7738	9.4143	2.9578	0.9147
Apr.	32	274	5.5227	0.8618	2.7904	9.3927	2.9868	0.9198
May	29	157	5.5377	0.9133	2.7962	10.4559	3.0756	0.9476
Jun.	11	43	2.6587	0.8339	1.6973	4.7768	1.9996	0.8405
Jul.	5	42	1.0701	0.7675	1.0910	1.4788	1.2353	0.6643
Aug.	8	67	1.6648	0.6308	1.1673	2.3689	1.3117	0.6268
Sept.	22	114	4.4339	0.8288	2.2995	8.1139	2.5619	0.8995
Oct.	29	161	5.5102	0.8625	2.6407	10.3234	2.9046	0.9249
Nov.	34	216	6.1392	0.8618	2.8012	11.3407	3.0392	0.9369
Dec.	34	305	5.7689	0.8781	2.9074	9.7993	3.0965	0.9425
Jan.09	34	256	5.9511	0.8187	2.6869	10.5189	2.8873	0.9188

Table 10.2: Monthly Variation in the Species Diversity and Richness of Molluscan Fauna at Study Station 2

	S	N	d	J'	Brillouin	Fisher	H'(loge)	1-Lambda'
Jan. 08	16	83	3.3945	0.8794	2.1741	5.8976	2.4383	0.9068
Feb.	25	122	4.9958	0.8506	2.4631	9.5216	2.7379	0.9248
Mar.	31	162	5.8966	0.8603	2.6849	11.3828	2.9543	0.9367
Apr.	29	185	5.3636	0.8593	2.6631	9.6543	2.8937	0.9356
May	20	86	4.2654	0.8381	2.2169	8.1878	2.5108	0.9075
Jun.	5	30	1.1760	0.8718	1.3504	1.7133	1.5641	0.8091
Jul.	4	21	0.9853	0.7259	0.8320	1.4652	1.0063	0.6142
Aug.	2	33	0.2860	0.9940	0.6290	0.4685	0.6890	0.5113
Sept.	11	51	2.5433	0.7577	1.5671	4.3104	1.8169	0.7921
Oct.	17	74	3.7174	0.8665	2.1550	6.9092	2.4552	0.9033
Nov.	22	102	4.5405	0.8750	2.4103	8.6209	2.7048	0.9225
Dec.	32	170	6.0360	0.8569	2.7030	11.6500	2.9698	0.9365
Jan. 09	20	121	3.9618	0.8261	2.2248	6.8261	2.4749	0.8968

The distribution and abundance of Molluscan fauna in the mangrove ecosystem may be attributed to the availability of food, shelter, and oviposit ion sites, the water bodies rich in organic and silt matter in the soil are known to support thriving population of these macrobenthic invertebrates because of the reduction in water current and as such the substratum tends to make molluscs to indistinguishable from their lentic habitat (Whitton, 1975). In the present study, the molluscan fauna

Table 10.3: Monthly Variation in the Species Diversity and Richness of Molluscan Fauna at Study Station 3

	S	N	d	J'	Brillouin	Fisher	H'(loge)	1-Lambda'
Jan. 08	17	40	4.3373	0.8673	1.9970	11.1701	2.4572	0.8974
Feb.	16	84	3.3853	0.8687	2.1438	5.8608	2.4087	0.8941
Mar.	19	93	3.9712	0.8668	2.2737	7.2245	2.5524	0.9088
Apr.	24	116	4.8384	0.8449	2.4038	9.1890	2.6852	0.9104
May	23	99	4.7876	0.8735	2.4298	9.4103	2.7388	0.9268
Jun.	8	34	1.9850	0.8721	1.5372	3.2981	1.8135	0.8395
Jul.	7	26	1.8415	0.9253	1.4885	3.1433	1.8005	0.8492
Aug.	8	30	2.0580	0.8808	1.5202	3.5695	1.8317	0.8252
Sept.	11	60	2.4423	0.8944	1.8926	3.9509	2.1447	0.8757
Oct.	15	73	3.2630	0.9183	2.1995	5.7215	2.4868	0.9178
Nov.	17	102	3.4594	0.8951	2.2919	5.8253	2.5360	0.9161
Dec.	20	113	4.0191	0.8448	2.2832	7.0574	2.5308	0.9031
Jan. 09	19	73	4.1953	0.8737	2.3912	8.3436	2.5727	0.9132

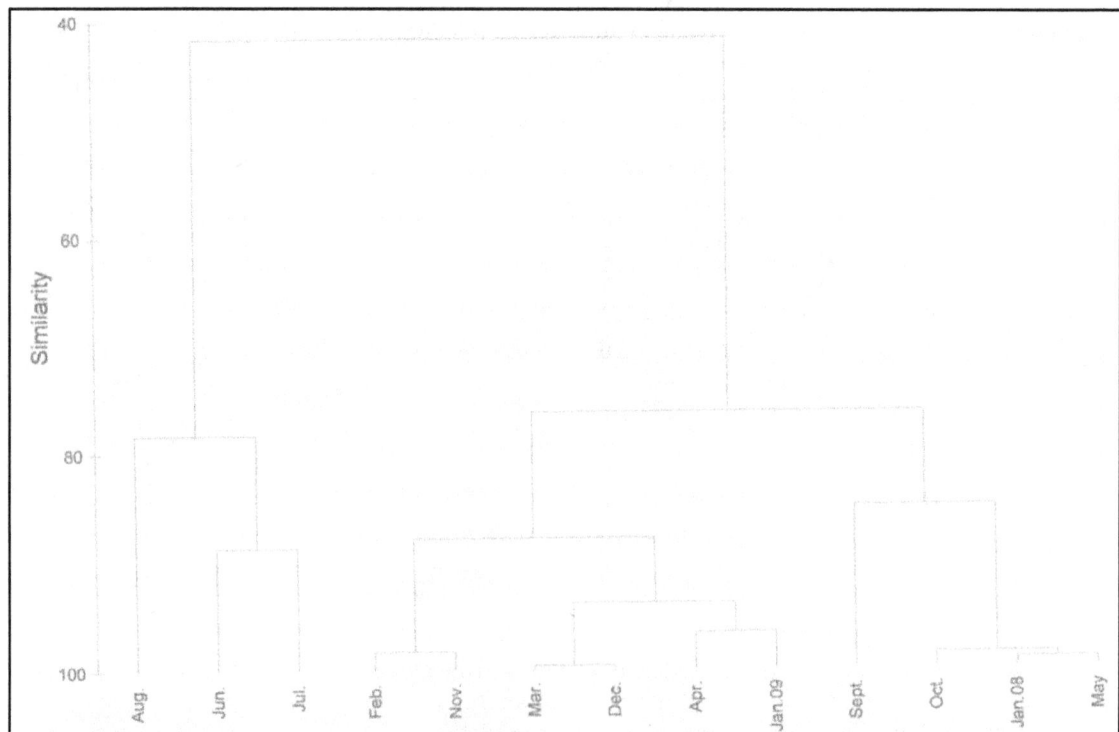

Figure 10.6: Dendrogram Showing Similarity of Mollusc Density at Station 1.

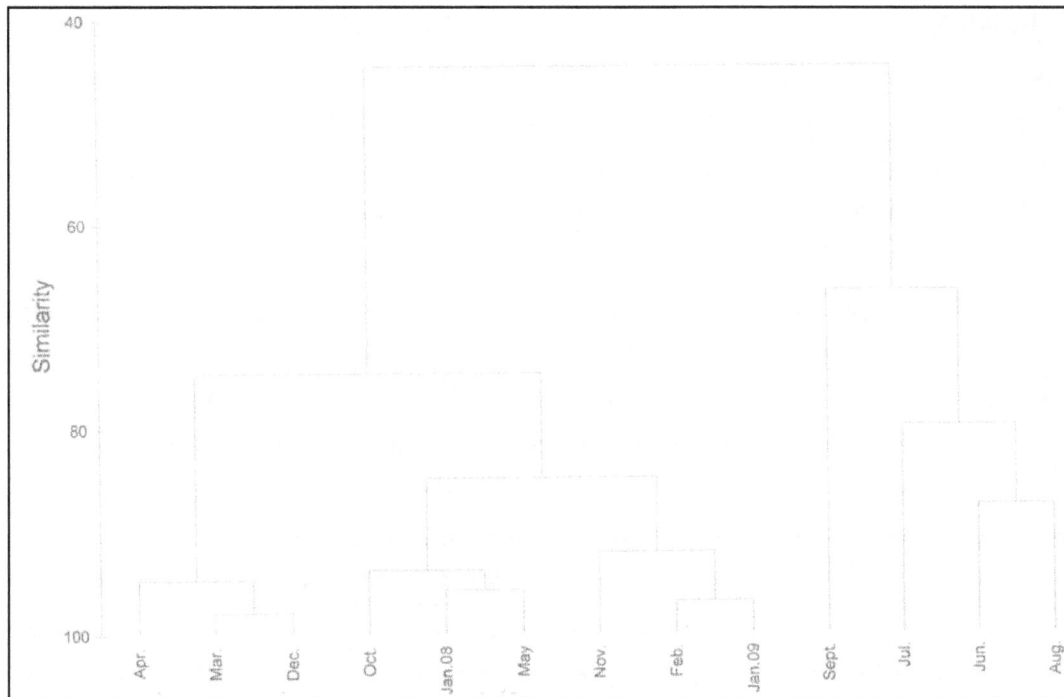

Figure 10.7: Dendrogram Showing Similarity of Mollusc Density at Station 2.

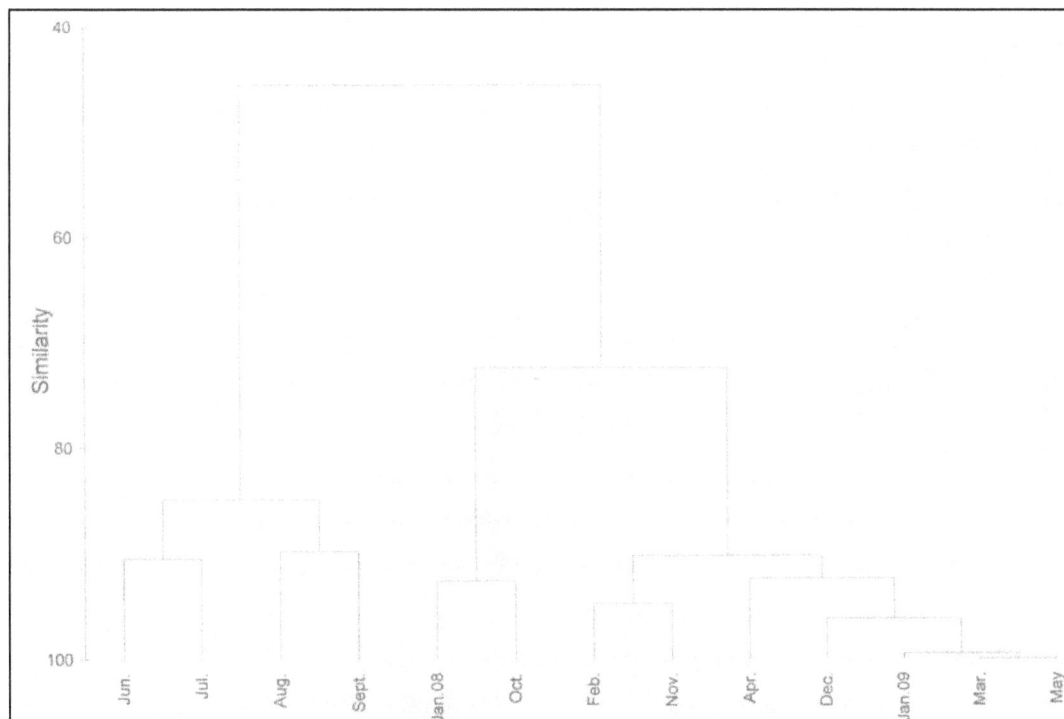

Figure 10.8: Dendrogram Showing Similarity of Mollusc Density at Station 3.

were comprised by bivalves and gastropods. The rich density of molluscan fauna in this mangroves ecosystem is mainly due to the virgin sediment of mangrove ecosystem which supports high density of molluscs were as the low density of faunal group was observed in the denuded or patchy mangroves (Schrijvers *et al.,* 1995). Based on these, composition of molluscan assemblage, the pollution effect in the mangroves forest can also be determined. The ecological conditions such as salinity, temperature and tidal amplitude are favourable which flourishes the molluscan population which provides them food and escalate their density. Hence, the molluscs are used as an indicator for fishery purposes. (Katherisan and Qasim, 2005). Gastropod and bivalves are the most diversed and dominant groups in the benthic water body.

Comparatively a lower density of molluscan fauna was recorded at three study sites, but within these study locales the study site 1, the density was higher (Figure 10.2) in June – Aug months. The lower density recorded during these months was probably due to the higher sand proportion (55.63 - 74.82 per cent) with the results, low organic carbon was found, the higher (2.20-3.06 per cent) organic carbon and silt were found higher during pre and post monsoon in these study stations anad this could be the causative factors held responsible for increase in this faunal community. This observation coincides the finding of earlier workers (Bhat, 1984; Philipose, 1981; Schrijivers *et al.,* 1995 and Katherisan and Qasim, 2005) in various estuary of India. Among the different groups of molluscan community, the Veneridae, Littorinidae and Osteridae were found comparatively higher in all study stations and the population of these bivalves and gastropods were rich in these environment, this could be probably due to the availability of sufficient plankton and detrital matter as they form the main diet of these animals, beside this the mangroves environment which provides an ideal niche for the animals because of the calm environment and little water movement (current), soft substratum and of course the presence of only few predatory organisms (Katherisan and Qasim, 2005). The molluscs were found abundant at stations 1 and 2 and their abundance might distributed to the presence of vegetation in the shallow depth and the nature of the sediment texture. As Guptha (1976) and Manoharan *et al.* (2006) states that during the post monsoon period, it forms a good feed, leading to their multiplication in the mangrove habitat the higher density of gastropods and bivalves recorded in the study site may be either due to the effect of reproduction of these macrobenthic invertebrates, as small size molluscs were observed in the collection during this period. It could be even due to the maximum abundance of decomposer settled organic matter and macrophyte on the bottom of the water body and increased water temperature. Activating the process of decomposition of organic sediment (Dutta and Malhotra, 1986 and Malhotra *et al.,* 1996).

The minimum density of these fauna recorded during south west monsoon season may be due to the aestivation (Singh and Munshi, 1992). Among these 19 groups, the maximum density was contributed by Veneridae (24.8–35.56 per cent), Littorinidae (16.08–22.48 per cent), Osteridae (7.05–31.92 per cent) and Trochidae (7.09–8.36 per cent) respectively.

Some of these gastropods and bivalves acts as a bioindicator of water, such high diversity of mollusc fauna may be attributed to the availability of suitable habitat (Wadaan, 2007) and organically enrich soft bottom (Sing, 1984) and slow water current (Shawhney, 2008).

Among dominant groups, the Littorinidae family- *Littorina scabra* (517/m²) alone constituted maximum share to the total density of molluscan fauna where as in Veneridae group (the first dominant groups), the *Meritrix meritrix* (75/m²), *Paphia malbarica* (119/m²) and *S. solandrii* (118/m²) were contributed much to the total density at stations 1 and 2.

At 3rd station, the *Telescopium telescopium* (210/m^2) was found in higher density followed by *Littorina scabra* (190/m^2) and *Paphia malabarica* (120/m^2). These species were dominant, in these study stations. The numerical abundance of *L. ccabra, T. telescopium, M. casta* and *P. malabarica* may be due to the reason that it is among the hardest in molluscs and reason it covers mainly due to its mode of reproduction (round the year) and can occupy great diversity of habitat (Shrama and Chowdary, 2012), besides this, these faunal community can tolerate high nutrient level and were found to be positively co-related with phosphate and nitrite and were found to be highly associated with the macrophyte floral stretch. The numerical abundance of bivalves indicated a greater nutrient concentration and are used as a bioindicator of water quality.

In all these study locales, the diversity index values were found comparatively higher (>2) except during certain months of the monsoon period, which indicate the habitat highly productive in terms of molluscan faunal population. The richness and evenness of species among stations and between the months indicates that, the higher values resulted due to the availability of nutrients and the texture of sediment might have escalated the higher densities in species richness and evenness in this habitat. The increase in species evenness could be due to the presence of some community in which abundance and distribution were more homogenous.

The present study indicates that, the molluscan population in these mangroves ecosystem may be playing central role in supporting both local and ecosystem level of biodiversity. The ultimate extirpation and extinction of such molluscan population may therefore have profound effects on the wider ecosystem the results emphasis the importance of conserving the hard shelled molluscan faunal population and invasion of non-local biota. These benthic macro invertebrates being wide spread and sensitive to any change in the environment and these groups of organisms are more often used for assessment of mangrove habitat quality. Application of these fauna as an bioindicator may helpful to improve the environment and an awareness and conservation of such biotic community has to be initiated and to safe guard such a versatile ecosystem in future.

References

Bhat, U. G., 1984. Studies on benthos of Kali estuary, Karwar. *Ph.D. Thesis,* Karnatak University, Dharwad.

Clarke, K.R. and Gorley, R.N., 2001. *PRIMER–v5* User manual/Tutorial (p.91). Plymouth *PRIMER–E* Ltd.

CMFRI Bulletin, 2000. *Taxonomy – Molluscs of India.*

Dutta, S.P.S. and Malhotra, Y.R., 1986. Seasonal variations in the macrobenthic fauna of Gadigarh stream (Miran Sahib) Jammu. *Indian J. Ecol.,* 113(1): 138–145.

Dutta, S.P.S., Malhotra, Y.R., Sharma, K.K. and Sinha, K., 2000. Diel variations in physico-chemical parameters of water in a relation to macarobenthic invertebrate in some pool adjacent to the River Tawi Nagrota Bye pass, Jammu. *Him. J. Env. Zool.,* 14: 13–24.

Gupta, S.D., 1976. Macrobenthic fauna of Loni reservoir. *J. Inland Fish. Soc., India,* 8: 49–59.

Harakantra, S.N., 1975. Benthos of Kali estuary, Karwar. *Mahasagar Bull. of NIO,* 8: 53–58.

Kathiresan, K. and Qasim, S.Z., 2005. *Bioldiversity of Mangrove Ecosystems.* Hindustan Publishing Corporation (India), New Delhi, pp. 251.

Murugesan, S.V.S. and Palaniswamy, R., 2006. Numerical abundance of macroinvertebrates in selected reservoirs of Tamil Nadu. *J. Inland Fish. Soc. India*, 38(1): 54–59.

Nelakantan, B., Bhat, U.G., Naik, U.G., Philipose, K.K. and Kusuma, M.S., 1983. On the molluscan resources of Uttara Kannada, Karnataka. *Harvest and Post Harvest Tech. Fish.* C. I. F. T. India.

Philipose, K.K., 1980. Studies on some aspects of the biology of green mussel; *Perna viridis* (Linneaus) and the backwater clam *Meretrix casta* (Chemnitz) from Karwar waters. *M.Sc. Dissertation Thesis*, Karnatak University, Dharwad.

Rane, Kavita 2010. Studies on the macrobethic production in Kanasgiri backwater, Kali estuary, Karwar. *M. Sc. Dissertation*, Karnatak University, Dharwad.

Schrijvers, J., J. P. Okondo, M. Steyaert and M. Vincx, 1995. The influence of epibenthos on the meiobenthos of *Ceriops tagal* mangrove sediment at Gazi Bay, Kenya. *Marine Ecology Progress Series* 128: 247–259.

Sawhney, N., 2008. Biomonitoring of river Tawi in the vicinity of Jammu city. *Ph.D. Thesis*, Univeristy of Jammu, Jammu.

Sharma, K. K. and Chowdhary, Samita 2012. Diveristy of molluscan fauna inhabited by River Chenab-fed stream (Gho–Manhasan). *Indian Journal of Ecology*, **39**(1): 48–51.

Singh, R. and Munshi, J. S. D., 1992. Molluscan diversity and role of certain abiotic factors on the density of gastropodas *Pila globosa* and *Belllamya bengalensis* in a tank at Jamlapur. *J. Freshwater Biol.*, **4**(2): 135–140.

Singh, R., 1984. Hydrobiological investigations of Neeru Nulla (Bhaderwah) with reference to the benthic macroinvertebrates. *M.Phil. Dissertation*, University of Jammu, Jammu.

Smrutha, Phal, 2010. Studies on the meiobethic production in Kanasgiri backwater, Kali estuary, Karwar. *M.Sc. Dissertation*, Karnatak University, Dharwad.

Wadaan, A. M., 2007. The freshwater growing snail *Physa acuta*: A suitable bioindicator for testing Cadmium toxicity. *Saudi J. Biological Sciences*, 14(2): 185–190.

Witton, B. A., 1975. Zooplankton and microinvertebrates. In: *Studies in river ecology*, (Ed.) B.A. Whitton. Baker Publisher Ltd., London, 2: 87–118.

Chapter 11

Seasonal Variation and Abundance of Microzooplankton in the Coastal Waters of Palk Strait Tamil Nadu

☆ *M. Kalaiarasi, C. Latha Sumathi, A. Priya,*
D. Chellaiyan and C. Stella

ABSTRACT

The present investigation has been made to study the distribution and abundance of microzooplankton in the coastal waters of S. P. Pattinam and Manamelkudi along the Palk bay of Tamil Nadu during July 2005 to June 2006. From the qualitative analysis, the microzooplankton comprised of Foraminiferans, Acantharians, Tintinnids and Metozoans like copepod nauplii, cirripede nauplii, decapod nauplii, veligers of Gastropods and Bivalves were observed during the study period. It included 4 species of foraminiferans, 2 species of Acantharians, 25 species of Tintinnids, large number of nauplii of copepode and cirripede, veligers of Gastropods and Bivalves. Higher density of microzooplankton was observed during summer and premonsoon seasons and lower density was observed during monsoon season. In microzooplankton, tintinnids formed the major group next to copepods and it comprised of 25 species belonged to 9 genera. Among these, the *Tintinnopsis* with 10 species are largely contributed to the total tintinnid population. Copepod nauplii were found the most dominant group of microzooplakton population throughout the period of study at both stations. The cyclopoid nauplii occurred more in number than harpacticoid nauplii. Their maximum occurrence was observed during summer and premonsoon months at both stations and they reached their peak during summer season at station–I and II. Gastropod and bivalve veligers were encountered throughout the study period in both stations.

Keywords: Zooplankton, Microzooplankton, Species composition, Succession, Density.

Introduction

Zooplankton are the vital link in the trophic tier of aquatic environment. In addition it is well known that the abundance and availability of plankton has direct influence on the fish population in any particular area. The study of zooplankton both in quality (Species) and quantity (biomass) reveals the productivity and also the health of the marine ecosystem. Hence it is necessary to widen our knowledge about plankton to assess the fertility of water as well as the management of the fishery resources. The role of microzooplankton in the 'food web' of aquatic environment is being increasingly realized as they are found to be fed by organisms of higher trophic levels (Santhanam *et al.,* 1975). It acts as an intermediate link between the primary producers, the phytoplankton (bacteria, naked flagellates, coccolithophores, peridinians and diatoms) and the macrozooplankton. Information on species diversity, richness, evenness and dominance evaluation on the biological components of the eco-system is essential to understand detrimental changes in environments or deterioration of water quality (Krishnamoorthy and Subramanian, 1999). Species diversity is a basic measure of community structure and organization and the most important parameter to understand the healthy status of the ecosystem. The diversity index gives a measure of how the individuals in a community are distributed. In the context of global loss of thousands of species as a result of pollution and habitat destruction, assessments of species diversity and richness are highly needed (May, 1986). Such studies assist the environment biologists to predict where and how many species go extinct so that certain effective measures may be taken to conserve them (Reise and Bartsch, 1990). While the information available on these aspects from the coast (Pillai *et al.,* 1973; Madhupratap, 1979; Goswami, 1982; Madhupratap, 1983; Goswami and Padmavathi, 1996 and Paulinose *et al.,* 1998) is fairly good, it is meager from the east coast of India (Srinivasan and Santhanam, 1991; Krishnamoorthy and Subramanian, 1999). How ever no accounts of microzooplankton were reported particularly in the waters of on the S. P. Pattinam and Manamelkudi along the Palk bay area. Hence the present study has been made to report the seasonal variation and abundance of microzooplankton in the coastal waters of Palk strait.

Description of the Study Area

The Palk Strait is a strait that lies between the Tamil Nadu state of India and the island nation of Sri Lanka. It connects the Bay of Bengal to the northeast with the Gulf of Mannar to the south. The strait is 40 to 85 miles (64-137 km) wide (Figure 11.1). Several rivers flow into it. The Palk Bay (Lat.9"40'N, Long. 79"20'E) and the Gulf of Mannar biosphere reserve (Lat.8"35'–9"25'N, long. 78"08'–79"30'E) situated along the southeast coast of India are separated by the Rameswaram Island. Sundarapandian Pattinam is commonly known as S. P. Pattinam. This village is located along the Palk Strait in the east coast of South Tamil Nadu. This small fishing village in Ramanathapuram District is known for its fishery resources. Manamelkudi is a village Panchayat in Avudaiyarkoil Taluk of Pudukkottai District. It extended over an area of 1135.24 hect. It is situated along the way of Palk Strait. Both stations are being influenced mainly by the northeast monsoon and has strong potential of both living and non living resources. These two stations are situated in the Palk Strait region of Tamil Nadu.

Materials and Methods

Zooplankton sampling was carried out once in every month. The microzooplankton collections were made by using a No. 32 (Mesh aperture size 54 µm) bolting silk plankton nets for the period of one year from July 2005–June 2006. For qualitative analysis, the plankton net was towed for 10 minutes. For quantitative analysis 140 litres of surface water was strained through No. 32 (Microzooplankton) bag nets. A country boat was used for plankton collection purpose maintaining a set speed and keeping the net dragging distance same. The gear was a conical international coarse silk net made up

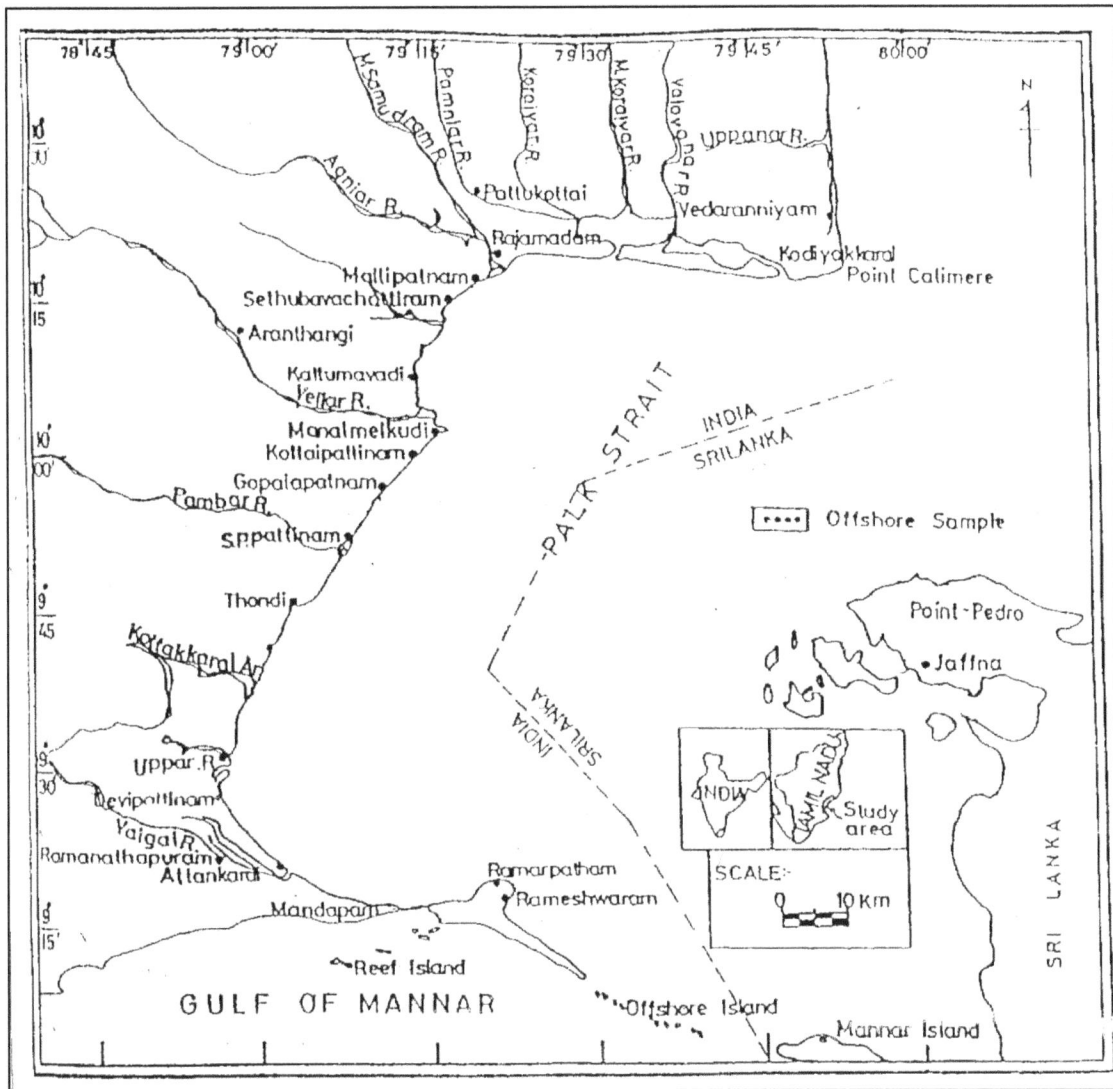

Figure 11.1: Map Showing the Palk Strait Area.

of blue bolting silk cloth No.10, the diameter of the net ring was 57 cm and the side length of the net was 2m. The construction and design of the net was similar to the net used by Russell and Colman (1931) in the Reef Lagoon and Wickstead (1961, 1963, and 1968) in the Singapore Strait, Zanzibar area and off Plymouth. The plankton collections were made by surface hauling (horizontal-towing) method after reaching suitable locations maintaining a constant speed of the boat. All the samples were preserved in 5 per cent formaldehyde and analyzed qualitatively and quantitatively using Sedgwick Rafter plankton counting cell and the number of organisms computed as individuals/1.

Results

Species Composition

Microzooplankton comprised of Foraminiferans, Acantharians, Tintinnids and Metazoans like Copepod nauplii, Cirripede nauplii, Decapod nauplii, veligers of Gastropods and Bivalves during the present study. Their percentage composition is presented in Tables 11.1 and 11.2. 4 species of Foraminiferans, 2 species of Acantharians, 25 species of Tintinnids, large number of nauplii of copepods and cirripede, veligers of Gastropods and Bivalves were the contributors for the total microzooplankton population. In the present study variations of microzooplankton were observed in different months at stations–I and II (Figures 11.2 and 11.3). Four species *Globigerina quinqueloba, G. pava, G. triculinoides, G. rubescens* of foraminifera were observed in the collections. They occurred rarely and present in certain months of the year. At station–I, they were recorded in summer and postmonsoon months while at station–II they were encountered during postmonsoon, summer months, and in the month of July. Two species of *Acantharians* were present few in number only at station–I. The observed species were *Acanthochiasma fusiforme* and *Acanthometron* sp.

In microzooplankton, tintinnids formed the major group next to copepods in most of the months and it comprised of 25 species belonged to 9 genera. Among these, the *Tintinnopsis* with 10 species largely contributed to the total tintinnid population. The other genera *Codonellopsis, Dictyocysta, Leprotintinnus, Tintinnidium* and *Favella* represented with one or two species each. The twenty five species were *Tintinnopsis nordquisti, T. cylindrica, T. tocantinensis, T. directa, T. beroidea, T. tubulosa, T. strigosa, T. acuminata, T. campanula, T. lobancoi, Codonellopsis ostenfeldi, C. pussila, C. ecaudata, Dicytocysta seshaiyai, D. elegans, Favella philippinensis, F. ehrenbergii, Leprotintinnus nordquisti, L. pellucidus, Hellicostomella longa, Rhabdonella* sp., *Metacyclis jorgenseni, Tintinnidium primitivum* and *T. incertum.* 9 species were *recorded* at both station–I and II. *Hellicostomella longa, Rhabdonella* sp., *Metacyclis jorgenseni, Codonellopsis pussila, C. ecaudata, Dicytocysta elegans, Leprotintinnus pellucides, Tintinnopsis campanula* were recorded only in station–I. *Tintinnopsis tocantinensis, Tintinnidium* sp., *Favella ehrenbergii, Dicytocysta seshaiyai* and *Codonellopsis ostenfeldi* were recorded only at station–II. *Favella ehrenbergii* and *F. philippinensis* were recorded almost throughout the study period in large number.

At station–I *Tintinnopsis tubulosa, T. directa, T. beroidea, Dicytocysta elegans, Codonellopsis pussila, C. ecaudata, Favella philippinensis* and *Hellicostomella longa* were found more abundant in most of the months, whereas *Rhabdonella* sp., *Tintinnopsis campanula, Leprotintinis pellucidus* and *Metacylis jorgenseni* were observed in two or three months. At station–II *Tintinnopsis cylindrica, T. strigosa* and *Favella philippinensis* were abundantly present among the Tintinnids. Next to these three species *Tintinnopsis beroidea, T. directa* and *Codonellopsis ostenfeldi* were dominated in number. *Tintinnopsis tocantinensis,* was present few in number and in few months of the year and other species rarely showed their occurrence. The most dominant microzooplakton population observed was copepod nauplii and found to present throughout the period of study at both the stations. The cyclopoid nauplii occurred more than harpacticoid nauplii. Eventhough they encountered throughout the period, their maximum occurrence was during summer and premonsoon months at both the stations and they reached their peak during summer season at stations–I and II. The cirripede nauplii were found to be occurring throughout the period, but their contribution was very less when compared to copepod nauplii. However, they showed their maximum number during premonsoon and monsoon seasons and occurred very few in number during summer season at station–I, while they reached their peak during monsoon season, especially in the month of December 2005 at station–II.

Figure 11.2: Variations in different Groups of Microzooplankton at Station–I.

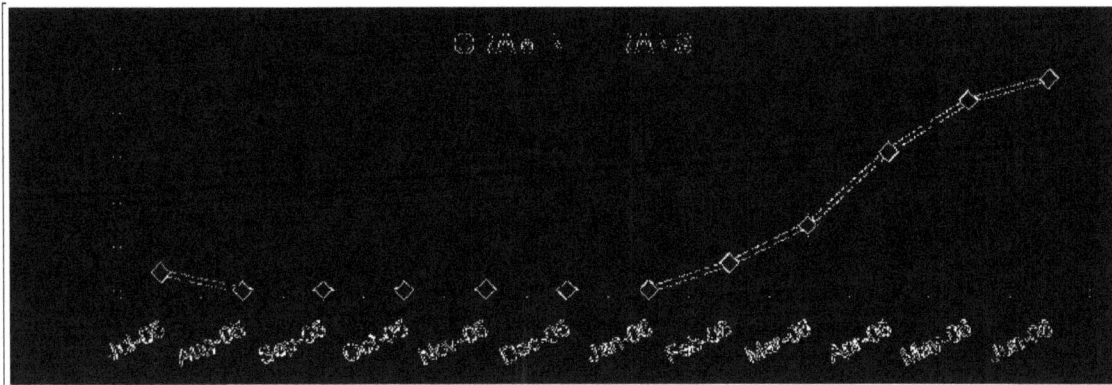

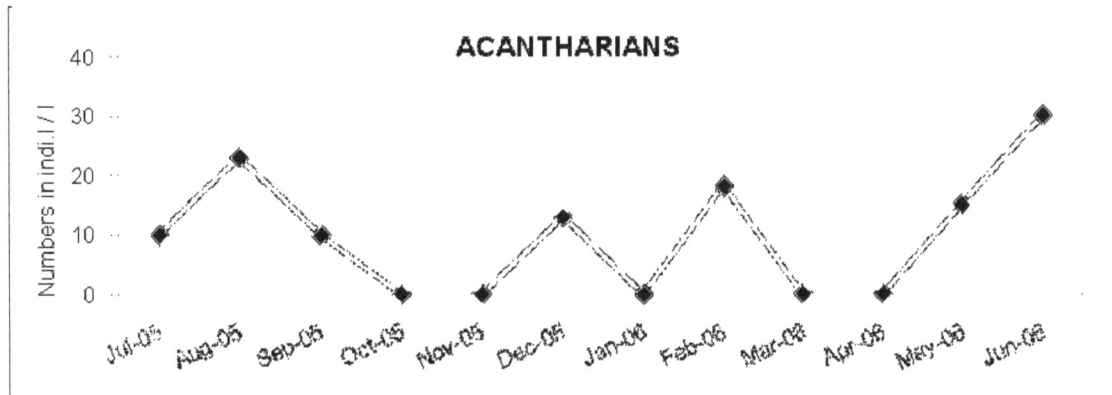

ACANTHARIANS

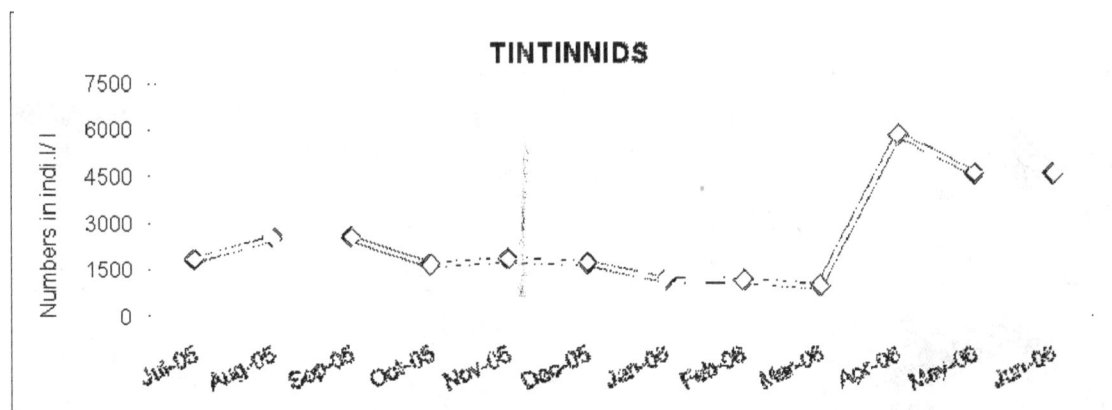

TINTINNIDS

Contd...

Figure 11.2–*Contd...*

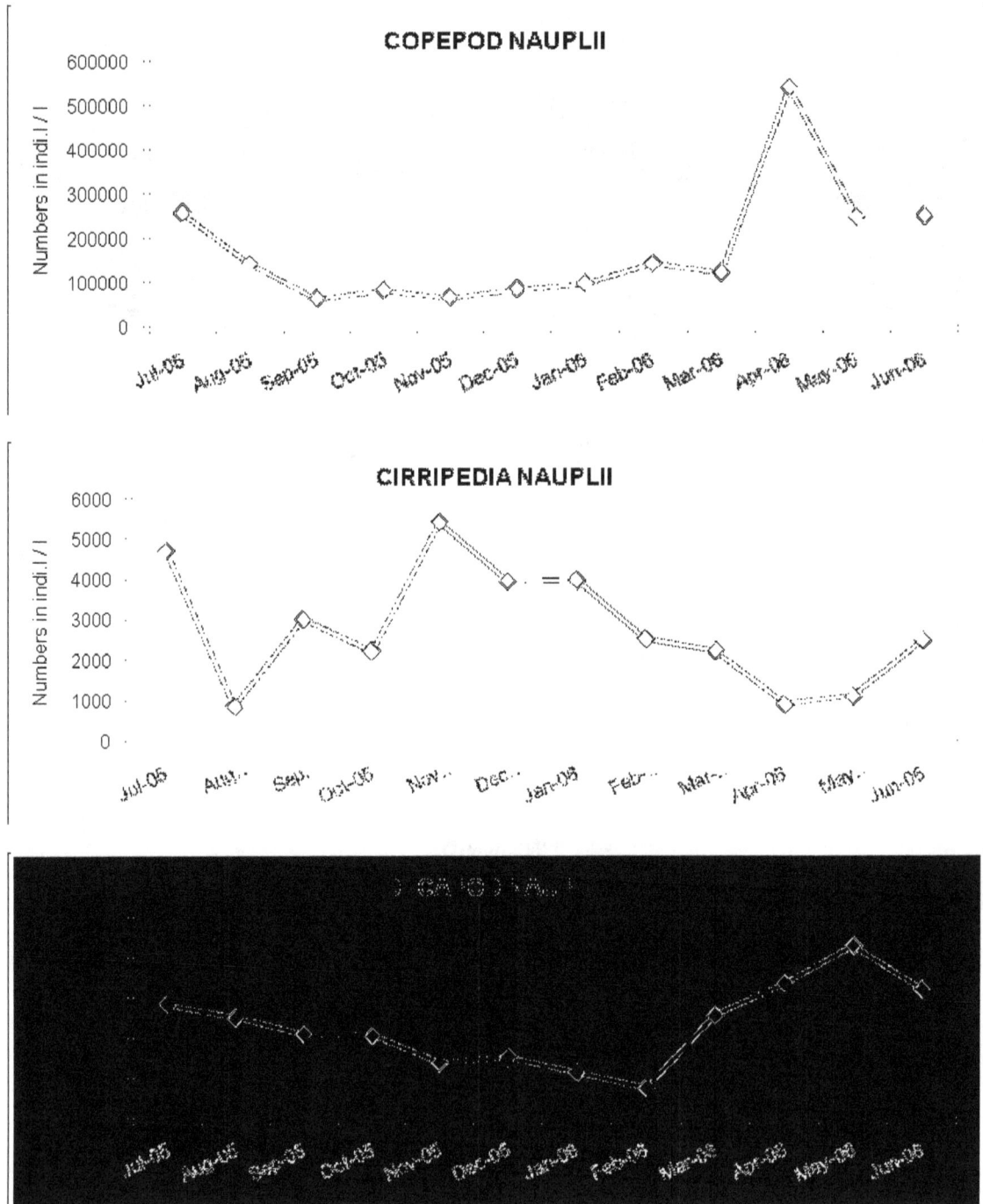

COPEPOD NAUPLII

CIRRIPEDIA NAUPLII

Contd...

Figure 11.2–*Contd...*

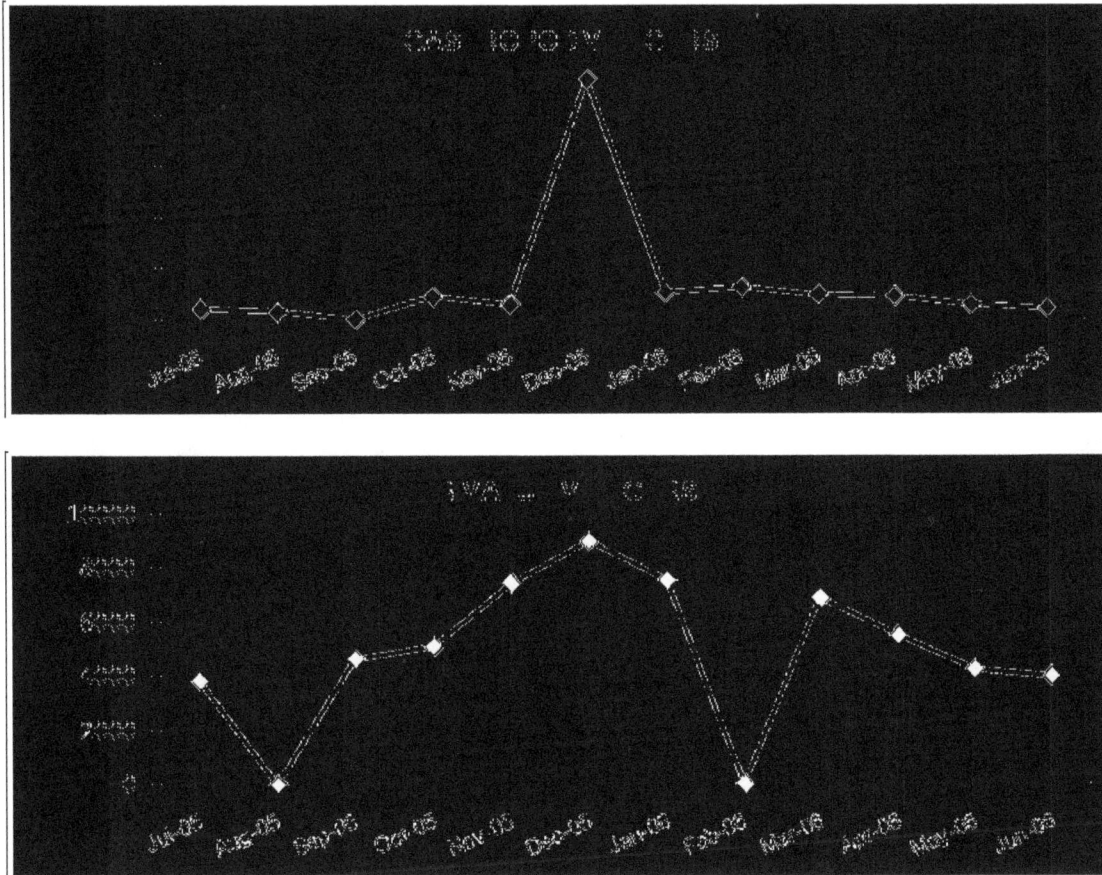

Decapod nauplii were observed large in number throughout the study period at both stations–I and II. They gradually decreased in their number from premonsoon period to postmonsoon period and there was gradual increase from post monsoon to summer season and the peak was noted during the month of May 2006 and again it was started to decline at station–I. At station–II *Penaeus* nauplii reached their maximum number during premonsoon season (September 2005) and gradually decreased from October 2005 to January 2006 and a sudden peak was noted during summer months. Gastropod and bivalve veligers were encountered throughout the study period in both the stations. The gastropod veligers were more in number than bivalve veligers. Their maximum number was observed in monsoon season at station–I. A sudden peak was recorded during the month of December 2005 at station–I, and not much variation was observed at station–II throughout the study period. Bivalve veligers at station–I showed much variation in their occurrence, and were not recorded during the months of August 2005 and February 2006. They reached their maximum number during monsoon season. At station–II they were abundantly observed throughout the study period and they were found high in number during the month of April 2006.

Figure 11.3: Variations in different Groups of Microzooplankton at Station–II.

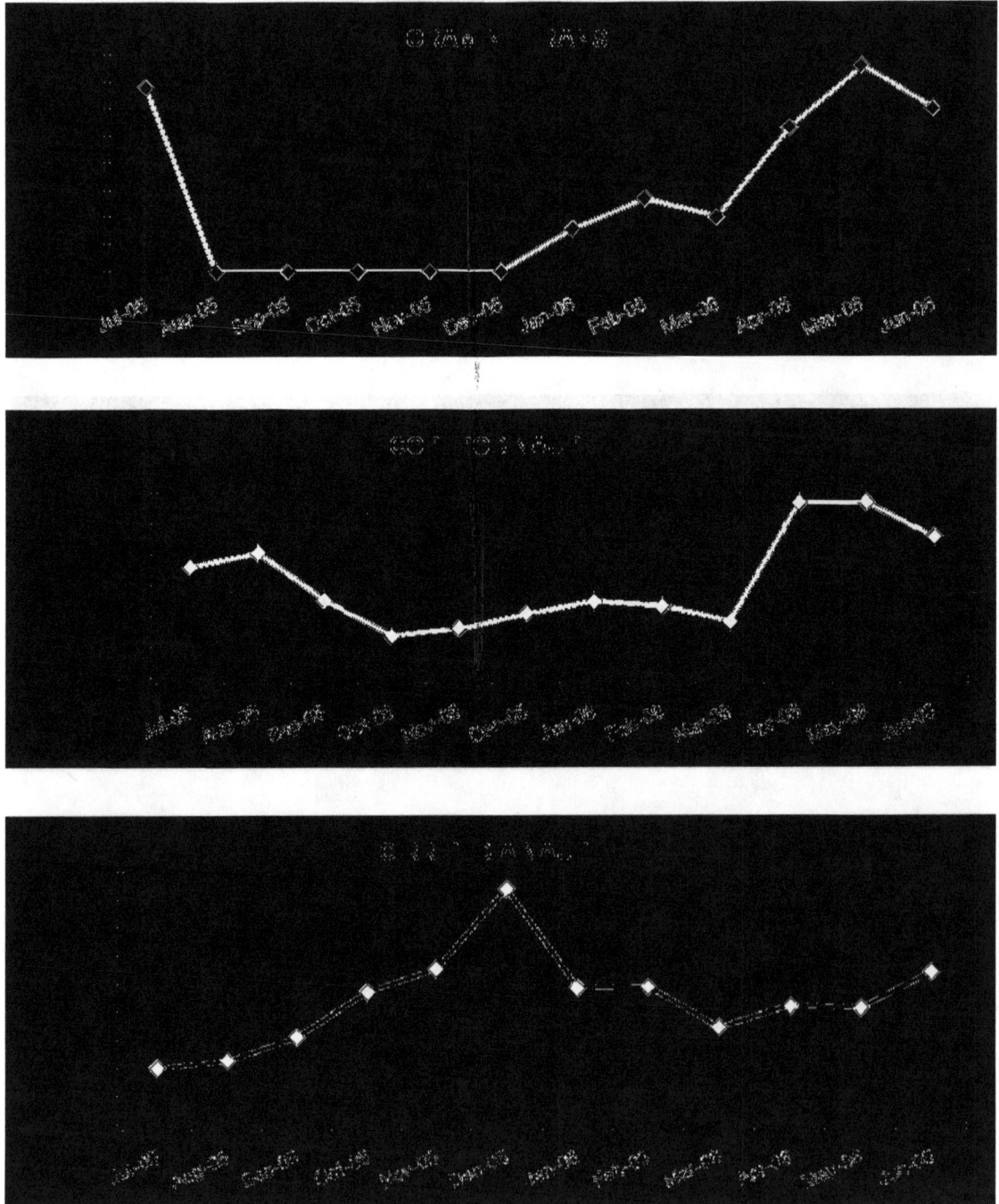

Contd...

Figure 11.3–*Contd...*

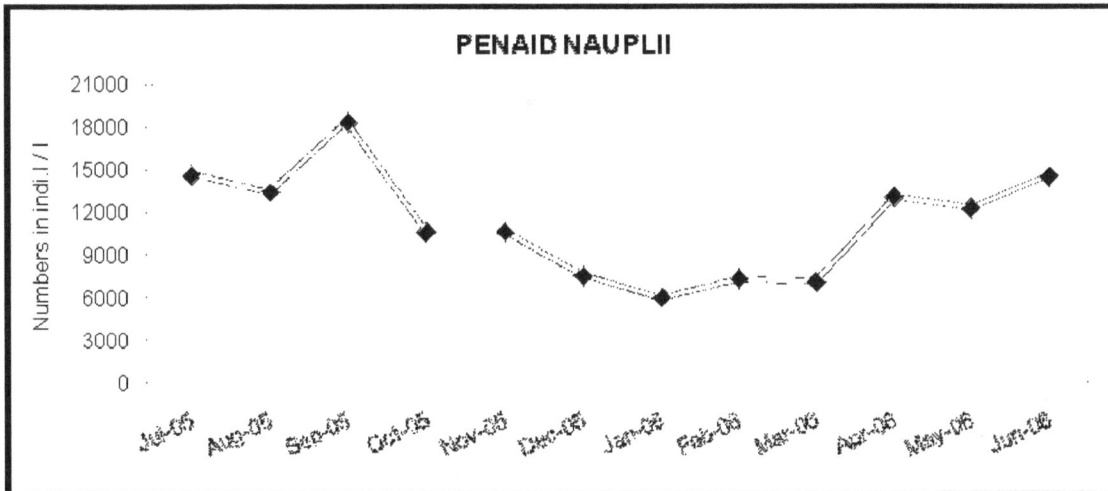

PENAID NAUPLII

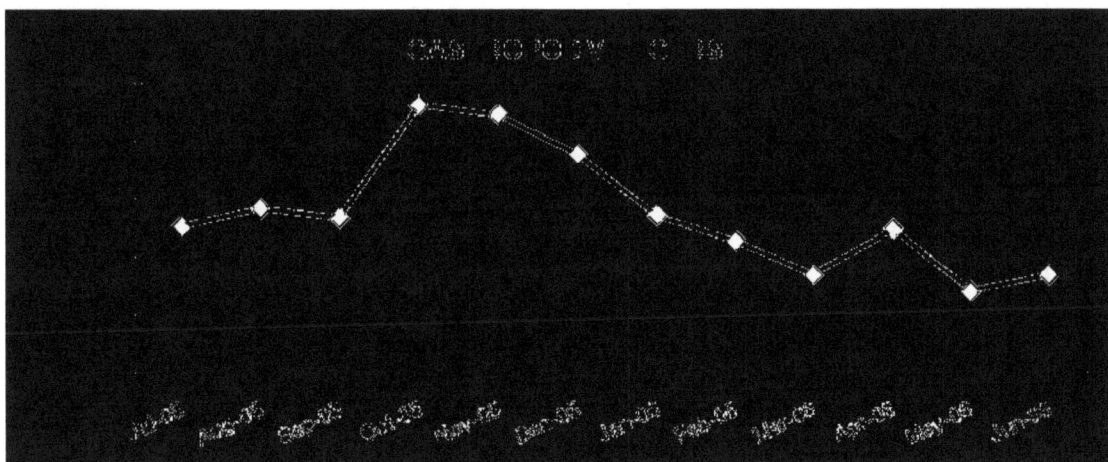

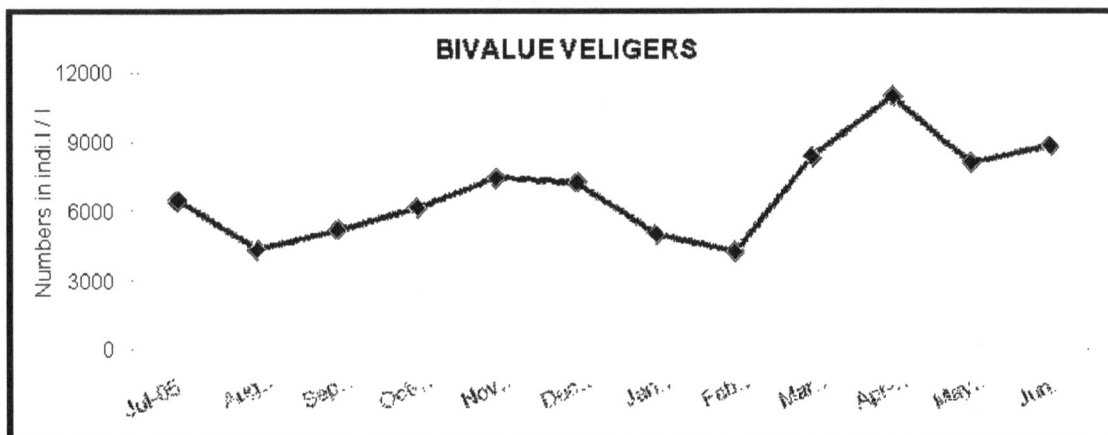

BIVALUE VELIGERS

Table 11.1: Percentage Composition of different Groups of Microzooplankton at Station–I

Percentage	Jul-05	Aug-05	Sep-05	Oct-05	Nov-05	Dec-05	Jan-06	Feb-06	Mar-06	Apr-06	May-06	Jun-06
Foraminiferans	0.0000%	0.0000%	0.0000%	0.0000%	0.0000%	0.0000%	0.0000%	0.0001%	0.0002%	0.0001%	0.0003%	0.0003%
Acantharians	0.0000%	0.0001%	0.0001%	0.0000%	0.0000%	0.0001%	0.0000%	0.0001%	0.0000%	0.0000%	0.0001%	0.0001%
Tintinnids	0.0065%	0.0161%	0.0322%	0.0150%	0.0197%	0.0069%	0.0090%	0.0068%	0.0063%	0.0101%	0.0164%	0.0165%
Copepod nauplii	0.9121%	0.9035%	0.7924%	0.7472%	0.7085%	0.3492%	0.7562%	0.8542%	0.7920%	0.9380%	0.8877%	0.9018%
Cirripedia nauplii	0.0167%	0.0054%	0.0381%	0.0202%	0.0580%	0.0162%	0.0306%	0.0151%	0.0145%	0.0016%	0.0039%	0.0089%
Decapod nauplii	0.0299%	0.0474%	0.0787%	0.0552%	0.0439%	0.0183%	0.0258%	0.0132%	0.0494%	0.0173%	0.0457%	0.0344%
Gastropod veligers	0.0212%	0.0275%	0.0000%	0.1165%	0.0904%	0.5725%	0.1203%	0.1105%	0.0931%	0.0233%	0.0308%	0.0235%
Bivalve veligers	0.0136%	0.0000%	0.0584%	0.0460%	0.0795%	0.0368%	0.0580%	0.0000%	0.0446%	0.0095%	0.0152%	0.0145%

Table 11.2: Percentage Composition of different Groups of Microzooplankton at Station–II

Percentage	Jul-05	Aug-05	Sep-05	Oct-05	Nov-05	Dec-05	Jan-06	Feb-06	Mar-06	Apr-06	May-06	Jun-06
Foraminiferans	0.0003%	0.0000%	0.0000%	0.0000%	0.0000%	0.0000%	0.0001%	0.0002%	0.0002%	0.0002%	0.0002%	0.0002%
Tintinnids	0.0230%	0.0081%	0.0236%	0.0202%	0.0263%	0.0374%	0.0281%	0.0244%	0.0211%	0.0227%	0.0259%	0.0344%
Copepod nauplii	0.8388%	0.8742%	0.7745%	0.6619%	0.6852%	0.7459%	0.8282%	0.8280%	0.7874%	0.8739%	0.8905%	0.8498%
Cirripedia nauplii	0.0047%	0.0049%	0.0095%	0.0227%	0.0233%	0.0300%	0.0162%	0.0172%	0.0148%	0.0064%	0.0064%	0.0100%
Penaid nauplii	0.0681%	0.0577%	0.1119%	0.1021%	0.0897%	0.0542%	0.0400%	0.0519%	0.0611%	0.0402%	0.0384%	0.0539%
Gastropod veligers	0.0352%	0.0363%	0.0487%	0.1337%	0.1126%	0.0806%	0.0540%	0.0481%	0.0434%	0.0228%	0.0132%	0.0188%
Bivalve veligers	0.0299%	0.0187%	0.0318%	0.0594%	0.0628%	0.0519%	0.0333%	0.0302%	0.0720%	0.0338%	0.0254%	0.0328%

Species Succession of Microzooplankton and Macrozooplankton at Stations–I and II (July 2005–June 2006) Showing Dominant and Sub-dominant Species with Density in Parenthesis

Premonsoon Season (July-September) - No. of Organisms (individuals/l)

During premonsoon seasons the Species succession of microzooplankton were observed in station 1, it was recorded *Parapontella brevicornis* (40,567), *Labidocera pava* (37,360), *Pontellinaplumata* (35,924), *Temoralongicornis* (34,127), *Centropagesfurcatus* (30,579), *Cyclopina longicornis* (27,736), *Metacalanus aurivilli* (22,290), *Microcalanus pusillus* (22,723), *Paracalanus parvus* (20,889) and *Pseudocalanus elongates* (20, 606) and in station 11 it was recorded *Penaid nauplii* (46,509), *Pontella danae* (40,227) *Pseudocalanus elongates* (36,838), *Paracalanus parvus* (34,105), *Isias tropica* (32,238), *Labidocera pava* (27,589), *Metacalanus aurivilli* (26,593), *Temora longicornis* (25,465), *Microcalanus pusillus* (24,076), *Gastropod veligers* (24,044) and *Bivalve veligers*.

Monsoon season (October-December)

During monsoon season in station 1 the Gastropod veligers (1,61,328), *Pontella danae* (25,444), Bivalve veligers (21,516), *Pontellina plumata* (16,062), *Metacalanus aurivilli* (14,710), *Microcalanus pusillis* (14,691), *Decapod nauplii* (14,672), *Oithona halgolandica* (14,671), *O. brevicornis* (14,238), *Labidocera pava* (12,832), *Pseudocalanus elongates* (12,369) were reported and the Gastropod veligers were the dominant and the peak was noted in December 2005 and in station 11 the Gastropod veligers (38,438), *Penaid nauplii* (28,772), *Bivalveveligers* (20,838) *Labidocera pava* (20,194), *Pontella danae* (18,662), *Paracalanus parvus* (16,796), *Parapontella brevicornis* (15,904), *Cyclopina longicornis* (13,340), *Centropages typicus* (12,091) and *Acartia* sp. (8,537) were reported.

Postmonsoon Season (January-March)

During Postmonson season, in station 1 it was recorded *Centropages furcatus* (54,050) Gastropod veliger (48,435), *Isias tropica* (20,350) *Paracalanus parvus* (20,251), *Pontellina plumata* (19,292), *Labidocera pava* (18,543), *Microcalanus pusillus*, *Metacalanus aurivilli*, *Pseudocalanus elongates*, *Eurytemora hirundoides* and *Bivalve veligers*. In the month of January 2006 the outnumbered species was *Pontella dana* and in station–II *Pontella danae* (31,441), *Parapontella brevicornis* (26,791), *Inachus dorsettensis* (24,764), *Paracalanus parvus* (23,506), *Labidocera pava* (23,106), *Cyclopina longicornis* (21,421), *Penaid nauplii* (20,448), *Gastropod veligers* (19,950), *Bivalve veligers* (17,639), *Temora longicornis* (15,359) and *Pseudocalanus elongates* (15,303) it was recorded.

Summer Season

During summer among tintinnids the *Favella philippinensis* was dominant in number and among copepods, the following species *Paracalanus parvus* (66,909), *Pseudocalanus elongates* (60,066), *Pontellinaplumata* (60,118), *Metacalanusaurivilli* (58,862), *Isiastropica* (54,091), *Pontella danae* (52,473), *Temora longicornis* (49,245), *Isias clavipes*, *Labidocera pava*, *parapontella brevicornis*, *Decapod nauplii* and *Acartia* sp. were dominated during summer season and in station–II Next to copepod nauplii, the decapod nauplii, *Metacalanus aurivilli* (59,663) *Paracalanus parvus* (57,442), *Labidocera pava* (52,711), *Cyclopina longicornis* (51,630), *Pseudocalanus elongatus* (46,920) *Pontella danae* (45,464), *Penaid nauplii* (39,849) and *Temora longicornis* (36,707) were dominated during summer season.

Population Density

The variation of population density at stations–I and II are given in Figures 11.4 and 11.5. At station–I population density variation ranged from 79,188/l (September 2005) to 57,7122/l (April 2005).

STATION I

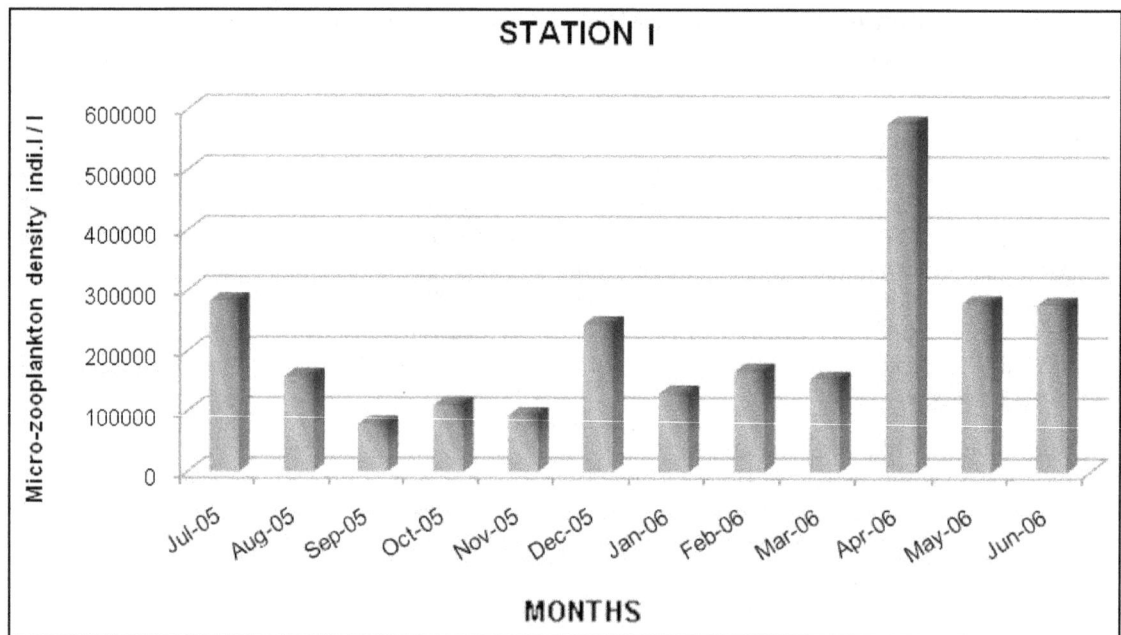

Figure 11.4: Variations in Microzooplankton Density at Station–I.

STATION II

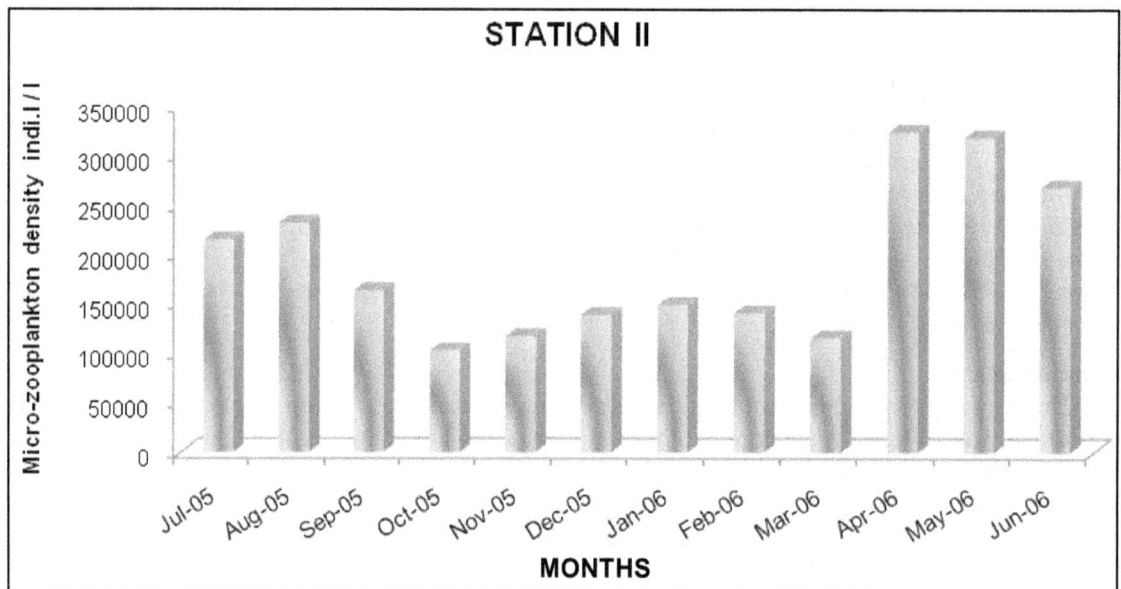

Figure 11.5: Variations in Microzooplankton Density at Station–II.

Density of Foraminiferans fluctuated from 8 organisms/l (July 2005) to 95 organisms/l (June 2006) during the study period. Acantharians showed fluctuation from 10 organisms/l (July and September 2005) to 30/l (June 2006). Tintinnids contributed from 966/l (March 2006) to 5841/l (April 2006). Population density of copepods varied from 62,748/l (September 2005) to 54,1365/l (April

2006). Cirripedia nauplii population density varied from 850/1 (August 2005) to 5,423/1 (November). Decapod nauplii showed variation from 2,200/1 (February) to 12,803/1 (May 2006). Population density of Gastropod veligers fluctuated between 4,346/1 (August 2005) and 1,40,000/1 (December 2005). The density of Bivalve veligers ranged from 3,850/1 (July 2005) to 9000/1 (December 2005). Higher population density was mainly contributed by more number of copepod nauplii, decapod nauplii, gastropod veligers, and tintinnids.

The population density of microzooplankton at station–II ranged from 1,03,662 organisms/1 (October 2005) to 3,25,218 organisms/1 (April 2006). Foraminiferans varied from 16/1 (January 2006) to 76/1 (May 2006). The population density of Tintinnids at station–II fluctuated from 1899/1 (August 2005) to 9267/1 (June 2006). Copepod nauplii was outnumbered when compared with all other organisms, their population density varied between 68614/1 (October 2005) and 284805/1 (May 2006). Cirripedia nauplii showed low density 1006/1 during the month of July 2005 and the high density 4193/1 was showed during December 2005. Gastropod veligers population varied from 4228/1 (May 2006) to 13858/1 (October 2005). Density variation of Bivalve veligers ranged from 4268/1 (February 2006) to 11000/1 (April 2006). In general the minimum value at station–I and station–II was observed during monsoon season, but the maximum value was observed during summer and premonsoon seasons.

Discussion

In the present study the higher density of microzooplankton was observed during summer and premonsoon seasons and lower density was observed during monsoon season. This similar observation was made by Damodara Naidu (1980) and Chandran (1982) in the south west coast of India. Prasad (1969) reported that the plankton biomass of the whole of the Arabian sea showed richer zooplankton biomass during the southwest monsoon. Reasonable extent of information on the seasonal distribution of zooplankton from Parangipettai coastal waters are also available (Subbaraju, 1970, Krishnamurthy and Santhanam, 1975; Santhanam *et al.*, 1975 and Kumar, 1993). Monsoon played a major role in the distribution and abundance of zooplankton population in tropical waters. Station–I and II are much influenced by the northeast monsoon. Salinity, tidal amplitude, circulation patterns, strength of the water currents besides predation, competition, size and external morphology are known to play a vital role in determining the quality and quantity of the zooplankton (Subbaraju and Krishnamurthy, 1972; Santhanam *et al.*, 1975; Madhupratap, 1976, 1978, 1980; Goswami and Selvakumar 1977; Peter *et al.*, 1977; Stephen, 1977; Mohamed and Rahaman, 1987).

In the present study higher density of foraminiferans was recorded during summer season at station–I and in station–II it was recorded in summer and early premonsoon (July 2005). Changes in the foraminifera population in the waters from time to time generally reflect the climatic conditions. A study of planktonic foraminiferans of south west coast of India during summer by Kameswara Rao and Jayalakshmy (1997), supported the present study. The minimum quantity of Tintinnoidea and foraminifera were present in the plankton populatiion almost throughout the year in the two stations. In the present study the appearance of *Globigerina* sp., and *Acanthometran* sp. were observed and their peak were noticed during summer season and this similar observation has been made by Maruthanayagam *et al.* (2001) in Palk Bay and Gulf of Mannar. Margalef (1963) Beers and Stewart (1969 a and b) showed that tintinnid numbers were approximately 20 per cent of the total ciliates. In the present study, tintinnid numbers were contributing 4.0 per cent to the microzooplankton population.

In Tintinnids, *Tintinnopsis* species formed the major component and occurred more throughout the period of study. In earlier reports Santhanam *et al.* (1975) mentioned the major occurrence of

Tintinnopsis species throughout the period of study in Vellar estuary. Occurrence of *Tintinnopsis nordquisti, T. tocantinensis, T. cylindrica* and *Leprotintinnus* sp. during summer in the present study coincides with the findings of Maruthanayagam *et al.* (2001) in Palk Bay and Gulf of Mannar. Tintinnids occurred throughout the year and the maximum numerical abundance was found in April 2006 at station–I, and in summer season at station–II. Foraminiferans showed regular occurrence. Peak occurrence was recorded in summer at station–I and II. Parsons and Kessler (1986) and Mangesh Gauns *et al.* (1996) reported the higher population of microzooplankton (protozoan) in coastal waters of eastern Arabian Sea during summer seasons Copepod nauplii were abundantly observed throughout the study period at both the stations due to the continuous breeding nature of these species and this similar observation has been made by Kumar, (1993). Maximum numerical counts of the copepod population have been reported when the temperature and salinity were high in the environment (Rajasegar, 1998). A positive trend was noted in two stations between the number of copepods and the number of nauplii and there was a direct relationship between the number of nauplii and the number of adults.

Gastropod veligers were more than the bivalve veligers in both the stations throughout the study period. Hopkins (1966) recorded abundant of mollusc larvae both gastropod veligers and pelecypod veligers in west Bay of St. Andrew Bay. Generally higher density of gastropod veligers reported during post monsoon season due to their peak breeding activity. This observation was supported by George (1958), Nair and Tranter (1972), Silas and Parameswaran Pilai (1975) in Cochin waters, Chandran (1982) in velar estuary. Gastropods differed completely from copepods in reaching their peak of dominance. In the present study the population density of Gastropod veligers was very high during monsoon season at both the stations. This may be due to the influence of environmental factors. The high zooplankton population density during the summer season could be related to the stable hydrographical condition while the low density was observed during the monsoon season was attributed to heavy flood and freshwater inflow (Santhanan *et al.*, 1975; Shanmugam *et al.*, 1986; Kumar, 1993 and Rjasegar, 1998). In the present study the density of the microzooplankton was comparatively higher at station–II than at station–I.

References

Beers, J. R. and Stewart, G. L., 1969a. The vertical distribution of microzooplankton and some ecological observations. *J. Cons. Inst. Explor. Mer.*, 33(1): 30–44.

Beers, J. R. and Stewart, G. L., 1969b. Micro–zooplankton and its abundance relative to the larger zooplankton and other season components. *Mar. Biol.*, 4: 182–189.

Chandran, R., 1982. Hydro biological studies in the gradient zne of the Vellar estuary. *Ph.D. Thesis*, Annamalai University, India, 195 pp.

Damodara Naidu, W., 1980. Studies on Tintinnids (Protozoa : Ciliata) of Porto Novo region. S. India. Ph. D., Thesis, Annamalai University, India, 281.

George, M. J., 1958. Observations on the plankton of the Cochin backwater. *Indian J. Fish.*, 5: 375–401.

Goswami, S. C and Padmavathi, 1996. Zooplankton production composition and diversity in the coastal waters of Goa. *Indian J. Mar. Sci.*, 25: 91–97.

Goswami, S. C. and Selvakumar, R. A., 1977. Plankton studies in the estuarine system of Goa. Proc. Symp. *Warm water Zoopl., Spl. Publ.* UNESCO/NIO, Goa: 226–241.

Goswami, S. C., 1982. Distribution and diversity of copepods in the Mandovi–Zuari estuarine system, Goa. *Indian J. Mar. Sci.*, 11: 292–295.

Hopkins, T. L., 1966. The plankton of the St. Andrew Bay system, Florida. Publs. *Inst. Mar. Sci. Univ. Tex.*, 11 : 12–64.

Kameswara Rao K. and Jayalakshmy, K.V., 1997. An overview of the planktonic foraminiferal fauna in waters off the kerala coast, south–west India during summer. *J. mar. biol. Ass. India*, 39 (1 and 2): 59–68.

Krishnamurthy K. and Santhanam, R., 1975. Ecology of tintinnids (Protozoa : Ciliata) in Porto Novo region. *Indian J. Mar. Sci.*, 4 : 181–184.

Krishnamoorthy P. and Subramanian, P., 1997. Meroplankton production in the Gulf of Mannar and Palk Bay on the south east coast of India. *J. Mar. Biol Ass. India*, 39: 44–48.

Kumar K., 1993. Studies on copepods occurring in coastal waters of. Parangipettai. *Ph.D. Thesis*, Annamalai University, India, 166 pp.

Madhupratap, M., 1976. Studies on the ecology of zooplankton of Cochin backwater (A tropical estuary). *Ph.D. Thesis*, Cochin University.

Madhupratap, M., 1978. Studies on the ecology of zooplankton of Chochin Backwaters. *Mahasagar–Bull. Natn. Inst. Oceanography.*, 11(19–2): 45–56.

Madhupratap, M., 1979. Distribution, community structure and species succession of copepods from Cochin back wares. *Indian J. Mar. Sci.*, 8: 1–8.

Madhupratap, M., 1980. Ecological relations of coexisting copepod species from Cochin backwaters, *Mahasagar, Bull. Natn. Inst. Oceanogr.*, 13(1): 45–52

Madhupratap, M., 1983. Zooplankton standing stock and diversity along an oceanic track in the western Indian Ocean. *Mahasagar. Bull. Nat. Inst. Oceanogr.*, 16(4): 463–467.

Mangesh Gauns, R. Mohan Raju and Madhupratap, M., 1996. Studies on the microzooplankton from the central and eastern Arabian Sea. *Ibid.*, 71: 874–877.

Margalef, R., 1963. Role des cilies dans le cycle de la vie pleagique en Mediterrance. *Rapp. Proc. Verb. C. I.E. S. M. M.*, 17(2): 511–512.

Maruthanayagam C., C. Senthil Kumar and P. Subramanian. (2001). *J. Mar. Biol. Ass. India*, 43 (1 and 2): 186–189.

May, R. M 1986. How many species are there? *Nature*, 324: 514–515.

Mohamed T. A. M. H. and A. A. Rahaman 1987. Seasonal distribution of Plankton in Agniar estuary. *J. Mar. Biol. Ass. India*, 29: 273–279.

Nair, K. K. C. and Tranter, D. J., 1972. Zooplankton distribution along salinity gradient in the Cochin, backwater before and after monsoon. *J. Mar. Biol. Ass. India*, p. 203–210.

Parsons T. and Kessler, T. A., 1986. In the role of freshwater outflow in coastal marine ecosystem (Ed.) *Skrestet Berlin*, p. 161–182.

Paulinose. V. T. C. H. I. Desai, V. R. Nair, N. Ramaiah and Gajbhiye, S. N., 1998. Zooplankton standing stock and diversity in the Gulf of Kachchh with special reference to larvae of decapods and pisces. *Indian J. Mar. Sci.*, 27 : 340–345.

Peter K. J., H. K. Iyer, S. John and Radhakrishnan, E.V., 1977. Distribution and coexistence of planktonic organisms along the south–west coast India. *Proc. Symp. warm Water Zoopl. Spl., Publ.,* UNESCO/NIO : 80–86.

Pillai, P., Parameswaran and Pillai, M. Ayyappan, 1973. Tidal influence on the diel variations of zooplankton with special reference to copepods in the Cochin Backwater. *Ibid.,* 15 (1) : 411–417.

Prasad R. R., 1969. Zooplankton in the Arabian Sea and the Bay of Bengal with the discussion on the fisheries of the regions. *Proc. Natn. Inst.* Sci; 35 (5) 8: 399–437.

Rajasegar, M., 1998. Environmental inventory on velar estuary (South coast of India) in relation to shrimp farming. *Ph.D.* Thesis, Annamalai University, India, 95 pp.

Reise K. and Bartsch, I., 1990. Inshore and offshore diversity of ephibenthos dredge in the North Sea. *Neth. J. Ses. Res.,* 25 : 175–179.

Russell F. S. and Colman, J. S., 1931. The zooplankton. 1. Gear, Methods and station lists. *Scient. Rept. Gt. Barrier Reef. Expd.,* 2 : 5–36.

Santhanam, R., K. Krishnamoorthy and Subbaraju, R. C., 1975. Zooplankton of Porto Novo, south India. *Bull. Dept. Mar. Sci. Univ. Cochin.,* 7 : 899–911.

Shanmugam, A., R. Kasinathan and Maruthamuthu, S., 1986. Biomass and composition of zooplankton from Pitchavaram mangroves south east coast of India. *Indian J. Mar. Sci.,* 15(2): 111–113.

Silas, E.G. and Pillai, Parameswaran, 1975. Dynamics of zooplankton in a tropical estuary (Cochin backwater) with a revies on the plankton fauna of the environment. *Bull. Dept. Mar. Sci. Univ. Cohin.* 7(2): 329–355.

Srinivasan A. and Santhanam, R., 1991. Tidal and seasonal variation in zooplankton of Pullavazhi brackish water, southeast coast of India. *Indian J. Mar. Sci.,* 20(3) 182–186.

Stephen R., 1977. Calanoid copepods from the shelf and slope waters off Cochin: Distribution, biomass and species diversity. *Proc. Symp. Warm Water Zoopl. Spl. Pub.* UNESCO/NIO : 21–27.

Subbaraju, R. C. and Krishnamurthy, K., 1972. Ecological Aspects of plankton production. *Mar. Biol.,* 14: 25–31.

Subbaraju, R. C., 1970. Studies on planktonic copepod of estuarine and inshore waters at Porto (S. India). *Ph.D. Thesis,* Annamalai University, India, 225 pp.

Wickstead J. H. 1961. A quantitative and qualitative survey of some Indo–West Pacific plankton. *Fish. Publs. Lond.,* 16 : 1–200.

Wickstead J. H. 1963. Estimates of total zooplankton in the Zanzibar area of the Indian Ocean with a comparison of the results with two different nets. *Proc. Zool. Soc. Lond.,* 141 : 577–608.

Wickstead J. H. 1968. Temperate and tropical plankton a quantitative comparison. *J. Zool. Lond.,* 155 : 253–269.

Chapter 12

Seasonal Changes in Proximate Composition of Muricids Species of *Chicoreus virgineus ponderosus* and *Siratus virgineus ponderosus* from Palk Bay in Tamil Nadu

☆ *R. Ravichandran, C. Latha Sumathi, D. Chellaiyan,*
A. Priya and C. Stella

ABSTRACT

The proximate composition of protein, carbohydrate and lipid are the yardstick to measure and assess the nutritional quality of food sources. This present investigation has been made to evaluate the seasonal variation of the proximate composition of two species of *Chicoreus virgineus ponderosus* and *Siratus virgineus ponderosus*. These proximate composition increased from July onwards due to the maturation of gonads and reached the maximum maturation during October-January. When the gonads became fully matured, the protein, carbohydrate and lipid decreased slowly from February to June. The fluctuations in the proximate composition of both the species are largely attributable to their reproductive and feeding activities.

Keywords: Proximate composition, Protein, Carbohydrate, Lipid, Muricidae, Chicoreus species, Palk bay.

Introduction

Molluscs are highly esteemed delicious sea foods and are considered as next in importance only to fishes and prawns. Proteins are fundamental bio molecules in all aspects of cell structure and function. An increasing demand for good quality of animal protein for the exploding population has led to effective and increasing exploitation of the aquatic resources. Carbohydrates are major sources of energy in all human diets. The ratio of carbohydrate was less when compare to the other nutrients such as proteins and lipids in animal tissues, especially in aquatic animals. Lipids can be defined as substances such as a fat, oil or wax that dissolves in alcohol but not in water. Fatty acids are the principal components in lipids. The knowledge of the Proximate Composition of any edible organisms is extremely important since the nutritive value is reflected in its biochemical contents (Nagabhushanam and Mane, 1975). The immense number of marine gastropods holds promise as a potential cheap and nutritive source. Investigations on the major Proximate Composition as protein, carbohydrate and lipid in Muricids are very much limited. These economically important edible marine Muricid gastropods are found along the South east coast of India. Studies were conducted on other gastropods like *Bullia vittata* (Thilaga, 1985); *Hemifusus pugilinus* (Anandakumar,1986); *Pythia plicata* (Shanmugam, 1987) and *Thais* sp. (Tagore, 1989) are estimate the Proximate Composition.

Seasonal variations in Proximate Composition of some tropical and temperate Muricids have been reported by Stickle (1970 a and b) in *Thais lamellosa*; Stickle Duerr (1970) in *Thais lamellose*; Stickle (1973) in *Thais lamellosa*; Kapeer *et al.* (1985) in *Thais haemastoma*; Murugan *et al.* (1991) in *Chicoreus ramosus*. Giese (1969) has reported a new approach to the proximate composition of the molluscan body. Suryanarayanan and Nair (1976) have studied the seasonal variation on the proximate composition of *Cellana radiata*. Lombard (1980) has examined the seasonal variations in energy and proximate composition of an edible gastropod, *Turbo sarmaticus*. Usmanghani *et al.*, (1989) have studied the composition of fatty acids of gastropod *Xancus pyrum*. Nirmal (1995) has studied the proximate composition prosobranchian gastropods *Babylonia zeylonica* (Neogastropods: Buccinidae: Fasciolariidae). Rajkumar (1995) has studied in *Rapana rapiformis* from the Parangipettai coastal waters. Stella (1995) has studied the taxonomy and eco biology of *Chicoreus* species from Parangipettai waters. Xavier (1996) has studied the biochemical and processing of edible meat of muricid gastropods *Chicoreus virgineus* and *Rapana rapiformis*.

The proximate composition is the yardstick to measure and assess the nutritional quality of food sources. Proximate composition varies during different stages in species and so it is more appropriate to assess the Proximate Composition of seafood. Since no such information was available in the species of *Chicoreus virgineus ponderosus* and *Siratus virgineus ponderosus*, Hence the present study was undertaken to elucidate the edible value of Indian Muricids *Chicoreus virgineus ponderosus* and *Siratus virgineus ponderosus*.

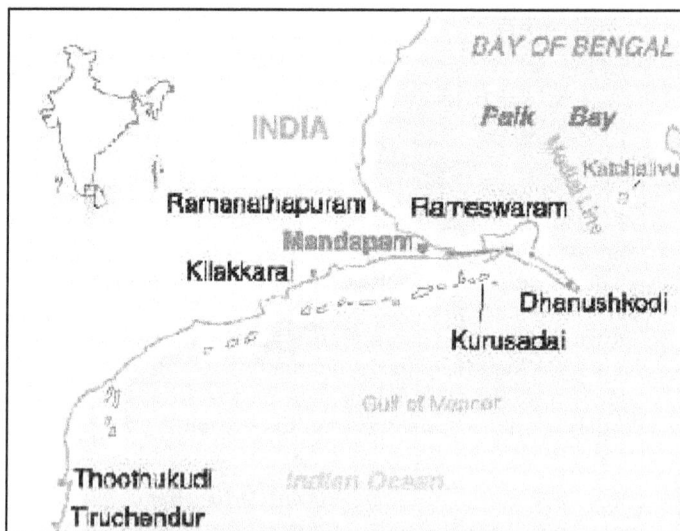

Map 12.1: Study Area–Palk Bay.

Materials and Methods

In the present study, a regular survey was conducted at Thondi coast in Palk Bay area (Lat. 9° and10° and Long 79° and 80°) Map 12.1. The species of *Chicoreus virgineus ponderosus* and *Siratus virgineus ponderosus* are exclusively marine in distribution. The specimens were collected from the trawlers. Specimens of *Chicoreus virgineus ponderosus* and *Siratus virgineus ponderosus* were collected monthly from the trawlers operated in Thondi Coastal waters were brought to the laboratory. The animals of the size group 7.5-9 cm in length were selected for biochemical studies. The animals were first washed in tap water. The outer hard shells were broken with a hammer. Care was taken not to damage the soft parts. The animals were removed from the shell and the meat was weighed separately. They were dissected to remove different body parts *viz.* mantle, foot, digestive gland and gonad. Different body parts were weighed and kept in an oven at 60°C for 24 hours. The dried components were brought to constant weight, after which different Proximate Composition were estimated. Four estimations were made in each case and the average was taken into consideration.

Protein

Protein was estimated by Biuret method as modified by Raymond *et al.* (1964) 0.5mg of dried material was taken and homogenized in a hand homogenizer with 1 ml of glass distilled water, 2ml of biuret reagent was added two times and the tissue grinded was cleaned before being transferred to the centrifuge tube. After 30 minutes, this sample was centrifuged for ten minutes and the supernatant fluid was transferred into another tube. Then the reading of the supernatant fluid was measured by using UV-VIS double beam spectrophotometer (UVD-2960) at the wave length of 540 um against the blank reading and then the percentage of protein was calculated.

$$\text{Percentage of Protein} = \frac{\text{Standard value} \times \text{OD}}{\text{Weight of tissue}} \times 100$$

Carbohydrates

For the estimation of the total carbohydrate content, the procedure of Dubios *et al.* (1956) was followed. 20 mg of dried tissue powder was taken and to this 1.0ml of the glass distilled water followed by 1.0 ml of 4 per cent phenol solution and 5 ml of concentrated sulphuric acid was added. After 30 minutes, reading was taken in UV-VIS double beam spectrophotometer (UVD-2960) at the wave length of 490 um against the blank reading and then the percentage of carbohydrate was calculated.

Lipid

The chloroform methanol extraction procedure of Folch *et al.* (1956) was adapted for extracting lipid from the tissue. 50 mg of finely powdered tissue was taken into a test tube to which 5 ml of chloroform methanol (3: 1) mixture was added and covered with aluminum foil and allowed to stand for overnight digestion. The mixture was then filtered by a micro filter and the filtrate was dried in an oven in a beaker. Percentage of lipid was calculated by using the following formula

$$\text{Percentage of Lipid} = \frac{\text{Weight of lipid}}{\text{Weight of tissue}} \times 100$$

Results and Discussion

Giese *et al.* (1967) have opined that the biochemical constituents fluctuate in relation to reproductive cycle, feeding and age of the organism. A pronounced seasonal variation in the biochemical constituents were noted in various gastropods (Krishnakumari 1985; Shanmugam 1987; Thivakaran 1988; Maruthamuthu 1988). In the present investigation an attempt has been made to evaluate the seasonal variation of the proximate composition of two Muricid species namely *Chicoreus virgineus ponderosus* and *Siratus virgineus ponderosus*. The biochemical constituents increased from July onwards due to the maturation of gonads and reached the maximum maturation during October-January. When the gonads became fully matured the biochemical constituents decreased slowly from February to June.

Protein

Male (Figures 12.1 and 12.2)

In *Chicoreus virgineus ponderosus*, the high value of protein was recorded in gonad (83.3 per cent) and the lower value was recorded in mantle (40.3 per cent) when it compared to other organs. In all the body parts except digestive gland, low values were found during summer and high values during monsoon (October-December 2006). The maximum protein value in whole animal was 78.7 per cent recorded during monsoon season and the minimum value 52.3 per cent was recorded during summer season.

In *Siratus virgineus ponderosus*, the high value of protein was recorded in gonad (80.3 per cent), and the low value of protein was recorded in mantle (44.3 per cent). In all the body parts except digestive gland, low value was found in summer and high value was found in monsoon season (October-December 2006). In whole animal the maximum protein value was 66.5 per cent recorded during monsoon season and the minimum value was 49.4 per cent recorded during summer season. In two species the gonad was found to contain higher amount of protein. When it compared to all other body parts.

Female (Figures 12.3 and 12.4)

In *Chicoreus virgineus ponderosus*, the high value of protein was recorded in gonad 89.4 per cent, digestive gland 80.34 per cent, foot 72.3 per cent, adductor muscle 69.3 per cent and mantle 60.7 per cent and the lower values of protein was recorded in different parts of gonad 60.5 per cent, digestive

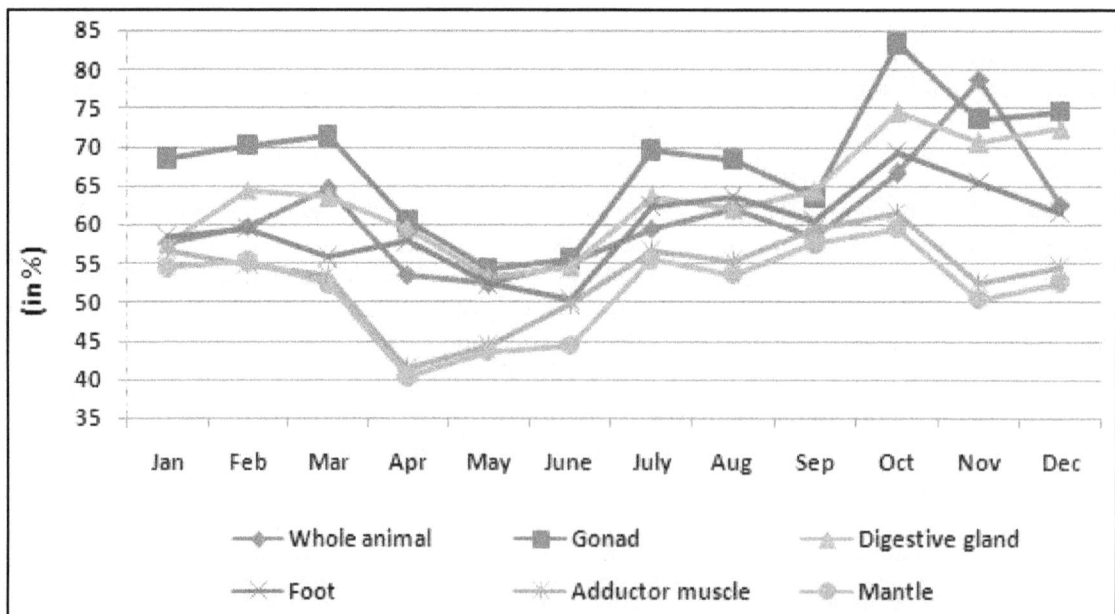

Figure 12.1 *: Chicoreus virgineus ponderosus* – Male – Protein

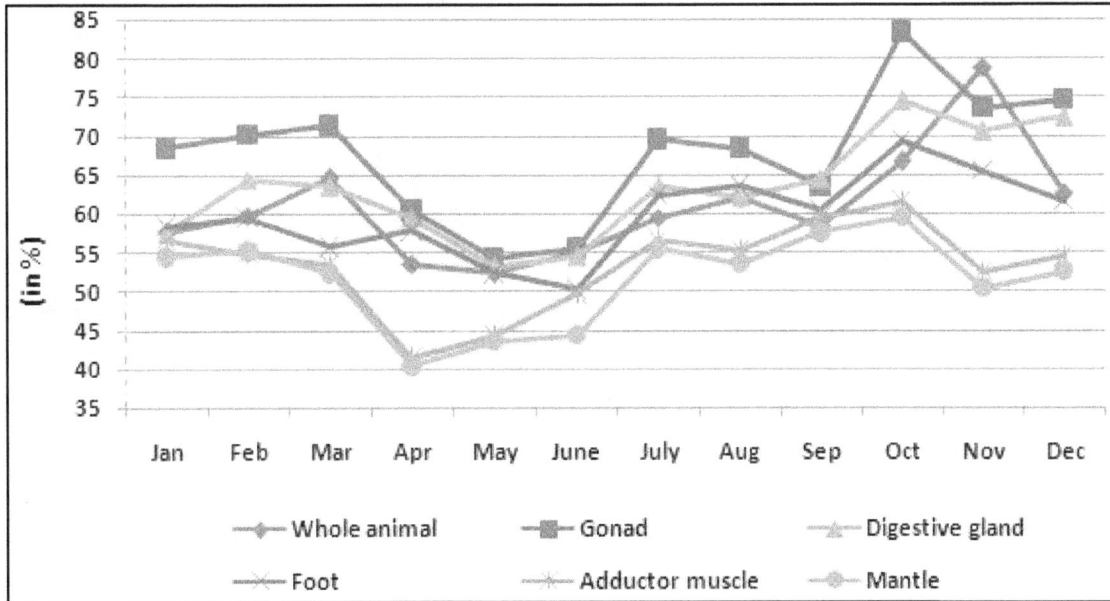

Figure 12.2: *Siratus virgineus ponderosus* – Male – Protein

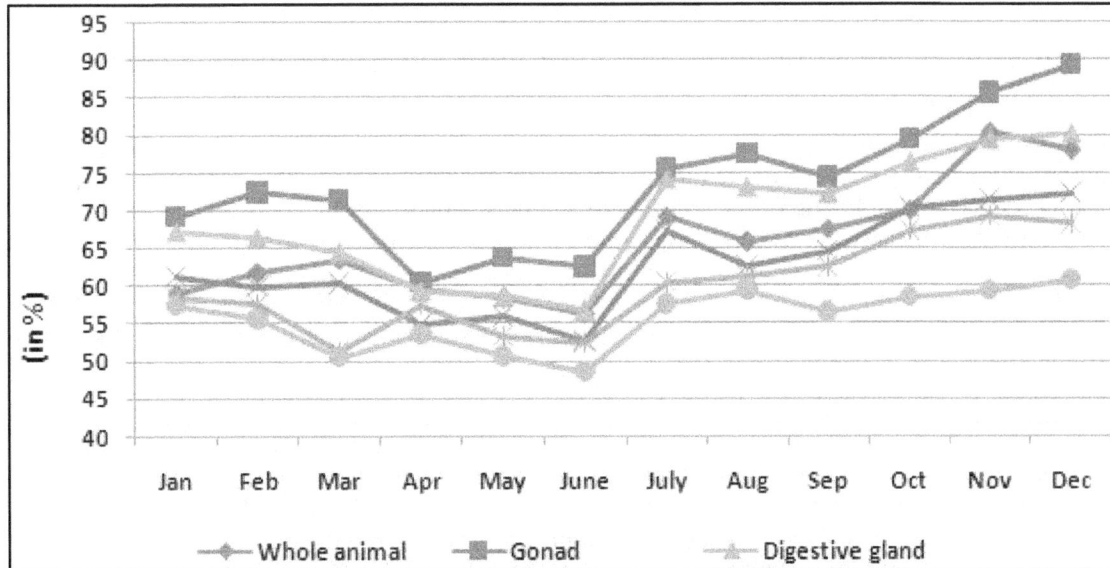

Figure 12.3: *Chicoreus virgineus ponderosus* – Female – Protein

gland 56.7 per cent, foot 52.7 per cent, adductor muscle 52.5 per cent and mantle 48.5 per cent. In all the body parts except digestive gland, low values were found during summer and high values during monsoon (October-December 2006). The maximum protein value 80.5 per cent in whole animal was recorded during monsoon season and the minimum value 56.3 per cent was recorded during summer season.

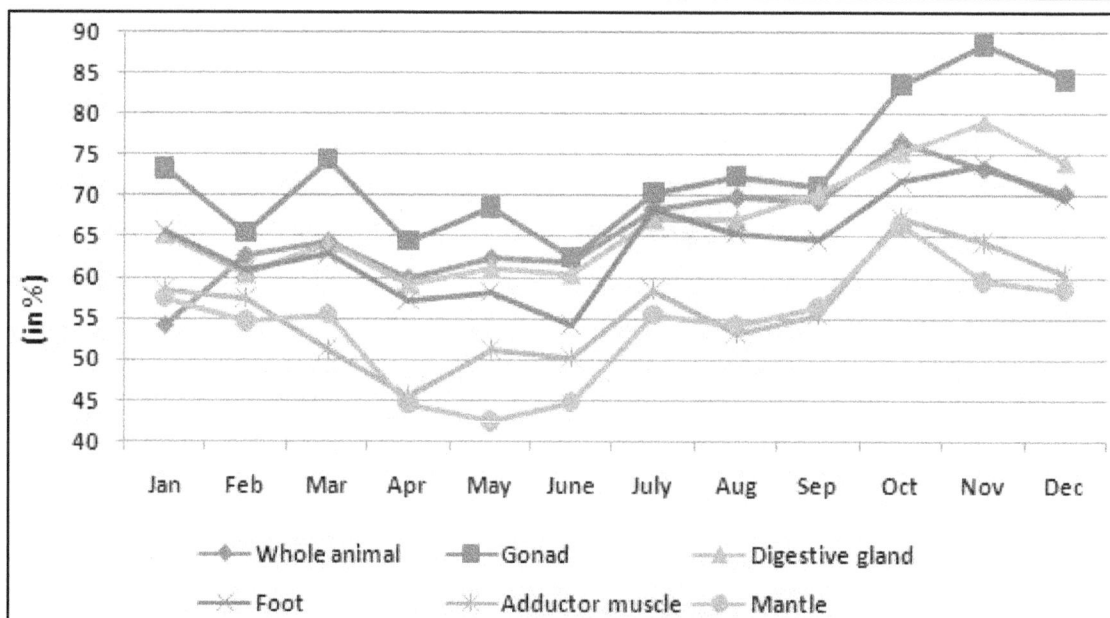

Figure 12.4: *Siratus virgineus ponderosus* – Female – Protein

In *Siratus virgineus ponderosus*, Protein values was high during monsoon and low during summer. The higher values of protein was recorded in gonad 88.5 per cent, digestive gland 79.2 per cent, foot 73.9 per cent, adductor muscle 67.4 per cent and mantle 66.6 per cent and the lower values of protein were recorded in different parts of gonad 62.4 per cent, digestive gland 59.3 per cent, foot 54.3 per cent, adductor muscle 45.4 per cent and mantle 42.4 per cent. In all the body parts except digestive gland, low values was found during summer and high values during monsoon (October-December 2006). The maximum protein value 76.5 per cent in whole animal was recorded during monsoon season and the minimum value 60.0 per cent was recorded during summer season. Of all the body parts analysed the gonad was found to contain higher amount of protein in these two species. Higher values of protein were recorded in *Chicoreus virgineus ponderosus* than in *Siratus virgineus ponderosus*. Protein was generally higher in females that in males of two species.

The results of the present study the Protein values are compared with those of other species of gastropods. In *Cellona radiate* the protein value ranged from 60.92 per cent–86.08 per cent in male and 61.98 per cent–85.75 per cent in female has been reported by Suryanarayanan and Nair, (1976). In *Cellona rota* the protein values were ranged from 42.33 per cent–82.45 per cent has been reported by Patil and Mane, (1982). In *Pythia plicata*, the protein values were ranged from 15.71 per cent to 29.80 per cent has been reported by Shanmugam, (1987). Thivakaran (1988) has reported that the protein values of *Litorina quadricentus* and *Nodilittorina pyramidalis* were 35.95 per cent and 35.63 per cent respectively. The maximum and minimum values of protein in the present investigation are comparatively higher than the earlier reports.

Presently the gonad and digestive gland showed higher protein values when compared to the other organs. Thus the gonad seems to serve as a storage organ of protein in *Chicoreus virgineus ponderosus* and *Siratus virgineus ponderosus*. In the digestive gland, the protein content was much higher than the other constituents and it exhibited a distinct seasonal cycle. The protein values were high during monsoon season could be due to intense proliferation of gonad and the low protein value in summer may be due to the spawning activity. Similar observation has been made in *Thais* sp. by Tagore, (1989). Lambert and Dehnel (1974) have reported that the protein value was high in *Thais lamellose*. Giese *et al.* (1967) also have reported that the higher organic constituent of Molluscs was protein.

Carbohydrate

Male (Figures 12.5 and 12.6)

In *Chicoreus virgineus ponderosus,* the higher values of carbohydrate was recorded in gonad 16.56 per cent, digestive gland 15.81 per cent, foot 14.51 per cent, adductor muscle 11.34 per cent and mantle 8.81 per cent and the lower values of carbohydrate was recorded in different parts of gonad 8.51 per cent, digestive gland 6.24 per cent, foot 6.41 per cent, adductor muscle 5.35 per cent and mantle 4.04 per cent. In all the body parts except digestive gland, low values were found during summer and high values during monsoon (October-December 2006). The maximum carbohydrate value 17.62 per cent in whole animal was recorded during monsoon season and the minimum value 8.10 per cent was recorded during summer season.

In *Siratus virgineus ponderosus,* protein value was high during monsoon and low during summer. The higher values of carbohydrate were recorded in gonad 17.52 per cent, digestive gland 15.80 per cent, foot 14.24 per cent, adductor muscle 12.81 per cent and mantle 10.50 per cent and the lower values of carbohydrate were recorded in different parts of gonad 7.62 per cent, digestive gland 7.25 per cent, foot 7.44 per cent, adductor muscle 7.21 per cent and mantle 6.54 per cent. In all the body parts except digestive gland, low values were found during summer and high values during monsoon (October-December 2006). The maximum carbohydrate value 16.62 per cent in whole animal was recorded during monsoon season and the minimum value 6.20 per cent was recorded during summer

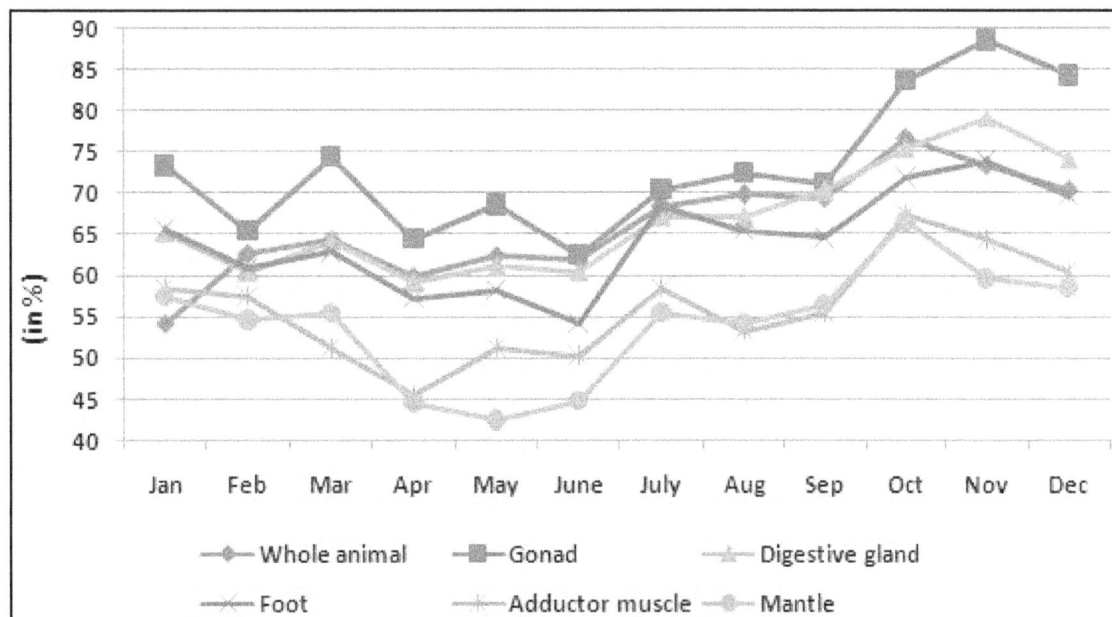

Figure 12.5: *Chicoreus virgineus ponderosus* – Male – Carbohydrate

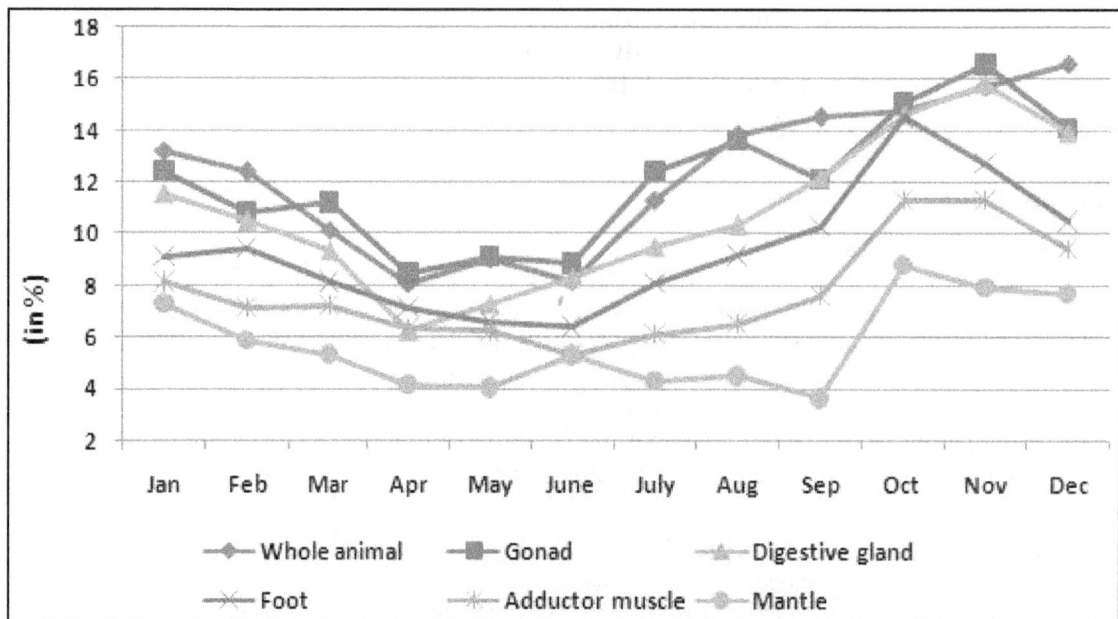

Figure 12.6: *Siratus virgineus ponderosus* – Male – Carbohydrate

season. Of all the body parts analysed the gonad was found to contain higher amount of carbohydrate in two species. Two species the maximum content of carbohydrate was observed in gonads when compared to other body parts.

Female (Figures 12.7 and 12.8)

In *Chicoreus virgineus ponderosus*, the higher values of carbohydrate were recorded in gonad 18.34 per cent, digestive gland 15.61 per cent, foot 14.56 per cent, adductor muscle 13.56 per cent and mantle 11.23 per cent and the lower values of carbohydrate were recorded in different parts : gonad 10.61 per cent, digestive gland 10.14 per cent, foot 10.0 per cent, adductor muscle 9.23 per cent and mantle 7.21 per cent. In all the body parts except digestive gland, low values were found during summer and high values during monsoon (October-December 2006). The maximum carbohydrate value 17.60 per cent in whole animal was recorded during monsoon season and the minimum value 10.52 per cent was recorded during summer season.

In *Siratus virgineus ponderosus*, carbohydrate value was high during monsoon and low during summer. The higher values of carbohydrate were recorded in gonad 18.71 per cent, digestive gland 17.45 per cent, foot 15.04 per cent, adductor muscle 14.53 per cent and mantle 12.44 per cent and the lower values of carbohydrate were recorded in different parts : gonad 14.42 per cent, digestive gland 12.0 per cent, foot 10.11 per cent, adductor muscle 8.24 per cent and mantle 6.29 per cent. In all the body parts except digestive gland, low values were found during summer and high values during monsoon (October-December 2006). The maximum carbohydrate value 18.6 per cent in whole animal was recorded during monsoon season and the minimum value 12.28 per cent was recorded during summer season. Of all the body parts analysed the gonad was found to contain higher amount of carbohydrate in two species. Higher values of carbohydrate were recorded in *Chicoreus virgineus ponderous* than in *Siratus virgineus ponderous*. Carbohydrates values are higher in females than in the males of two these species.

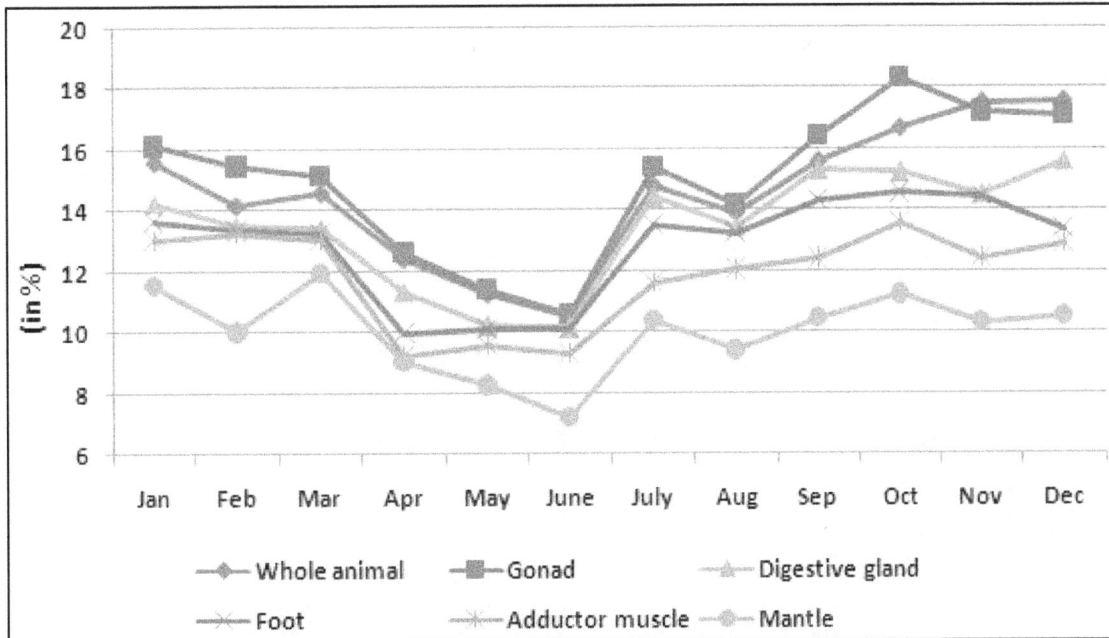

Figure 12.7: *Chicoreus virgineus ponderosus* – Female – Carbohydrate

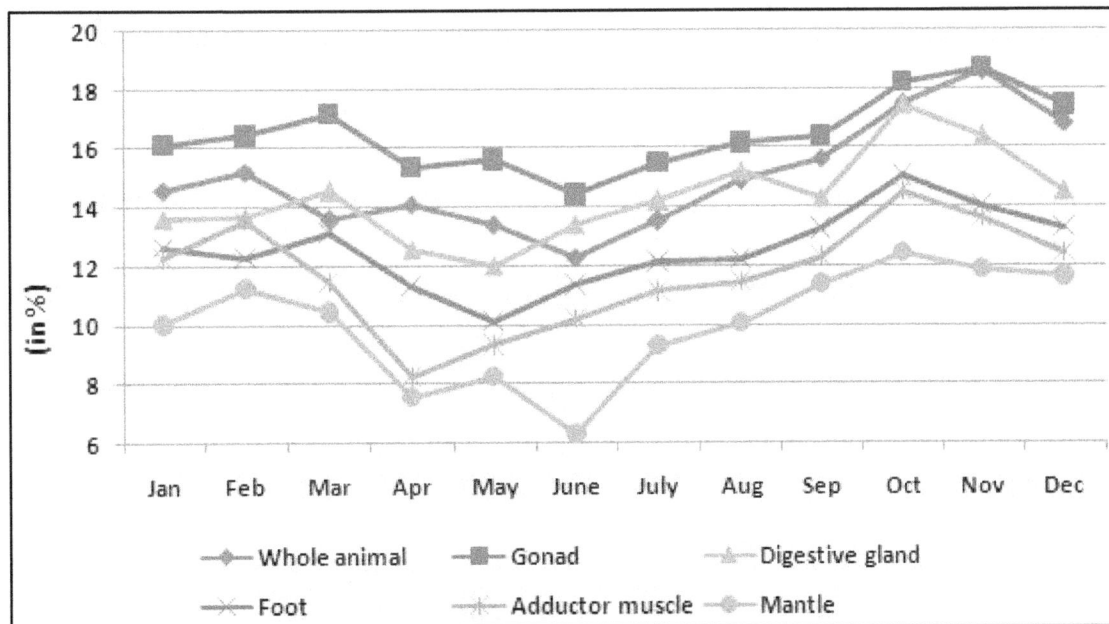

Figure 12.8: *Siratus virgineus ponderosus* – Female – Carbohydrate

In the present study carbohydrate values are observed to be high in the gonad during monsoon and very low in summer when the gonad was in ripe stage. Generally carbohydrate values were higher than lipid values. Krishnakumari (1985) has reported that the carbohydrate level in *Cerithium rubus* varied between 4.85 and 11.89 per cent. Shanmugam (1987) has reported that it was varied from 0.84 to 3.04 in *Pythia plicata*. The maximum carbohydrate levels was 5.31 per cent in *Littorina quadicentus* and 13.6 per cent to 15.2 per cent in *Thais bufo* has been reported by Tagore (1989). In both species the carbohydrate content was high when compared to other gastropods. Ansell *et al.* (1973) have reported that the wide variation in the carbohydrate content in molluscan tissues especially in gonad may be act as the storage organ and utilized under favourable condition.

Lipid

Male (Figure 12.9 and 12.10)

In *Chicoreus virgineus ponderosus*, the higher values of lipid were recorded in gonad 6.82 per cent, digestive gland 5.43 per cent, foot 3.72 per cent, adductor muscle 3.63 per cent and mantle 3.49 per cent and the lower values of lipid were recorded in different parts : gonad 3.42 per cent, digestive gland 3.40 per cent, foot 3.24 per cent, adductor muscle 2.28 per cent and mantle 2.13 per cent. In all the body parts except digestive gland, low values were found during summer and high values during monsoon (October-December 2006). The maximum lipid value 6.45 per cent in whole animal was recorded during monsoon season and the minimum value 3.13 per cent was recorded during summer season.

In *Siratus virgineus ponderosus*, lipid values were high during monsoon and low during summer. The higher values of lipid were recorded in gonad 5.32 per cent, digestive gland 3.56 per cent, foot 3.63 per cent, adductor muscle 3.62 per cent and mantle 2.90 per cent and the lower values of lipid were recorded in different parts : gonad 2.52 per cent, digestive gland 2.23 per cent, foot 2.28 per cent, adductor muscle 2.50 per cent and mantle 2.59 per cent. In all the body parts except digestive gland, low values were found during summer and high values during monsoon (October-December 2006). The maximum lipid value 5.45 per cent in whole animal was recorded during monsoon season and

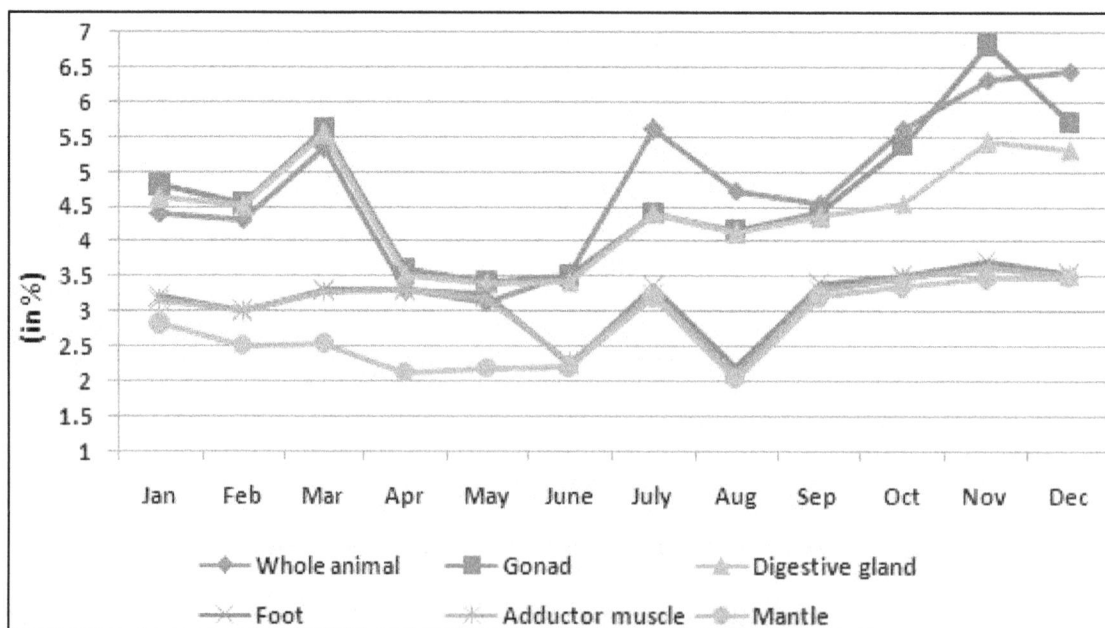

Figure 12.9: *Chicoreus virgineus ponderosus* – Male – Lipid

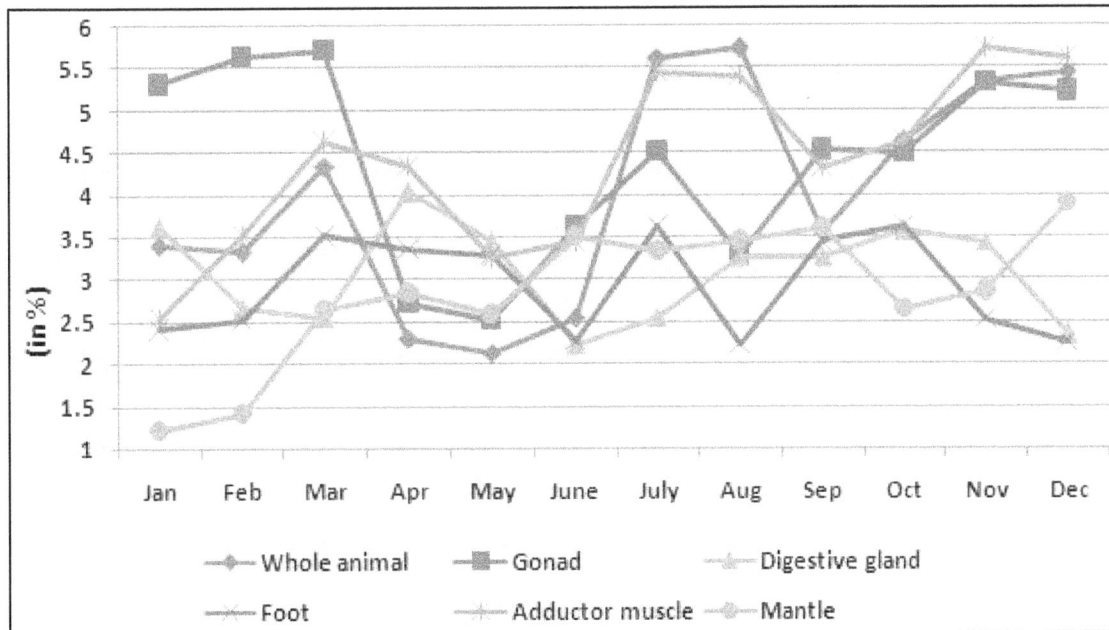

Figure 12.10: *Siratus virgineus ponderosus* – Male – Lipid

the minimum value 2.13 per cent was recorded during summer season. Of all the body parts analysed the gonad was found to contain higher amount of lipid in two species. Higher value of lipid was recorded in *Chicoreus virgineus ponderosus* than in *Siratus virgineus ponderosus*. Lipid was generally higher in females than in males of two these species.

Female (Figures 12.11 and 12.12)

In *Chicoreus virgineus ponderosus*, the higher value of lipid was recorded in gonad 5.34 per cent, digestive gland 4.81 per cent, foot 4.34 per cent, adductor muscle 3.41 per cent and mantle 2.85 per cent and the lower values of lipid were recorded in different parts of gonad 3.61 per cent, digestive gland 3.60 per cent, foot 3.24 per cent, adductor muscle 2.21 per cent and mantle 1.29 per cent. In all the body parts except digestive gland, low value was found during summer and high values during monsoon (October-December 2006). The maximum lipid value 5.72 per cent in whole animal was recorded during monsoon season and the minimum value 3.22 per cent were recorded during summer season.

In *Siratus virgineus ponderosus*, lipid values were high during monsoon and low during summer. The higher values of lipid were recorded in gonad 5.85 per cent, digestive gland 5.80 per cent, foot 3.61 per cent, adductor muscle 3.29 per cent and mantle 2.89 per cent and the lower values of lipid were recorded in different parts : gonad 3.42 per cent, digestive gland 3.35 per cent, foot 2.34 per cent, adductor muscle 2.13 per cent and mantle 1.45 per cent. In all the body parts except digestive gland, low values were found during summer and high values during monsoon (October-December 2006). The maximum lipid value 5.47 per cent in whole animal was recorded during monsoon season and the minimum value 3.51 per cent was recorded during summer season. Of all the body parts analyzed the gonad was found to contain higher amount of lipid in two species. Higher value of lipid was recorded in *Chicoreus virgineus ponderosus* than in *Siratus virgineus ponderosus*. Lipid was generally higher in females than in males of these two species.

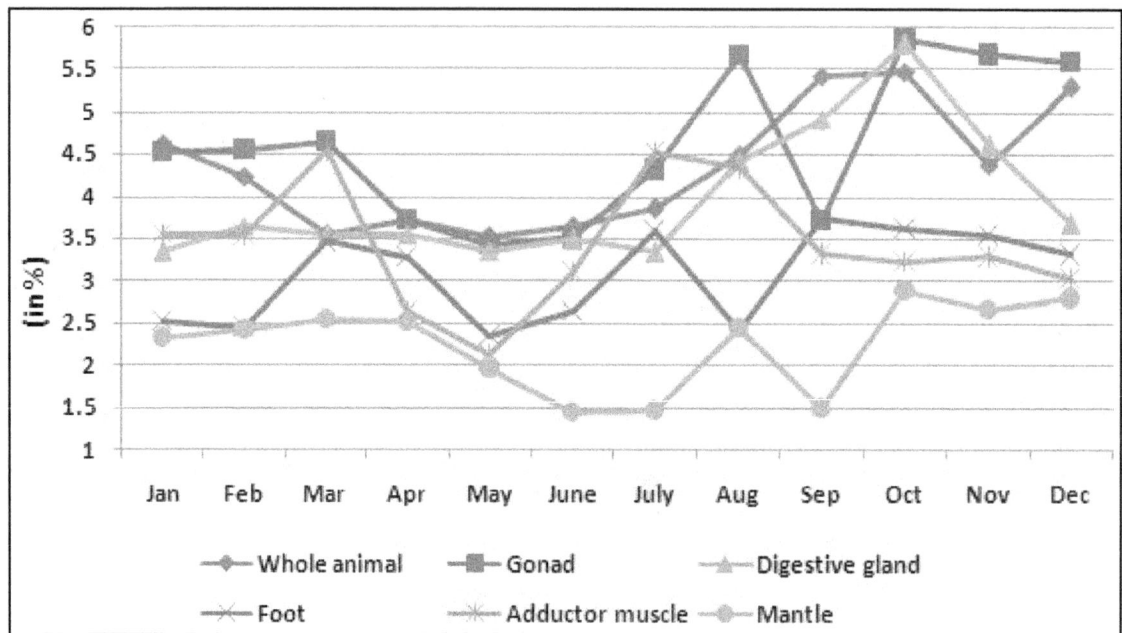

Figure 12.11: *Chicoreus virgineus ponderosus* – Female – Lipid

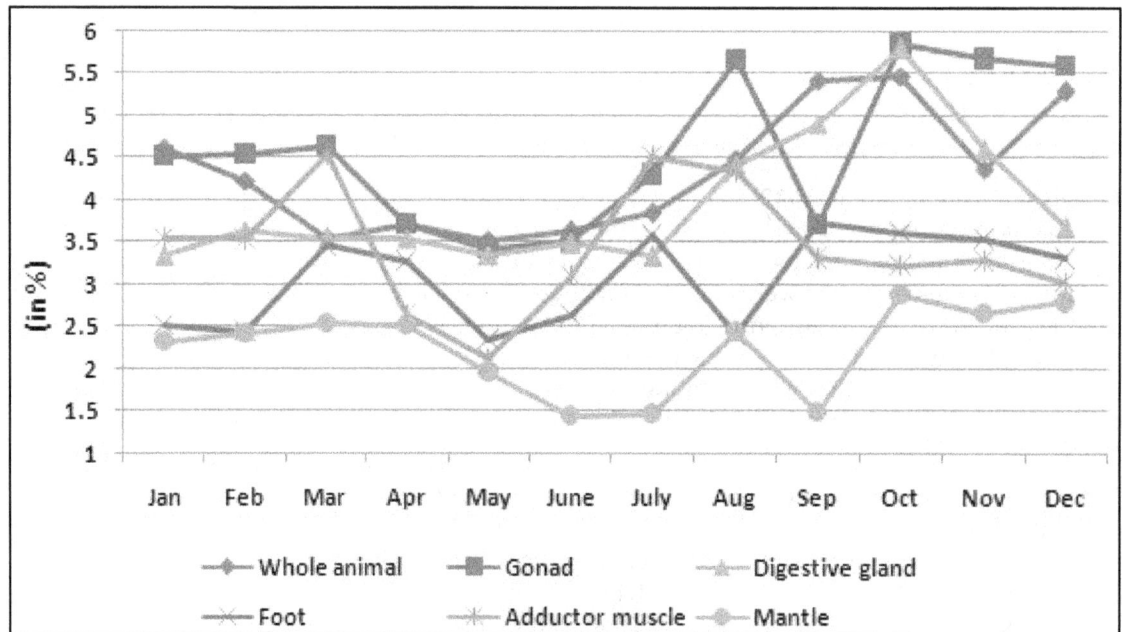

Figure 12.12: *Siratus virgineus ponderosus* – Female – Lipid

The maximum value of lipid content (5.72) was recorded in females of *Chicoreus virgineus ponderosus*. The fluctuations in lipid values closely followed by Protein and carbohydrate levels. Thivakaran

(1988) has reported that the variation in lipid in *Littorina quadricentus* was (0.79 per cent). Krishnakumari (1985) has reported that the lipid values ranged from 5.10 to 22.93 per cent in males and from 4.90 per cent to 24.19 per cent in females of *Cerithium rubus*. In *Cellona rota* the lipid values ranged from 0.8 to 10.75 per cent (Suryanarayanan and Nair, 1976). Tagore (1989) has reported that the lipid values ranged from 0.80 to 2.42 per cent in males and from 0.84 to 2.68 per cent in females of *Thais biseralis* and *Thais bufo*. In the present study in both species the maximum lipid content was found in digestive gland. Similar observation has been made in *Thais lamellose* (Lambert and Dehnel, 1974). In most of the molluscs, digestive gland acts at the storage organ (Owen, 1966; John, 1980). Tagore (1989) has reported that the digestive gland acts as the storage site in *Thais biserialis* and *Thais bufo*. In both *Chicoreus virgineus ponderosus* and *Siratus virgineus ponderosus,* biochemical constituents were high during the spawning period. The levels of protein, carbohydrate and lipid vary during the feeding and spawning periods. The biochemical constituents were high during monsoon month when there was active feeding and reduced spawning. During the summer the level will go to the minimum when the animals were actively spawning.

Generally the variation in the Proximate Composition are associated with the reproductive activity and food availability (William, 1970). Umadevi (1983) has reported progressive increase and decrease in the levels of Proximate Composition are attributed the maturation and spawning activity of *Morula granulata* for the variation. Similar observation has been made in *Pythia plicata* by Shanmugam, (1987) and *Littoriana quadricentus* and in *Nodilittorina pyramidalis* by Thivakaran (1988). In the present study the fluctuations in the proximate composition of both the species are largely attributable to their reproductive and feeding activities.

References

Anandakumar, S., 1986. Studies on *Hemifusus pugilius* (Bron) (Mollusca: Gastropoda: Volemidae) from Portonovo waters. *M. Phil. Thesis,* Annamalai University, pp. 177.

Ansell, A. D., Sivadas, P. and Narayanan, B., 1973. The Ecology of two sandy beaches in Southwest India. IV. The biochemical composition of four common invertebrates. *Spl. Publ. J. Mar. Biol. Ass. India,* p. 333–348.

Dubios, M., Giller, K. A., Hamilton, J. K., Roberts, R. A. and Smith, F.1956. Colorimetric method for determination of sugars and related substances. *Analyt. Chem.,* 28: 350–356.

Folch, J., Lees, M. and Solane Stantley, G. H.1956. A simple method for the isolation and purification of total lipids from the animal tissues. *J. Biol. Chem.,* 826: 497–509.

Giese, A. C., Hart, M. A., Smith, A. M. and Cheang, M. A.1967. Seasonal changes in body component indices and chemical composition in Pismo clam *Tivela stultorum. Comp. Biol. Chem. Physio.,* 22: 549–561.

Giese, A. C., 1969. A new approach to the biochemical composition of the Mollusc body. *Oceanogr. Mar. Blo. Annu. Rev.,* 7: 175–229.

John, G., 1980. Studies on *Anadara rhombea* (Born) (Mullusca: Bivalvia) from Portonova coastal water. *Ph.D. Thesis,* Annamalai University, pp. 190.

Kapeer, M. A., Stickle, W. B. and Blakency, E., 1985. Volume regulation and nitrogen metabolism in the Murisid gastropod *Thais haemastoma. Biol. Bull.,* 169: 458–475.

Krishanakumari, L., 1985. Ecological and Biochemical studies with special reference to pollution on selected species of Molluscs from Bombay *Ph.D. Thesis,* University of Bombay, pp. 322.

Lambert, P. and Dehenel, P. A., 1974. Seasonal Variations in Biochemical composition during the reproductive cycle of the intertidal gastropod *Thais lamellose* (Gmelin) (Gastropoda: Prosobranchia). *Can. J. Zool.,* 52: 305–318.

Lombard, H. H., 1980. Seasonal variations in energy and biochemical components of an edible gastropod. *Turbo sarmaticus* (Turbinidae). *Aquaculture,* 19(2): 117–125.

Maruthamuthu, S., 1988. Studies on *Littorina littoraria* from Tranquebar and Mandapam, South east Coast of India. *Ph.D. Thesis,* Annamalai University.

Murugan, A., J. K. Patterson Edward and Ayyakkannu, K., 1991. Ecological, biochemical and antibacterial data from India. *Phuket Mar. Biol. Cent. Spec. Publ.,* 9: 108–110.

Nagabhushanam, R. and Mane, U. H., 1975. Seasonal variations in the biochemical composition of the clam *Katelysia opima. Mar, Biol.* 68(4): 279 –321.

Nirmal. A., 1995. Biochemical studies on prosobranchian gastropods *Babylonia zeylonica* (Neogastropods: Buccinidae: Fasciolariidae). *M.Sc., Dissertation,* Annamalai University, pp. 30.

Owen, G., 1966. Digestion, In: *Physiology of Mollusca.* Academic press. New York, 53–59.

Patil, P. J. and Mane, V. H., 1982. Levels of protein, fat and glycogen in the limpet *Cellona rota* (Chemnitz). *Indian J. Mar. Sci.,* 11(2): 182–183.

Rajkumar, T., 1995. Studies on *Rapana rapiformis* Born (Mollusca: Gastropoda: Muricidae: Rapaninae) from the Parangipettai Coastal waters, India. *Ph.D. Thesis,* Annamalai University, India. pp. 185.

Raymont, J. E.G., Austin, A. and Lingford, E. 1964. Biochemical studies on marine zooplankton. I. Biochemical composition of *Neomysis integer. J. Cons. Perm. Explor. Mer.,* 28: 354–363.

Shanmugam, A., 1987. Studies on *Pythia plicata* (Gray) (Gastropoda: Pulmonata: Ellobiidae) from Pitchavaram mangroves. *Ph.D. Thesis,* Annamali University, pp. 137.

Stella, C., 1995. Studies on the taxonomy and ecobiology of *Chicoreus* species from Parangipettai waters, South East Coast of India. *Ph.D. Thesis,* Annamalai University. pp. 195.

Stickle, W. B., 1970a. Some physiological aspects of the reproductive cycle of the intertidal prosobranch *Thais lamellosa* (Gmelin, 1792). *Ph.D. Thesis,* University of Sakatchewan Regina Campus.

Stickle, W. B., 1970b. The metabolic effects of starving *Thais lamellosa* immediately after spawning. *Comp. Biochem. Physiol.,* 40A: 627–634.

Stickle, W. B. and Duerr, F. G., 1970. The effects of starvation on the respiration and major nutrient stores of *Thais lamellosa. Comp. Biochem. Physiol.,* 33: 689–695.

Stickle, W. B., 1973. The reproductive physiology of the intertidal prosobranch *Thais lamellosa* (Gmelin) II. Seasonal changes in Biochemical composition. *Biol. Bull.,* 148: 448–460.

Suryanarayanan, H. and Nair, N. B., 1976. seasonal variation in the biochemical constituents *Cellona radiate* (Born) *Ind. J. Mar. Sci.,* p. 126–128.

Tagore, J., 1989. Studies on the Thaisids *Thais biserialis* and *Thais Bufo* from the Tranquebar rocky shore (South East Coast of India). *Ph.D. Thesis.* Annamalai University, pp. 212.

Thilaga, R. D., 1985. Studies on *Bullia vittata* (Linnaeus) (Mollusca: Gastropoda) from Portnovo waters. *M.Phil. Thesis,* Annamalai University, pp. 35.

Thivakaran, G. A., 1988. Studies on the littorinids *Littorina quadricentus* (Phillip) and *Nodilittorina pyramidalis* (Quoy and Gaimard, 1833) Gastropoda: *Prosobranchia littorinidae*) from the tranquebar rocky shore (South East Coast of India) *Ph.D. Thesis*, Annamalai University, pp.179.

Umadevi, V., 1983. Studies on the intertidal gastropod *Morula granulata* (Duclos) with special reference to respiration, starvation and seasonal changes in some biochemical constituents. *Ph.D. Thesis*, Andhra University, S. India.

Usmanghani. K., Shahida, S., Mansoor, A., Afzal, M. K., 1989. Composition of fatty acids of gastropod *xancus pyrum*, zoology. *J. Islamic Acad. Sci.*, 2(3): 165–167.

Williams, E. E., 1970. Seasonal variations in the biochemical composition of the edible winkle *Littoria littorea* (L) *Comp. Biochem. Physiol.*, 33: 655–661.

Xavier R. M., 1996. Studies on the biochemistry and processing of edible meat of Muricid gastropods *Chicoreus virgineus* (Roding, 1798) and *Rapana rapiformis* (Born, 1778). *Ph.D. Thesis*, Annamalai University, India. pp. 88.

Chapter 13

New Records of Bivalve Species from Sea Grass Beds of Palk Bay Area in Tamil Nadu

☆ *C. Stella, R. Ravichandran, M. Thayalan and D. Chellaiyan*

Introduction

Palk Bay areas is known for the economically important renewable resources like seaweeds, seagrasses, finfishes and shell fishes. In Palk bay area 11 species of seagrasses were recorded. The molluscan fauna associated with seagrass beds has received much attention in European seas. Along the Atlantic coasts, the molluscan fauna associated with *Zostera marina* and *Zostera noltii* is well known (Jacobs and Huisman, 1982). Ark clam is the common name of salt water marine bivalve molluscs in the family Arcidae. Ark calms vary both in shape and size. The shells of ark clams are often white or cream, but in certain species, the shell is striped with, tinted with or completely coloured with a rich brown. The shell of most species has a thick layer of brown Periostracum covering the harder calcareous part of the shell. In some genera such as Barbatia, the periostracum can be tufted at the end of the shell in to some thing that resembles a beard, hence the name Barbatia or bearded one. Dimyidae is a family of bivalve molluscs related to the scallops and oysters. Dimya is a genus of very small clams, marine bivalve molluscs in the family Dimyidae.

Satyanarayana Rao and Sundram (1972) studied the fauna distribution in the Intertidal zone of the Gulf of Mannar and Palk bay. Ganapati and Sarma (1972) reported the bivalves and gastropods species of the Indian seas. Seventy seven species of bivalves belonging to 2 subfamilies, 5 orders and 21 families were collected from Tuticorin coast of Gulf of Mannar (CMFRI Ann. Rep, 2003-2004). Sahul Hameed and Somasundaram (1998) reported that, out of 55 species of bivalve molluscs collected from Gulf of Mannar, 49 species belonging to two subclasses, four orders and eighteen families were identified and classified.

A survey was conducted as in Palk bay area (Map 13.1) from 2007-2008. The shells were collected manually from the intertidal area and fish landing centers of Palk bay and the specimens were transferred to the laboratory for identification. The specimens were stored in 10 per cent formalin and the colour of the shells was recorded. Identification was mainly based on the external morphology of the shells. All the seven species are under five families namely, *Scapharca cornea, Barbatia velata,* (Arcide) *Potamocorbula ustulata ustulata* (Corbulide), *Donax tranculus* (Donacidae), *Dimya japonica* (Anomiidae), *Gelonia bengalensis* (Gelonoiidae), *Glauconome chinensis* (Gluconomidae).

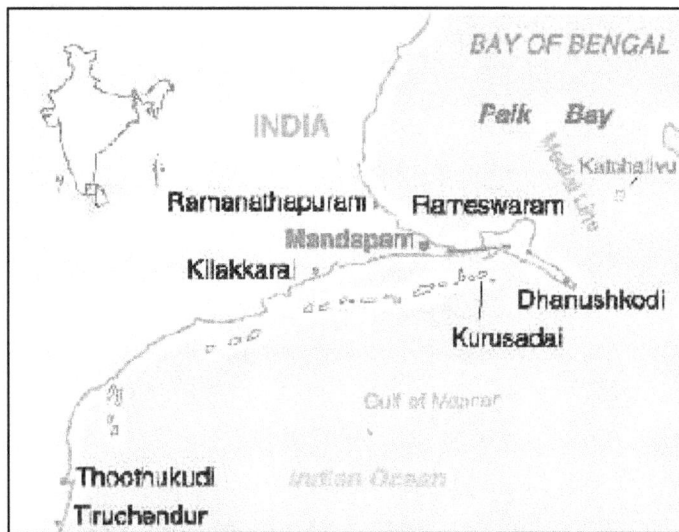

Map 13.1: Map of Palk Bay Study Area.

Family: Arcidae (Ark Shells)

1. *Barbatia velata*

Shells up to 5.2 cm in length, elongated, thick, oval or quadrilateral. Shell compressed with wide gap with the lower edge of the shell. Umbo round and elevated. Shell sculptured with numerous radial ribs crossed equally by numerous concentric grooves and covered by a coarse often tufted periostracum. Hinge line is long and straight with numerous teeth.

Common Name: Ark shells; **Locality**: Coral area; **Distribution**: India: Palk Strait and Gulf of Mannar; **Status**: Uncommon (Figure 13.1).

2. *Scapharca cornea*

Shell up to 7.8 cm in length, heavy, sturdy and small. Equal valve. Sculpture with elongate radial ribs, strong and smooth. Umbo broad and strong, facing each other. Hinge straight and elongate with numerous teeth. Ligament short with transversal striations. Periostracum thick. Colour of the shell brown. Inner surface of the shell pale brown in colour and smooth.

Common name: Burnt –End Ark; **Locality**: Intertidal and coral area; **Distribution**: India: Palk Strait and Gulf of Mannar; **Status**: Uncommon (Figure 13.2).

Family: Corbulide

3. *Potamocorbula ustulata ustulata* (Reeve, 1844)

Shell 6.8 cm in length, large, thick and triangularly ovate in shape. Strongly inflated. Umbo elevated and prominent. Gibbose appearance. Umbonal area smooth and polished. 2 cardinal teeth bifid in nature. Hinge strong. Ligament elongated and broad. Pallial sinus broad and deep. Inside the shell white in colour and outer surface of the shell creamy white.

Locality: Intertidal and coral area; **Distribution**: India: Palk Strait and Gulf of Mannar; **Status**: Common (Figure 13.3).

Figure 13.1: *Barbatia velata.*

Figure 13.2: *Scapharca cornea.*

Family: Donacidae

4. *Donax tranculus*

Shell 3.1 cm in length, thin, elongate moderately flatted, and triangular. Shell with rounded umbones, Ligament much longer than behind. Fine radial lines on surface. Pallial sinus large and deep. Colour of the shell grayish blue and violet. Often rayed. Inner surface of the shell smooth and usually violet.

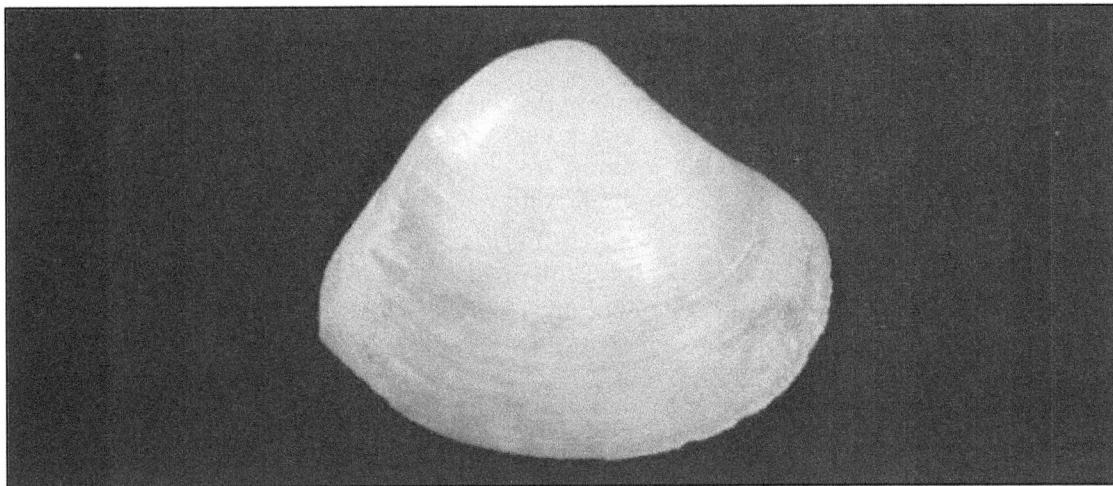

Figure 13.3: *Potamocorbula ustulata ustulata.*

Common name : Truncate donax; **Locality** : Intertidal and coral area; **Distribution**: India: Palk Strait and Gulf of Mannar; **Status**: Common (Figure 13.4).

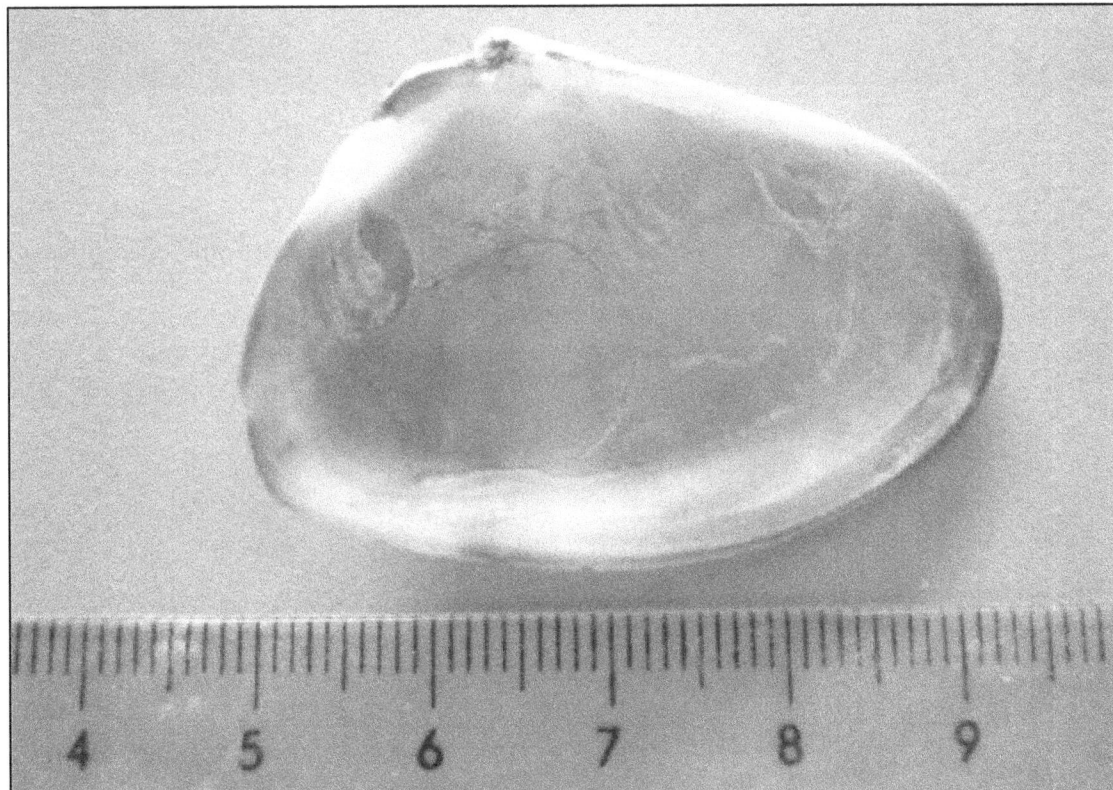

Figure 13.4: *Donax tranculus.*

Family: Dimyidae

5. *Dimya japonica*

Shell upto 8.8 cm in length, thin, fragile and flat. Moderately large, outer surface of the shell slightly rough, concentric lines widely developed. Shell almost transparent. Inner surface of the shell smooth and glossy. Umbo small, Adductor muscle scar present at center of valve. Dark brown in colour with black patches are present on both side. Shiny and silky appearance. Uncommon (Figure 13.5).

Common name : Saddle Oysters; **Locality** : Intertidal and coral area; **Distribution**: India: Palk Strait and Gulf of Mannar; **Status**: Uncommon.

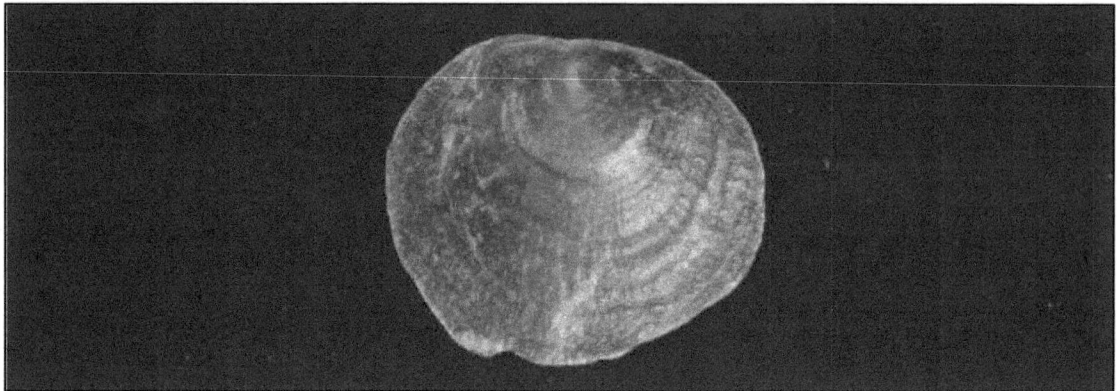

Figure 13.5: *Dimya japonica.*

Family: Geloniidae

6. *Gelonia bengalensis*

Shell up to 5.8 cm in length thick, solid and heavy. Umbo prominent and smooth. Outer surface of the shell bears strong, close set, crenulated ridges. Interstices bear radial ribs due to both radial and concentric ridges. Hinge thick with 3 cardinal teeth. Anterior teeth more elongated towards the posterior

Figure 13.6: *Gelonia bengalensis.*

teeth. Middle teeth small. Ligament elongated. Pallial sinus broad and U-shaped. Outer surface of the shell dark brown in colour. Inner surface white. Anterior and Posterior adductor scar present (Figure 13.6).

Locality : Intertidal and coral area; **Distribution**: India: Palk Strait and Gulf of Mannar; **Status**: Uncommon.

Family: Glauconomidae

7. *Glauconome chinensis* (Gray, 1828)

Shell up to 6.4 cm in length, large, heavy and thick. Umbo elevated and prominent. Gibbose appearance. Strong keel radiating from umbo. Two keels on hind end more pronounced than rest Anterior rim rounded. Surface bears thin concentric rings. Two cardinal teeth. Hinge small. Inner and outer surface of the shell white (Figure 13.7).

Locality : Intertidal and coral area; **Distribution**: India: Palk Strait and Gulf of Mannar; **Status**: Common.

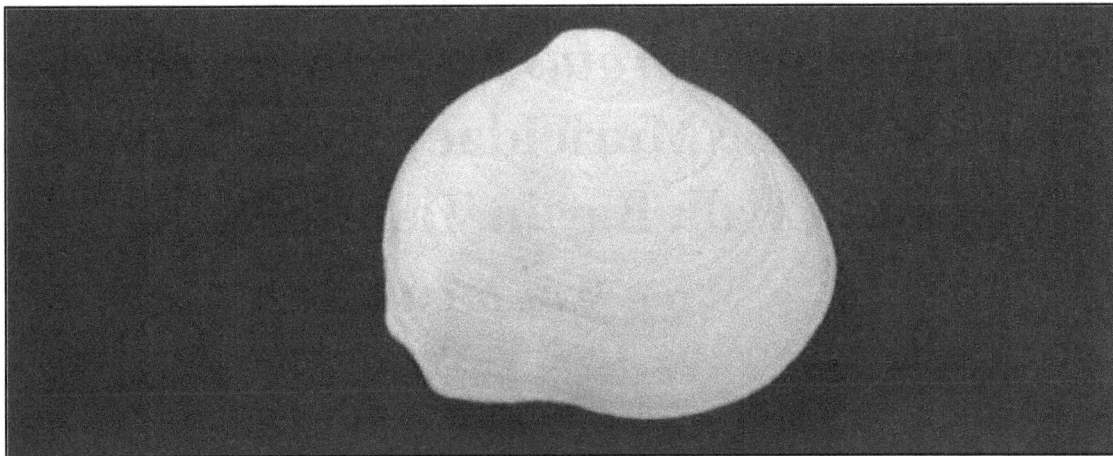

Figure 13.7: *Glauconome chinensis.*

References

CMFRI., 2004. Biodiversity of marine Molluscs. *CMFRI Ann. Rep.* (2003–2004) MOL/BIOD/01. pp. 55.

Ganapati, P.N and Sarma, A.L.N., 1972. Bivalve gastropods of the Indian seas. *Proc. Ind. Nat. Sci. Acad.,* 38B: 240–250.

http://microseashell. com/bbs/view.

Jacobs, R.P.W.M. and Huisman, W.H.T., 1982. Macro benthos of some *Zostera* beds in the vicinity of Roscoff (France) with special reference to relations with community structure and environmental factors. *P.K. Ned. Akad. Wetensc.,* 85: 335–356.

Sahul Hameed, P. and Somasundaram, S.S.N., 1998. A survey of bivalve mollusca in Gulf of Mannar, India. *Indian. J. Fish.,* 45(2): 177–181.

Satyanarayana Rao, K. and Sundaram, K.S., 1972. Ecology of intertidal molluscs of Gulf of Mannar and Palk Bay. In: *Proc. Indian Natl. Sci. Acad.,* Pt B. 38(5–6): 462–474.

Chapter 14

Length-Weight Relationship and Allometry of *Chicoreus virgineus ponderosus* and *Siratus vigineus ponderosus* (Gastropoda : Muricidae) from Thondi Coast, Palk Bay in Tamil Nadu

☆ *C. Stella, D. Chellaiyan, C. Latha Sumathi, A. Priya and R. Ravichandran*

ABSTRACT

In the present study, the changes in the constant allometry of length-weight relationship are associated with increase in size and sexual maturity. The lower 'b' values are obtained for *Chicoreus virgineus ponderosus* and *Siratus virgineus ponderosus* of male and females though looked strange, may be attributed to heavy shell weight with less meat weight observed in these species. The Correlation co-efficient 'r' of two species was found to be significant (P 0.001). Results of the analysis of variance for various parameters of two species of male and females were studied. It is evident that there are significant differences were observed between males and females of *Siratus virgineus ponderous* and there are significant differences were observed in the intercepts between males and females of *Chicoreus virgineus ponderous*. Changes in allometric relations were observed between different shell characters with respect to growth and total weight of the animal in both males and females of two species.

Keywords: Length-weight, Allometry, Muricidae, Chicoreus species, Siratus species, Palk bay.

Introduction

Length-weight relationship has been mathematically proven to be having a constant relationship between total length and weight of the individual. Length weight relationship is of great importance in fishery biology for the calculation of average weight at a certain length and the conversion of an equation of growth in terms of weight at a certain length and conversion of growth equation in terms of length and weight (Entsna–Mensah *et al.*, 1995; King, 1996; Bernandes *et al.*, 2000; Muto *et al.*, 2000). According to Gould (1966) knowledge of allometry in shell and soft body characters is essential to fully understand the growth of a species. The length-weight relationship has a number of applications in stock assessment. Among the applications, the estimation of standing stock biomass, calculating condition index and comparing the ontogeny of population from different regions are important. (Pauly and Munro, 1983). The growth rate of various organisms is not uniform. The growth of some dimension of a body component in relation to the whole body is termed as allometric growth. It is useful in knowing the variations from the expected weight for various length groups, as some organisms are known to change their form or shape during growth (Lacren, 1951). Length-weight relationship of gastropods is not well documented. Geldiay (1956) has made length-weight studies on *Ancylus fluviatilis*. Berrie (1966) has studied the growth and seasonal changes of *Lymnaea stagnalis*. Kemp and Bertness (1993) have studied the snail shape and growth rates evidence for plastic shell allometry in *Littorina littorea*. Thivakaran (1988) has observed the allometric relationships in *Umbonium vestiarium* and *Nassa stolatus*.

Age and growth and length-weight relationship in *Littorina scabra* have been observed by Maruthamuthu and Kasinathan (1985). Paninee (1991) has studied the length-weight relationship in *Chicoreus ramosus*. Stella *et al.*, (1992) have studied the analysis of size class distributions of *Chicoreus ramosus* collected from the Gulf of Mannar coast of India. Benny and Ayyakkannu (1992) have observed the length-weight relationship in *Chicoreus ramosus*. Stella (1995) has studied the length-weight relationship in *Chicoreus virgineus* and *Muricanthus virgineus* in Parangipettai waters. The change in the shape of growing animals at a point of time is due to the concurrent increase in size and weight respectively. Analysis of the length-weight relationship has become a standard practice in fishery studies (Ricker, 1975). Rao (1988) has reported that the knowledge of the length-weight relationship has a vital importance in fishery as it is not only helps in establishing the yield, but in converting one variable to another. Of the two, length is easier to measure and can be converted into weight in which the catch is invariably expressed (Bal and Rao, 1984).

A number of studies have been used for allometry and growth models to study the population structure, growth modeling, and shell morphometrics of a number of snails. Grossowicz *et al.* (2003) have examined the distribution of Melanopsis snail and studied the population assessment of commercial gastropods and the nature of gastropods fishery in Panglao Bay. Heller *et al.* (2002) have reported the systematics of Melanopsis from the coastal plain of Israel. Ismail and Elkarmi (2006) have reported the age, growth and shell morphometrics of *Monodonta dama* from the Gulf of Aqaba, Elkarmi and Ismail (2006) have studied the population structure and shell morphometrics of *Theodoxus macri* from Jordan. Tuberculata (Thiaridae) living in hot springs and freshwater pools by Joeppette *et al.*, (2007). Elkarmi and Ismail (2007) have studied the growth models and shell morphometrics of two populations of Melanoides Mohamed H. Yassien (2009) have studied the shellfish fishery in the North western part of the Red sea. Jaikumar *et al.*, (2011) have studied the length weight relationship of *Lambis lambis* from Tuticorin waters.

The standard comparison used for gastropods being total length against total weight. The length weight relationship formula besides providing a means for evaluating weight from length, a direct way of converting logarithmic growth rates calculated on length into growth rates. The length of animal increases with weight, showing that weight of animal is a function of length. As length is a linear measure and the weight a measure of volume, the relationship between length and weight can be expressed by the hypothetical cubic law $W = CL^3$. Where, W = weight, L = length and C = constant. But Le Cren (1951) has suggested that it is advisable to fit parabolic equation of the form $W = aL^b$. This formula expresses the relation between weight and length better than the cubic formula. If the form and specific gravity remains constant, the formula could be used to calculate the weight of known length and *vice versa*. The length-weight relationship can be expressed graphically by plotting the observed length and weight as a scattered diagram. The present study has been carried out in *Chicoreus virgineus ponderosus* and *Siratus virgineus ponderosus*, in order to understand the length-weight and other allometric relationships between various morphological characters.

Description of the Study Area

During the Quaternary period the Palk Strait must have originated introducing a close connection to the Southern Gulf of Mannar and to the northern Bay of Bengal with in the latitude of 90° and 10° N and longitude of 79° and 80° E. Northern boundary of the strait is of Kodiyakkarai (Map 14.1). The Southern one is restricted to the Adams Bridge and the Eastern limit is to the Sri Lanka and Thalai-mannar region. The Palk Bay is influenced mainly by the North east monsoon. The Bay has strong potential of living and non living resources. Thondi is a small village

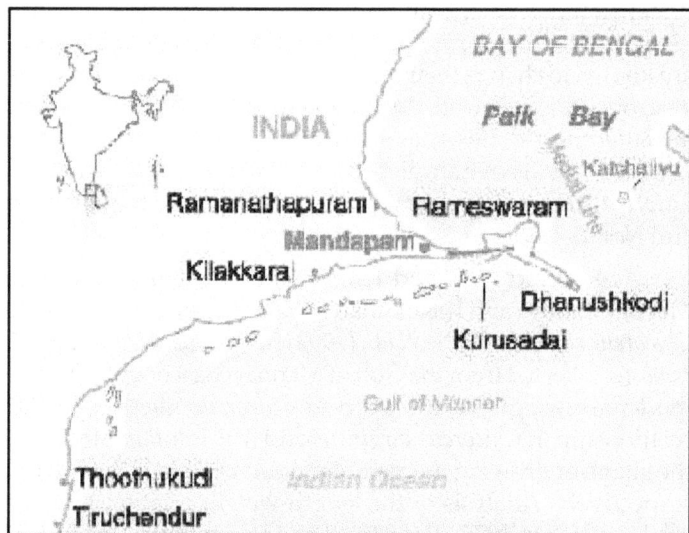

Map 14.1: Palk Strait Study Area Map.

situated in the Palk Bay region of Tamil Nadu. The study area lies in the latitude of 9°44'N and longitude of 79°19' E. The present study was carried out at Thondi coast, South east coast of India. The species of *Chicoreus virgineus ponderosus* and *Siratus virgineus ponderosus* are exclusively marine in distribution. The collection of these species was made from 10 to 15 fathom lines with muddy bottom. The species live in sandy mud benthic zones. The animals were collected from the fishing trawlers along with other benthic gastropods like *Rapana bulbosa, Babylonia spirata, Murex trunculus*, and *Conus* sps. etc.

Materials and Methods

Random samples of *Chicoreus virgineus ponderosus* and *Siratus virgineus ponderosus*, were collected once in a month from trawlers of Thondi coast for a period of one year (January 2006 to December 2006). Random samples of two species, ranging in size from 4.0 to 10.7 cm in *Chicoreus virgineus ponderosus* and from 4.0 to 12 cm in *Siratus virgineus ponderosus* were collected. The shells were measured with the help of Vernier Caliper to the nearest 0.1 mm. The shell length, width, aperture length, aperture width were measured using a Vernier Caliper.

Results and Discussion

The parabolic equation, $W = aL^n$, used in the study can be expressed in the logarithmic form as

$$\text{Log } W = \log_a + n \log_w$$

i.e. $Y = a + bx$, where, $a = \log$; $b = n$; $Y = \log_n$

$X = \log_L$ which is a linear relationship between X and Y.

Length-Weight Relationship (Figures 14.1–14.4)

The comparison of regression lines between male and female of two species is presented in Tables 14.1–14.3. Correlation co-efficient 'r' for two species was found to be significant (P 0.001). In the

Table 14.1: Regression Parameters and Results of Statistical Point from Length-Weight Relationship of *Chicoreus virgineus ponderosus* and *Siratus virgineus ponderosus*

	N	a	b	R	S.E.	T Value
		Chicoreus virgineus ponderosus				
Male	547	3.561586	3.324779	0.226402	0.612714	5.426317
Female	840	4.443329	3.271298	0.241852	0.453379	7.215376
Combined	1387	4.105817	3.290618	0.236092	0.363929	9.041928
		Siratus virgineus ponderosus				
Male	339	−58.2203	14.58857	0.892392	0.352895	41.33973
Female	725	−72.188	16.87561	0.921337	0.264824	63.72386
Combined	1164	−66.8104	16.01529	0.910625	0.2132	75.11857

N: Number of animals; a, b: Constants of regression equation; S.E.: Standard error; r: Correlation coefficient p<0.001.

*: Indicates the largely significance of p<0.001

Table 14.2: Comparison of Regression Lines between Males and Females of *Chicoreus virgineus ponderosus*

	df	X^2	Y^2	xy	Regression Coefficient	Deviation from Regression		
						df	ss	ms
Within								
Male	546	527.1000	113672.6300	1752.4900	3.3200	545	107846.0000	197.8826
Female	839	1007.1800	901391.2300	2837.4500	2.8200	838	173481.1000	207.0181
						1383	281327.1000	203.4180
Pooled w	1385	1534.2800	1015063.8600	4589.9400	2.9916	1384	289628.9891	209.2695
		Difference between slope				1	8301.8891	8301.8891
Between B	1	0.0050	435.7950	2.0600				
W+b	1386	1534.2850	1015500.0000	4592.0000		1385	306242.5891	
		Between adjacent error				1	16613.6000	16613.6000

Comparison of slope = f= 8301.8891/203.4180 = 40.8120 (df = 2772) N.S.

Comparison of elevation F = 16613.6000/209.2695 = 79.3885 (df= 2773) (significant test p<0.01).

present study, the changes in the constant allometry of length-weight relationship are associated with increase in size and sexual maturity as observed in the case of *Anadara rhombea* (John, 1980) and *Pythia plicata* (Shanmugam, 1987). The lower 'b' values obtained for both species *Chicoreus virgineus ponderosus* and *Siratus virgineus ponderosus* of male and females though looked strange, may be attributed to heavy shell weight with less meat weight observed in these species.

Table 14.3: Comparison of Regression Lines between Males and Females of
Siratus virgineus ponderosus

	df	X^2	Y^2	xy	Regression Coefficient	Deviation from Regression		
						df	ss	ms
Within								
Male	438	548.83	146673.47	8006.63	14.59	437	29868.1700	68.3482
Female	724	901.29	302374.71	15209.80	16.39	723	45700.0900	63.2090
						1160	281327.1000	65.1451
Pooled w	1162	1450.12	449048.18	23216.43	15.73	1161	468067.1000	403.1585
	Difference between slope					1	186739.9578	186739.9578
Between B	1	14.47	3960.08	239.39				
W+b	1163	1464.59	453008.26	23455.82		1162	843718.8000	
	Between adjacent error					1	375651.7000	375651.7000

Comparison of slope = f = 186739.9578/65.1451 = 2866.5235 (df = 2326) N.S.

Comparison of elevation F = 375651.7000/403.1585 = 931.7717 (df= 2327)(significant test p<0.01).

Allometric Relationship between Various Morphological Characters

The relationship between shell total length (TL), total width (TW), aperture length (AL), aperture width (AW) and weight of the whole animal (WA) in male and female of *Chicoreus virgineus ponderosus* were studied in all possible combinations using the linear regression and the results are presented in Tables 14.4 and 14.5. In males, the correlation between the variables, TL and TW (r = 0.2834), TL and Al (r = 0.8515), TL and AW (r = 0.8805) and TL and WA (r = 0.1019) and in females TL and TW (r = 0.2673), TL and AL (r = 0.8268), TL and AW (r = 0.8605) and TL and WA (r = 0.09189). It is evident that the correlation coefficient values for various combinations of both forms of *Chicoreus virgineus ponderosus* are statistically significant. In *Siratus virgineus ponderosus* in males, total length influences the other variables; TL and TW (r = 0.5071), TL and AL (r = 0.8350), TL and AW (r = 0.8640) and TL and WA (r = 0.4685) so also in females TL and TW (r = 0.5053), TL and AL (r = 0.8441), TL and AW (r = 0.8761) and TL and WA (r = 0.4810). It is evident that the correlation coefficient values for various combinations in both the sexes are highly significant.

Results of analysis of variance for various parameters of two species of male and females are presented in the Tables 14.6 and 14.7. It is evident that there are significant differences between males and females of *Siratus virgineus ponderosus* and there are significant differences in the intercepts between males and females of *Chicoreus virgineus ponderosus*. Changes in allometric relations between different shell characters with respect to growth were observed between shell characters and total weight of the

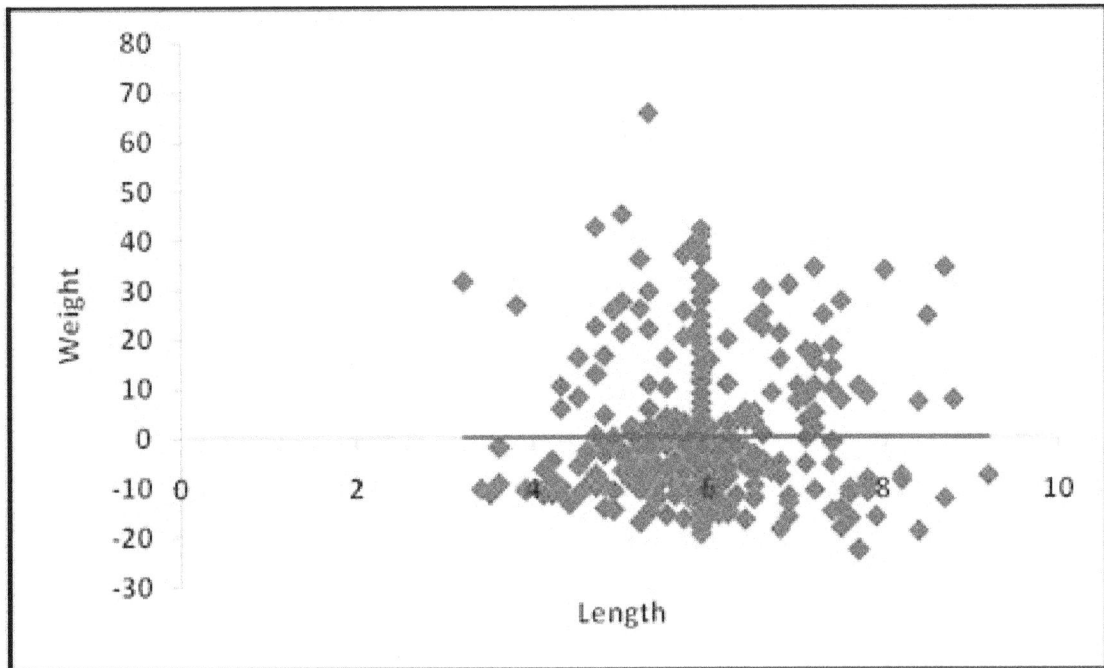

Figure 14.1: Length-Weight Relationship of *Chicoreus virgineus ponderosus*–Male.

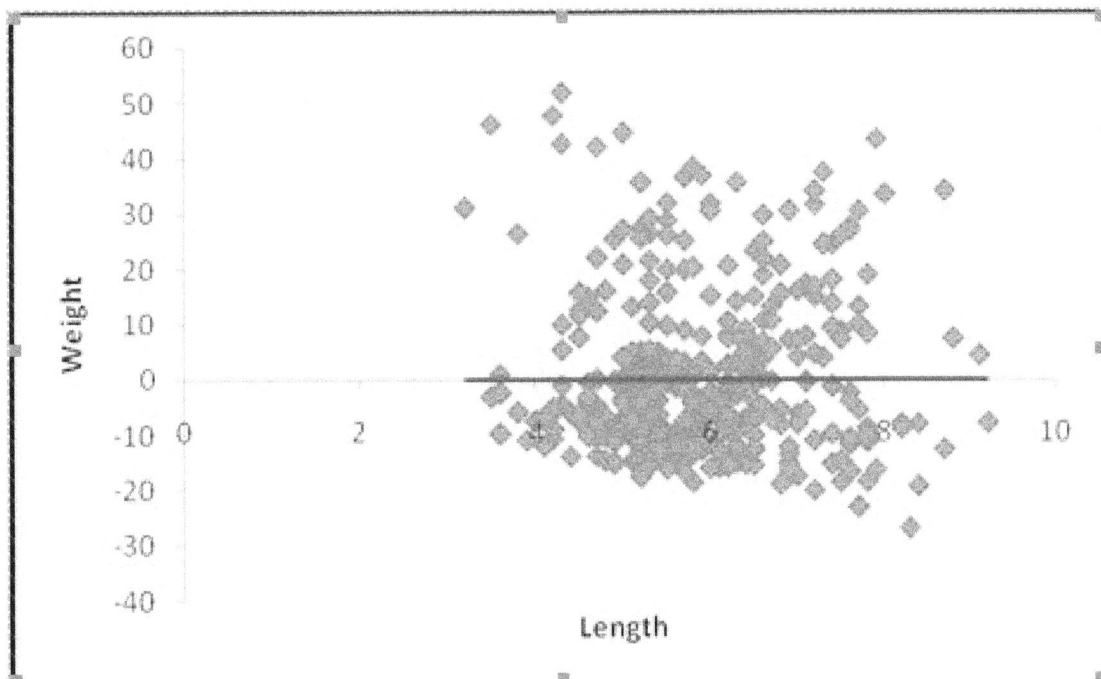

Figure 14.2: Length-Weight Relationship of *Chicoreus virgineus ponderosus*–Female.

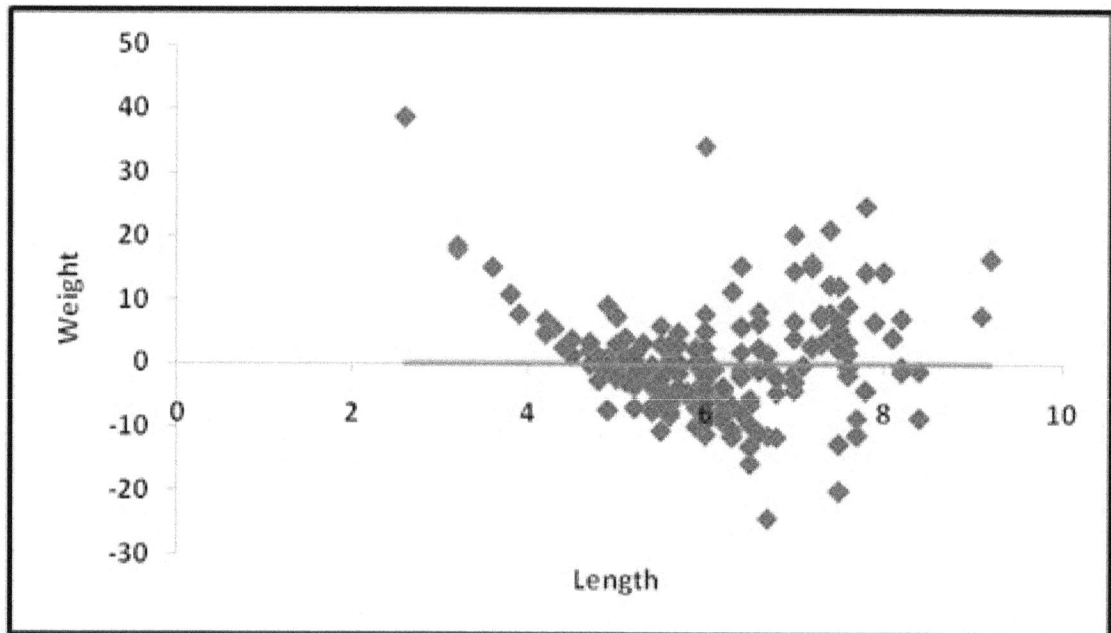

Figure 14.3: Length-Weight Relationship of *Siratus virgineus ponderosus*–Male.

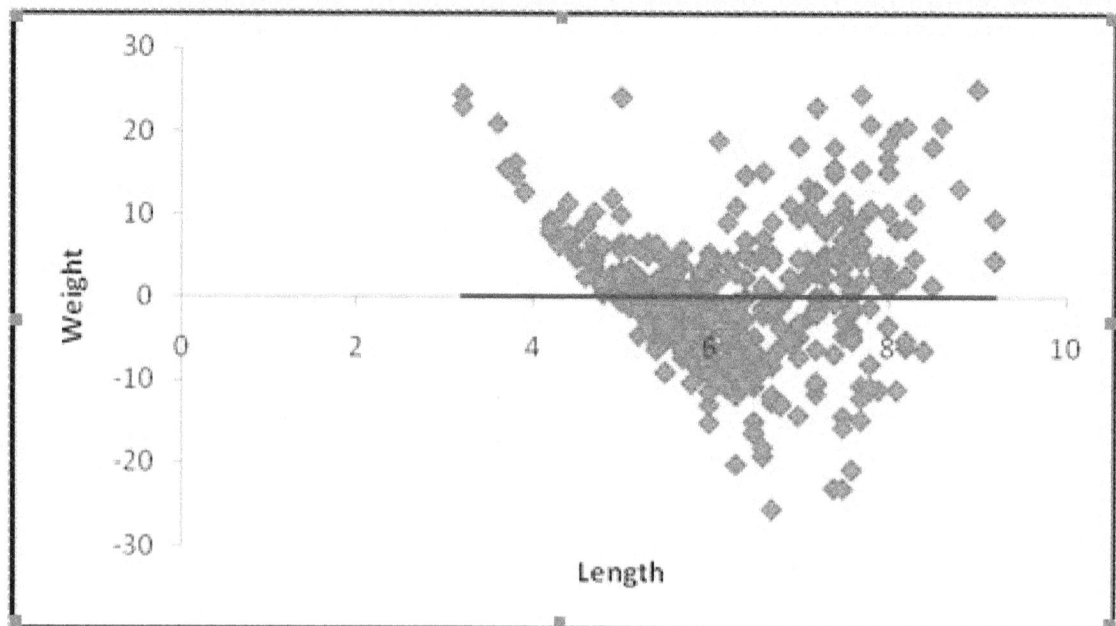

Figure 14.4: Length-Weight Relationship of *Siratus virgineus ponderosus*–Female.

Table 14.4: Correlation Coefficient and Results of *t*-test in Males and Females *Chicoreus virgineus ponderosus*

Variables	R	Coefficient	t	df
		Male		
TL=TW	0.2834	0.2635779	20.78	1092
TL=AL	0.8515	0.138098	79.13	1092
TL=AW	0.8805	0.098977	89.71	1092
TL=WT	0.1019	3.324779	11.13	1092
		Female		
TL=TW	0.2673	0.242401	24.74	1678
TL=AL	0.8268	0.123174	89.49	1678
TL=AW	0.8605	0.093165	101.7	1678
TL=WT	0.09189	2.81723	13.03	1678

Table 14.5: Correlation Coefficient and Results of *t*-test in Males and Females of *Siratus virgineus ponderosus*

Variables	R	Coefficient	t	df
		Male		
TL=TW	0.2834	0.2635779	20.78	1092
TL=AL	0.8515	0.138098	79.13	1092
TL=AW	0.8805	0.098977	89.71	1092
TL=WT	0.1019	3.324779	11.13	1092
		Female		
TL=TW	0.2673	0.242401	24.74	1678
TL=AL	0.8268	0.123174	89.49	1678
TL=AW	0.8605	0.093165	101.7	1678
TL=WT	0.09189	2.81723	13.03	1678

animal in both males and females of two species. The results were found to support variations in growth between males and females of *Chicoreus virgineus ponderosus*. Kasinathan *et al.*, (1987) also have reported significant differences in growth between the male and female of *Oliva oliva*. Even though positive allometric growth was observed between shell height and weight in *Chicoreus virgineus ponderosus*, the difference has been observed to be significant between these males and females forms (Males F = 124.0; Females F = 169.0) and in *Siratus virgineus pondersous* it was observed to be significant between these males and females forms (Males F = 772.1082; Females F = 1346.241). Maruthamuthu and Kasinathan (1985) have reported significant difference in length-weight relationship between male and female of *Littorina scabra* from Parangipettai waters. Kasinathan *et al.*, (1987) also have reported significant difference between length-weight and opined that the difference appears to be due to sexual maturity and also due to increase in size of both sexes. Kuenzler (1961) and Shafee (1976) also have reported changes in the allometry to length-weight relationship associated with sexual maturity in mussels. Stella (1995) has reported in *Chicoreus virgineus* and *Muricanthus virgineus*.

Table 14.6: Analysis of Variance among Various Morphometric Measurements of Males and Females of *Chicoreus virgineus ponderosus* (*** P< 0.001)

Variables	SS	DF	MS	F Ratio
Male				
TL=TW				
Between	1053.58	1	1053.58	412.3777
Within	2789.942	1092	2.554892	
TL=AL				
Between	3756.806	1	3756.806	5373.368
Within	763.475	1092	0.699153	
TL=AW				
Between	4731.75	1	4731.75	6889.727
Within	749.9676	1092	0.686783	
TL=Wt				
Between	94040	1	94040	124.0
Within	828300	1092	758.5	
Female				
TL=TW				
Between	1565.037	1	1565.037	612.2554
Within	4289.276	1678	2.556183	
TL=AL				
Between	5813.616	1	5813.616	8008.42
Within	1218.124	1678	0.725938	
TL=AW				
Between	7309.177	1	7309.177	10350.63
Within	1184.932	1678	0.706158	
TL=Wt				
Between	163700	1	163700	169.8
Within	1618000	1678	964.1	

Table 14.7: Analysis of Variance among Various Morphometric Measurements of Males and Females of *Siratus virgineus ponderosus*

Variables	SS	DF	MS	F Ratio
Male				
TL=TW				
Between	963.3805	1	963.3805	901.2963
Within	936.3417	876	1.068883	
TL=AL				
Between	3403.255	1	3403.255	4433.86
Within	672.3829	876	0.76756	

Contd...

Table 14.7–*Contd...*

Variables	SS	DF	MS	F Ratio
TL=AW				
Between	4220.091	1	4220.091	5564.652
Within	664.3363	876	0.758375	
TL=Wt				
Between	129762	1	129762	772.1082
Within	147222.3	876	168.062	
Female				
TL=TW				
Between	1671.037	1	1671.037	1484.43
Within	1630.028	1448	1.12571	
TL=AL				
Between	5973.681	1	5973.681	7854.832
Within	1101.219	1448	0.76051	
TL=AW				
Between	7468.073	1	7468.073	10252.36
Within	1054.759	1448	0.728425	
TL=Wt				
Between	281963	1	281963	1346.241
Within	303276	1448	209.4448	

References

Bal, D. V. and Rao, K. V., 1984. *Marine Fisheries of India*. Tata McGraw-Hill Publishing Company, New Delhi, 51–73 pp.

Benny, A. and Ayyakkannu, K., 1992. Length-weight relationship in *Chicoreus ramosus*. *Phuket Mar. Boil. Cent. Sec.*, 10: 199–201.

Bernandes, R. A. and Rossi-Wongts Chowski, C. L. D. B., 2000. Length-weight relationship of small pelagic fish species of the Southeast and South Brasilian exclusive economic Zone. *NAGA (ICLARM)*, 23(4): 30–32.

Berrie, A. D., 1966. Growth and seasonal changes in the reproductive organs of *Lymnae stagalis* (L). *Proc. Malac. Soc. Lond.*, 37, 83–92.

Elkarmi, A. and Ismail, N., 2006. Population structure and shell morphometrics of the gastropod *Theodoxus macri* Neritidae: Prosobranchia) from the Gulf of Aqaba, Red Sea. *Pak. J. Biol. Sci.*, 9: 843–847.

Elkarmi, A. and Ismail, N., 2007. Growth models and shell morphometrics of two populations of *Melanoides tuberculata* (Thiaridae) living in hot springs and freshwater pools. *J. Limnol.*, 66(2): 90–96.

Enstna-Mensah, M., Abunyewa, A. and Palomares, M.L.D., 1995. Length-weight relationships of fishes from tributaries of the Vata River, Cihana, Part I. Analysis of fold data sets. *Naga ICLARM,* 18(Y2): 36–38.

Geldiay, R., 1956. Studies on local populations of the freshwater limpet *Ancylus fluviatilis. Jour. Animal. Ecology,* 25: 389–402.

Gould, S.J., 1966. Allometry and size in ontogeny and phylogeny. *Biol. Rev.,* 41: 587–640.

Grossowicz, N., Sivan, N. and Heller, J., 2003. *Melanopsis* from the Pleistocene of the Hula Valley (Gastropoda: Cerithiodea). *Isr. J. Earth Sc.,* 52: 221–234.

Heller, J., Sivan, N. and Bea-Ami, F., 2002. Systematics of *Melanopsis* from the coastal plains of Israel (Gastropoda: Cerithiodea). *J. Conchology,* 37: 589–606.

Ismail, N.S. and Elkarmi, A.Z., 2006. Age, growth and shell morphometrics of the gastropod *Monodonta dama* (Neritidae: Prosobranchia) from Azraq Oasis, Jordan. *Pak. J. Biol. Sci.,* 9(3): 549–552.

Jaikumar, M.R., Gunalan, Ramkumar and Kangu, L., 2011. Length-weight relationship of *Lambis lambis* (Mollusc: Gastropoda) from Tuticorin coastal waters, Gulf of Mannar, South East Coast of India. *World Applied Science Journal,* 14(2): 207–209.

Joppette, J. Hermosilia and Narido, Charina I., 2007. Population assessment of commercial gastropods fishery in Panglao Bay, *Bohol, Philippines Journal,* Holy Name University, Tagbilaran city, Bohol., 18 (2).

John, G., 1980. Studies on *Anadara rhombea* (Born) (Mullusca: Bivalvia) from Portonova coastal water. *Ph.D. Thesis,* Annamalai University, pp. 190.

Kasinathan, R., Maruthamuthu, S. and Tagore, J., 1987. Allometric relationships *Oliva olive* from Portonovo waters. *Indian Jounal of Fisheries,* 34: 213–218.

Kemp, P. and Bertness, M.D., 1983. *Snail Shape and Growth Rates Ecological Implications of Body Size.* Cambridge University Press, N.Y., 329 pp.

King, R.D., 1996b. Length-weight relationship of the Nigerian coastal water fishes. *Naga, ICLARM,* 19(4): 53–58.

Kuenzler, E.J., 1961. Structure and energy flow of a mussel population in a Georgia Salt marsh. *Limnol. Oceanogr.,* 6: 400–415.

Lacren, C. D., 1951. The length-weight relationship and seasonal cycle in gonad weights and condition in the perch (*Perca Fluviatilus). J. Anim. Ecol.,* 20: 201–219.

Maruthamuthu, S. and Kasinathan, R., 1985. Age and growth and length-weight relationship in estuarine periwinkle *Littorina scabra* (Linne.) *Indian J. Mar. Sci.,* 14: 102–104.

Mohamed, H. Yassien, 2009. Shell fish fishery in the North Western part of the sea. *World Journal of Fish and Marine Sciences,* 1(2) : 97–104.

Muto, E.Y., Soares, L.S.H. and Rossi-Wongtschowski, C.L.D.B., 2000. Length-weight relationship of marine fish species of Sao Sebastion system, Sao Paulo, Southeastern Brazil. *Naga (ICLARM),* 23(4): 27–29.

Paninee, T., 1991. Basic data on *Chicoreus ramosus. Phuket. Mar. Boil. Cent. Spec. Oub.,* 9: 94–96.

Pauly, D., and Munro, J.L. 1984. Once more on the comparison of growth in fish and invertebrates. *ICLARM, Fishbyte,* 2(1): 21.

Rao, G.S., 1988. Biology of *Meretrix casta* (Chemnitz) and *Paphia Malabarica* (Chemnitz) from Mulky estuary, Dakshina Kannada. *Bull. Cent. Mar. Fish. Res. Inst.*, (42)1: 148–153.

Ricker, W.E., 1975. Computation and interpretation of biological statistics of fish populations. *Bull. Fish. Res. B. Can.*, 191: 382.

Shafee, M.S., 1976. Studies on the varies allometric relationships in the intertidal green mussel *Perna viridis.* (Linnaeus) of Ennore estuary, Madras. *Indian. J. Fish.*, 23 (1 and 2).

Shanmugam, A., 1987. Studies on *Pythia plicata* (Gray) (Gastropoda: Pulmonata: Ellobiidae) from Pitchavaram mangroves. *Ph.D. Thesis,* Annamalai University, pp. 137.

Stella, C., Rajkumar, T. and Ayyakkannu, K., 1992. Analysis of size class distributions of *Chicoreus ramosus* collected from the Gulf of Manner area, South East Coast of India. *Phuket Mar. Biol. Cent. Spec. Publ.*, 11: 91–93.

Stella. C., 1995. Studies on the taxonomy and ecobiology of *Chicoreus* species from Parangipettai waters, South East Coast of India. *Ph.D. Thesis,* Annamalai University, pp. 195.

Thivakaran, G.A., 1988. Studies on the littorinids *Littorina quadricentus* (Phillip) and *Nodilittorina pyramidalis* (Quoy and Gaimard, 1833) Gastropoda: prosobranchia littorinidae) from the tranquebar rocky shore (South East Coast of India) *Ph.D. Thesis,* Annamalai University, pp. 179.

Chapter 15

Avifauna of Mangrove Ecosystem of Karwar, West Coast of India

☆ *B. Vasanthkumar*

ABSTRACT

The present study was carried out in different mangrove ecosystem of Kali estuary, Karwar (14°50′21″ N and 74°10′05″E) for the period of thirteen months from January 2008 to January 2009. The study area lying within the grid of 14° 50′ 39″ N and 14° 51′ 49″ N and 74°07′ 44″ E and 74° 13′ 25″ E of was selected for the distribution and abundance of avifaunal community. The bird community on the exposed mangrove mud flat area was scanned at monthly intervals from an indigenous dugout canoe with an outboard motor engine moving at uniform speed of 10 km per hour for the period of one year. Sampling began at 7.00 AM and continued upto 9.00 AM and in the evening, bird sampling was done from 4.00 PM to 6.00 PM. In each habitat, minimum of 8 censuses were conducted. About four to eight censuses were carried out per month in each study site.

Keywords: Avifauna, Mangroves, Kali river, Karwar.

Introduction

Studies on avian community provide effective tools for monitoring forest in general and mangrove forest in particular. Evaluating bird communities of the mangrove forests to plan for biodiversity-friendly development is gaining significance. Ali and Ripley (1983) reported the avifauna of Indian subcontinent being represented by 2094 forms belonging to 1200 species of which 19.9 per cent (417) forms are wetland birds (Rao *et al.*, 1997). In recent past years, importance of bird diversity and its conservation has been emphasized and such studies are encouraged. Much work related to the avifaunal diversity has been done in temperate forests, while, a very limited data is available in the tropics in general and mangrove forest in particular.

In west coast of India, especially in mangrove ecosystems harboring a fairly rich faunal and floral wealth are relatively unexplored. Only a few reports are available pertaining to the mangrove forests. Not many studies have been made in the region of Uttara Kannada district of Karnataka. Bhagwat (2003) examined the effects of landscape modification on bird diversity in Kodagu district. Some reports have been recorded on several aspects of bird life in and around Uttara Kannada district. Other reports related to survey of birds was also carried out in Uttara Kannada district and adjoining areas and no serious studies exclusively on bird diversity across different landscapes are made in the western Ghat regions. From the review of the above literature, it was inferred that not much studies are made in this region especially on mangrove ecosystem and that studies on census, diversity and relationship with vegetation is needed. Hence, the present study was attempted on the birds of natural forests including mangrove forests.

Foot trails were conducted into the dense mangroves not accessible by boat and the birds therein were listed. The birds sighted were identified using relevant field guides (Woodcock, 1980; Ali and Ripley, 1983; Sonobe and Usui, 1993; Grewal, 1995' Ali, 1996; Grimett *et al.,* 1998). A combination of Total Count Method and Linear Transact count method standardized to meet the requirement of the site were employed. An binocular and 15–45 × 60 spotting scope was used for this purpose.

Materials and Methods

The present study was carried out in different mangrove ecosystem of Kali estuary, Karwar (14°50′21″ N and 74°10′05″E) for the period of thirteen months from January 2008 to January 2009. The study area lying within the grid of 14° 50′ 39″ N and 14° 51′ 49″ N and 74°07′ 44″ E and 74° 13′ 25″ E of was selected for the distribution and abundance of avifaunal community.

The bird community on the exposed mangrove mud flat area was scanned at monthly intervals from an indigenous dugout canoe with an outboard motor engine moving at uniform speed of 10 km per hour for the period of one year. Sampling began at 7.00 AM and continued upto 9.00 AM and in the evening, bird sampling was done from 4.00 PM to 6.00 PM. In each habitat, minimum of 8 censuses were conducted. About four to eight censuses were carried out per month in each study site. At each study site, all the birds that were seen were recorded by species with field binoculars and census was not conducted on rainy and heavy misty mornings. All birds that were heard while survey and were not counted because of the lack of information on the identity of species by bird call alone. The species identification was based on Ali (1996) and the nomenclature was based on Grimmett *et al.* (1999).

Results and Discussion

The climate of Karwar is strongly influenced by the southwest monsoon occurring normally from June to September. However, since the present study is confined to water birds ecology the seasons have been categorized as summer, extending from March to May; monsoon from June to September; post monsoon from October to November and winter from December to February. Table 15.1 depicts the check-list of avifauna in the mangrove ecosystem of Kali estuary.

Mangrove ecosystem harbored as 66 species of birds of which 23 species were of terrestrial birds. On the whole the cumulative biannual bird population in Kali River was1236 individuals. Although species diversity of terrestrial bird was relatively high, they contributed a mere 25.70 per cent to the total avifaunal population. The terrestrial birds dominating in the mangrove ecosystem included Warbler; Black capped Kingfisher, Blue eared Kingfisher and Collared Kingfisher. All three species of Kingfisher were recorded exclusively inside the mangrove ecosystem indicating that although the ecosystem is estuarine wetland, which provides an ideal habitat even for terrestrial birds that feed largely on insects. Similar findings were also reported from estuarine wetland of Goa by Sonali *et al.* (2005).

Table 15.1: List of Avifauna Recorded in the Mangrove Forests of Kali Estuary, Karwar

Scientific Name	Common Name	Scientific Name	Common Name
Spilornis cheela	Crested serpent eagle	*Heliastur Indus*	Brahminy kite
Heliaeetus leucogaster	White bellied sea eagle	*Sterna aurantia*	Indian River tern
S. hirunda	Common tern	*Aegithina tiphia*	Common Iora
Alcedo atthis	Common Kingfisher	*Halcyon capensis*	Stork-billed Kingfisher
H. smyrnensis	White-throated Kingfisher	*H. pileata*	Black-capped Kingfisher
Ceryle rudis	Pied Kingfisher	*Amaurornis phoenicurus*	White-breasted waterhen
Ardea alba	Large Egret	*A. cinerea*	Grey heron
Ardeola grayii	Pond heron or paddy bird	*Phalacrocorax niger*	Little Cormorant
Pluvialis fulva	Pacific golden plover	*Bubulcus ibis*	Cattle Egret
Butorides striatus	Little heron	*Casmerodius albus*	Great Egret
Centropus sinensis	Crow-pheasant	*Ceryle rudis*	Lesser pied Kingfisher
Charadrius dubius	Little ringed plover	*C. leschenaultia*	Large sand plover
C. mongolus	Lesser sand plover	*Larus brunnicephalus*	Brown headed gull
L. fuscus	Lesser black backed gull	*Chlidonius hybridus*	Whisked tern
Ephippirhynctus asiaticus	Black-necked stork	*Egretta garzetta*	Little Egret
Gelocheidon nilotica	Gull-billed tern	*Haematopus ostralegus*	Eurasian ovstercatcher
Halcyon capensis	Stork billed Kingfisher	*H. chloris*	White collared Kingfisher
H. smyrensis	White breasted Kingfisher	*Lanius schach*	Rufousbacked shrike
Merops orientslis	Green bee-eater	*Nectarinia asiatica*	Purple sunbird
Orilus xanthornus	Black hooded oriole	*Orthotomus sutorius*	Tailor bird
Passer domesticus	House sparrow	*Pericrocotus cinnamomeus*	Small minivet
Psitacula krameri	Roseringed parakeet	*Pycnonotus cafer*	Redevented bulbul
P. jocosus	Redwhiskered bulbul	*Streptopelia chinensis*	Spotted dove
Turdoides striatus	Jungle babbler	*Zosterops palpebrosa*	White eye
Acridotheris fuscus	Jungle myna	*Acridotheris tristis*	Common myna
Hierococcyx barius	Common hawk cuckoo	*Hirundo rustica*	Swallow
Eudynamys scolopacea	Asian koel	*Chrysocolaptes lucidus*	Greater flameback
Calmator jacobinus	Pied cuckoo	*Copsychus saularis*	Magpie-robin or dovel
Coriacias benghalensis	Indian roller	*Coracina macei*	Large cuckooshrike
Corvus splendens	House crow	*Cuculus micropterus*	Indian cuckoo
Cypsiurus parvus	Palm swift	*Dendrocitta bagabunda*	Indian tree pie
Dinopium benghalense	Lesser golden backed woodpecker	*Dendronanthus indicus*	Forest wagtail
Dicrurus adsimilis	Black drongo	*Dicrurus aenus*	Bronzed drongo

The waterfowl community of Mangrove ecosystem was constituted by 21 species including both residents and migrants, the latter far exceeding the former group. In terms of population the biannual cumulative total of the water birds alone was 1254. Anatids, ciconids, aradeids, charadrids and lards were the five dominant groups. While the ciconids recorded in the region, were migrants all the ardeids were residents. Large Egret *Casmerodius albus*, little Egret, *Egretta garzetta* and Cattle Egret, *Bubulcus ibis* were observed to breed in this ecosystem. Reef Egret, *E. gularis*, *Anastomus oscitans*, white-necked Storks, *Ciconia episcopus* were the wading birds inhabiting in the mangrove ecosystem of wetland area. Among the Anatids the migrant pintails, *Anas acuta* were observed actively foraging on the mudflats in extremely large flocks of more than 300 individuals were encountered thereby confirming the speculations of Walia (2000).

Summary

During the present study, totally 66 species of birds were recorded of which 23 species were terrestrial birds. On the whole the cumulative biannual bird population observed in the mangrove ecosystem of Kali estuary was 1236 individuals. The species diversity of terrestrial bird was relatively high (25.70 per cent) to the total avifaunal populations. The dominant species were Warbler, black capped Kingfisher, blue eared Kingfisher and collared Kingfisher were some of the dominant species of terrestrial birds. The water fowl community of mangrove ecosystem was constituted by 21 species including both resident and migrants. Seasonally these avifauna showed remarked variation in their population, with highest density recorded during the winter (1864 nos.) and lowest in southwest monsoon period (51 nos.) for water birds followed by waders (62–142 individuals) and terrestrial birds (123 – 318 (individuals)–Good correlation exists between avian fauna and macrobenthic fauna in the mangrove ecosystem.

References

Ali, S. and Ripley, S.D., 1983. *A Pictorial Guide to the Birds of the Indian Subcontinent.* Oxford University Press, Mumbai.

Ali, S., 1996. *The Book of Indian Birds.* Bombay Natural History Society and Oxford University Press.

Borges, Sonali D. and Shanbhag, A.B., 2005. Avifauna of the Dr. Salim Ali bird sanctuary at Chorao–an Estuarine wetland of Goa, India.

Grewal, B., 1995. *Birds of the Indian Subcontinent.* Guide Book Company Ltd. Hong Kong.

Grimmett, R., Inskipp, C. and Inskipp, T., 1998. *Birds of the Indian Subcontinent.* Oxford University Press, Delhi.

Moreira, F., 1997. The importance of shorebirds to energy fluxes in a food web of a south European estuary. *Est. Coast Shelf Sci.,* 44: 67–78.

Ramarao, K.V., Muley, E.V., Raghunathan, K. and Karmakar, A.K., 1985. Estuarine biological methods. *Zoological Survey of India,* Calcutta, pp.1–94.

Rao, V.V., Nagulu, V., Anjaniyulu, B.S., Srinivasulu, C. and Rao, Ramana 1997.

Shanbhag, A.B., Walia, R. and Borges, S.D., 2001. The impact of Konkan Railway Project on the avifauna of Carambolim Lake in Goa. *Zoos' Print Journal,* 16(6): 503–508.

Sonobe, K. and Usui, S. (Eds.), 1993. *A Field Guide to the Waterbirds of Asia.* Wildbird Society of Japan, Tokyo.

Swedmark, B., 1964. The interstitial fauna of marine sand. *Biol. Rev.* 39: 1–42.

Walia, R. and Shanbhag, A.B., 1996. Birds at Santa Monica Lake, Goa: An integrated ecological study. In: *Pan–Asian Ornithological Congress and XII Birdlife Asia Conference,* Coimbatore, India, 9–16, Nov.

Woodcock, M., 1989. *Collins Handguide to the Birds of the Indian Subcontinent.* Williams Collins Sons and Co. Ltd., London.

Chapter 16

Reservoir Fisheries of Satara District, Maharashtra

☆ *V.B. Sakhare*

Maharashtra, one of the largest states in the country in population and geographical area having a number of rivers like the Godavari, the Bhima, the Krishna, the Narmada, the Tapti and other several rivers and their tributaries having a total of 1600 km of river length. Reservoirs constitute the prime inland fishery resources of Maharashtra in terms of their vast area and enormous production propensity. These open water systems not only allow quick, yield enhancement at low capital/environment cost, but also fisheries development of these big water bodies directly benefits some of the weakest sections of our society. These benefits are accrued through increase in yield and thereby improve the quality and living standards of poor fishermen. Unlike culture fishery, where income is shared between limited investors, in reservoir fishery, the cream of augmented yield is more or less equitably distributed among the masses. This being a community based development process, has a direct bearing on our rural economy.

Sugunan (1995) mentions the total reservoir area in the state is 2,73,750 ha. However, according to Sreenivasan (1998), Maharashtra is endowed with 1,79,430 ha. of reservoir area and the state produced 7.83 kg/ha fish from its reservoirs. However. IIMA (1985) worked out 1,05,202 ha. reservoir area comprising 72 reservoirs. Pathak (1990) is of the view that, the area under reservoirs in the state of 20 ha and above estimated to about 2,36,157 hectares. No limnological studies or survey of fish fauna has been carried out so as to know the productivity of water and fish faunal diversity. Sugunan (1995), Sreenivasan (1991), Valsangkar (1993 and 1980) and Sakhare (1999 and 2001) documented information on fisheries of some of the reservoirs in the state.

Satara district is located in the western part of Maharashtra. It is bounded by Pune district to the north, Solapur district to the east, Sangli district to the south and Ratnagiri district to the west. Raigad district lies to its north-west. The district lies between 17.5 degree and 18.11 degree North latitude and between 73.33 degree and 74.54 degree East longitude. It covers 10,480 sq kms. Most of the central Satara district's area falls in the river Krishna basin and limited area falls in the river Bhima basin.

In summer, western part of the district experiences cool and pleasant climate and during the rainy season the climate is cold. In summer the central part of the district experiences hot climate. In winter the nights are colder and days are warm. The climate in eastern part is hot compared to western and central part. Change in season affects the climate in the district. On an average district receives average rainfall of 1426 mm. The atmospheric temperature varies from 11.6°C to 37.5°C. The climate of this district is on the whole agreeable. The year may be divided broadly into four seasons. The cold season is from December to about the middle of February. The hot season, which follows lasts till the end of May. June to September is the southwest monsoon season and the two months October and November form the post-monsoon or the retreating monsoon season. In the south-west monsoon months the air is highly humid but in the summer and the cold seasons the air is dry particularly in the afternoons. In the plains, the dryness is more marked than in the hills. Winds are strong particularly on the hills in the south-west monsoon season. In the rest of the year they are light to moderate.

The soil pattern in the district differs widely due to marked variations in the topography of the region. The soil in the hill slopes especially in the western part of the district is of low type, reddish in

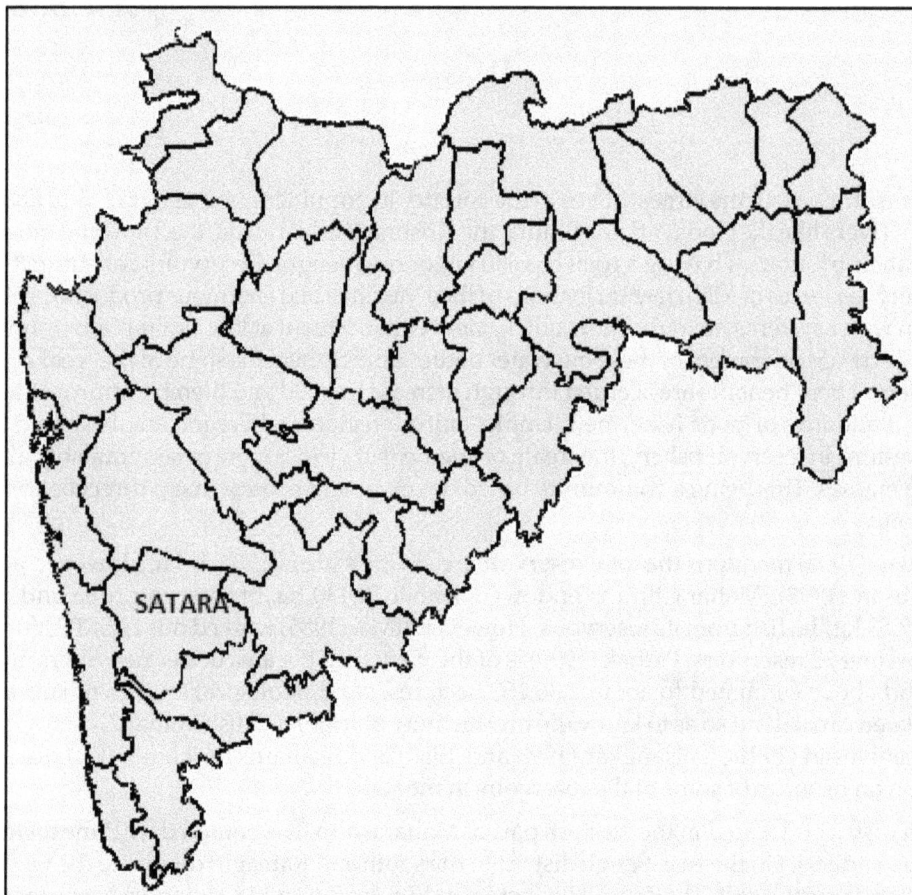

Figure 16.1: Map of Maharashtra.

colour. The soil in the parts surrounding rivers especially of Krishna and Koyana in central part of the district is black. The land is very fertile. Kas and nearby area, western part of Satara taluka has been declared as World Heritage Site. Here excellent flora and fauna is seen after month of September. The soil type also varies from block to block. It ranges from, light laterite type in hilly tracts of Mahabaleshwar, Jaoli, Patan to fertile cotton black cotton soils in some parts of the district *i.e.,* Phaltan and Karad.

The main source of supply of fish in the district is at present confined to the Urmodi, Vena, Krishna and Koyna river. The other rivers are not of much importance from the fisheries point of view as

Figure 16.2: Map of Satara District.

they get dry for most part of the year. The important reservoir in Satara district are Dhom, Urmodi, Veer, Kanher, Ghom Balakwadi and Koyana. Total fish production from the reservoirs is not known. Although 7 reservoirs in district are under regular stocking, latest fish production figures are available only in respect of two reservoirs *i.e.,* Dhom and Kanher. Few references are available on the fish fauna of this district (Pawar and Pawar, 2012, Jadhav *et al.,* 2011, Valsangkar, 1993). The present Chapter deals with the fisheries of important reservoirs in Satara district of Maharashtra.

Dhom Reservoir

The dam was constructed across the river Krishna near Wai in Satara district. It is an earthfill and gravity dam. The capacity of dam is 13.50 TMC. The reservoir is about 20 kms in length. It is an hydroelectric project with capacity of 4 mg electricity generation capacity. The main purpose of the dam is to supply the water to agriculture, industries and for the drinking to Wai, Pachgani, Mahabaleshwar and the surrounding villages on the bank of reservoir. The salient features of the reservoir are depicted in Table 16.1.

Table 16.1: Salient Features of Dhom Reservoir in Satara District

1. Name of reservoir	Dhom	2. Location	Wai
3. Year of construction	1976	4. River	Krishna
5. Type	Earthfill-gravity dam	6. Height of dam	160 feet
7. Length of dam	8130 feet	8. Volume	6335 km³
9. Capacity of dam	331,100 km³	10. Surface area	2498 km²

The reservoir is regularly stocked by the seed of Indian major carps. The details of the seed stocking and fish production in the reservoir are depicted in Table 16.2. The fish production for five years (2007-08 to 2011-12) varied from 13736 kg (in 2007-08) to 40061 kg (in 2011-12). The average fish production is calculated at 21.18 kg/ha/yr.

Table 16.2: Fisheries of Dhom Reservoir

Sl.No.	Year	Seed Stocking (Lakhs)	Seed Stocking (nos/ha/yr)	Fish Production (kg)	Fish Production (kg/ha/yr)
1.	2007-08	2.50	179.34	13736	9.85
2.	2008-09	4.00	286.94	23178.50	16.62
3.	2009-10	10.00	717.36	32520	23.32
4.	2010-11	13.725	984.57	38203	27.40
5.	2011-12	4.403	315.85	40061	28.73

Kanher Reservoir

It is an earthfill and gravity dam constructed across river Wenna near Satara. The dam is situated in Northwest of Medha Tahsil of Satara district. It is located at latitude 17°44" 16°02" North and longitude 73°53" 43°10" East. The height of the dam above lowest foundation is 165.2 feet while the length is about 6411 feet. The volume content is 6308 km^3 with gross capacity of 286,000 km^3. The main purpose of dam is to supply water for drinking, domestic purpose and irrigation as well as fishing practices are carried out under the supervision of district fishery development office, Satara. The morphometry of the reservoir is presented in Table 16.3.

Table 16.3: Morphometry of Kanher Reservoir in Satara District

1. Name of reservoir	Kanher	2. Location	Satara
3. Year of construction	1986	4. River	Wenna
5. Type	Earthfill- gravity dam	6. Height of dam	165.2 feet
7. Length of dam	6411 feet	8. Volume	6308 km^3
9. Capacity of dam	271,680 km^3	10. Surface area	18.63 km^2
11. Designed spillway capacity	3203 m^3/sec		

Table 16.4: Years-wise Seed Stocking and Fish Production in Kanher Reservoir

Sl.No.	Year	Seed Stocking (Lakhs)	Seed Stocking (nos/ha/yr)	Fish Production (kg)	Fish Production (kg/ha/yr)
1.	2007-08	5.00	445.23	76.50	0.068
2.	2008-09	3.00	267.14	12590	11.21
3.	2009-10	5.50	489.75	74960	66.74
4.	2010-11	NA	NA	60000	53.42
5.	2011-12	9.56	851.29	NA	NA

Figure 16.1: Sluice Gates of Koyana (Shivajisagar).

Figure 16.2: Fishes Sold in Local Fish Market.

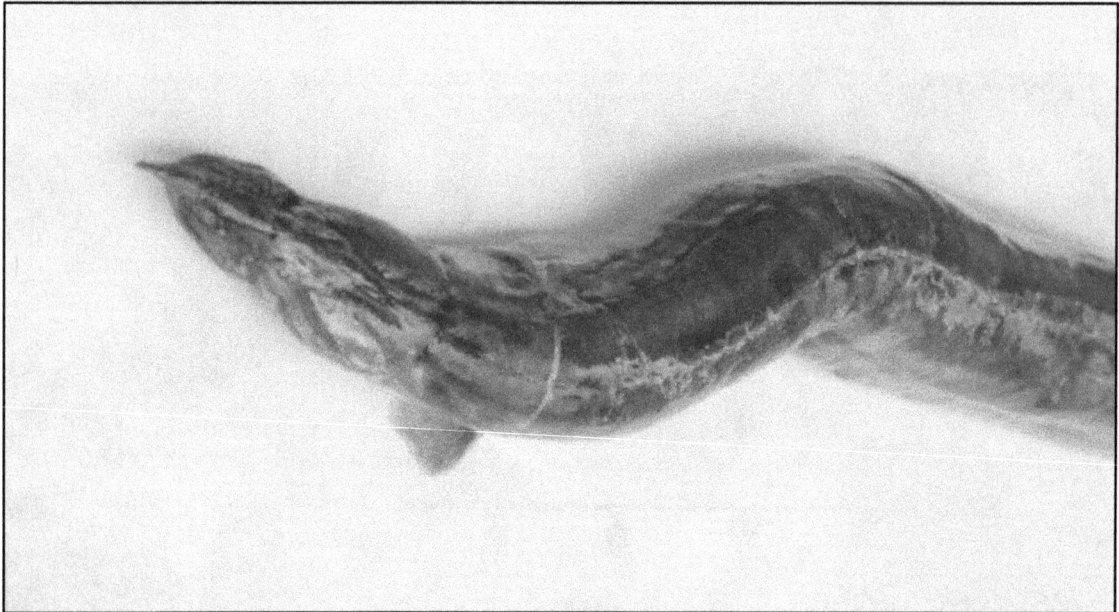

Figure 16.3: *Mastacembelus armatus.*

Figure 16.4: *Mystus seenghala.*

Figure 16.5: *Catla catla.*

Figure 16.6: *Cyprinus carpio.*

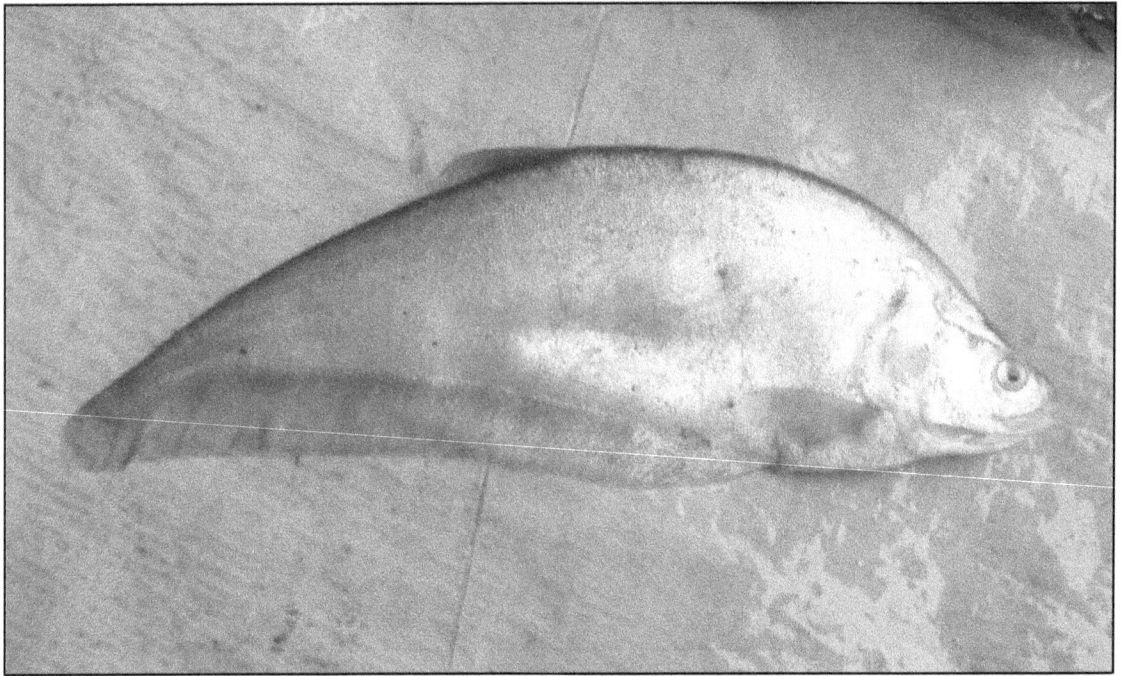

Figure 16.7: *Notopterus notopterus.*

Figure 16.8: *Channa marulius.*

Figure 16.9: Fishing in Progress.

Figure 16.10: Fisherman with Fish Catch.

Figure 16.11: Fisherman Using Hook and Line at the Bank of Kanher Reservoir.

Presently, there is no organized fishing in the reservoir. A few local people inhabiting along the reservoir margin are engaged in improvised fishing practices, using surface gill nets. Reservoir is stocked with seed of Indian major carps. Seed stocking varied from 267.14 to 851.29 nos/ha/yr (Table 16.4). The maximum seed was stocked in year 2011-12, while minimum seed stocking was done in year 2008-09. Details on seed stocking for year 2010-11 are not available.

The details of the fish production from the reservoir for the period of 2007-08 to 2011-12 is shown in Table 16.4. The study of the data reveals that the maximum fish production for the last five years (from 2007-08 to 2011-12) was recorded at 74960 kg during 2009-10. It gave an average fish production of 66.74 kg/ha/yr. The minimum production was during 2007-08, at 76.50 kg, giving an average fish production of 0.068 kg/ha/yr. The average fish production from reservoir is 32.85 kg/ha/yr.

Pawar and Pawar (2012) reported the occurrence of 50 fish species belonging to 5 orders from the Kanher reservoir. The fishes belonging to order cypriniformes were dominant with 40 species to be followed by fishes of order siluriformes with 5 species, while order perciformes was with 3 species only. The orders like synbranchiformes and anguilliformes were represented by one species each. They concluded that 10.32 per cent fish species were carnivorous, 60.43 per cent omnivorous and 38.02 per cent herbivorous.

Koyana (Shivajisagar) Reservoir

Koyana is one of the major tributaries of Krishna river system. Both the river and the tributary originate from Mahabaleshwar plateau, the famous hill resort of Maharashtra state, located in western ghats (locally called Sahyadri). They flow in opposite directions at their origin, Krishna to east and Koyana to west; both then turn southwards to meet at Karad, to form the famous confluence 'the preetisangam'. Both get the benefit of high rainfall of 6182 mm/yr at Mahabaleshwar, during southwest monsoon.

Table 16.5: Morphometric Features of Shivajisagar Reservoir

1. Location	Koyananagar	2. Coordinates	17°24'06"N73°45'08"E
3. Type of dam	Concrete dam	4. Construction began	1956
5. Catchment Area	871.78 sq. km.	6. Maximum submerged Area	11535 ha.
7. Minimum submerged Area	1435 ha.	8. Maximum depth of water column	78.8 m
9. Minimum depth of water column	30.5 m	10. Nature of submergence	Deep valley
11. Rainfall at catchment area	6182 mm/yr	12. Length of dam	807 m
13. Height of dam	80.35 m	14. Gross storage capacity	2797.4 mcm
15. Dead storage capacity	707.90 mcm	16. Maximum depth of dead storage	30.5 m
17. Number of radial gates	6 tender gates	18. Purpose of dam	Electricity generation
19. Number of generators	12	20. Generation capacity	1920 MW

The Koyna dam is one of the largest dams in Maharashtra constructed on Koyna river. It is located in Koyna Nagar of Satara district, nestled in the Western Ghats on the state highway between Chiplun and Karad. The main purpose of dam is to provide hydroelectricity with some irrigation in neighboring areas. Today the Koyna hydroelectric project is the largest completed hydroelectric power plant in India having a total installed capacity of 1,920 MW. Due to its electricity generating potential Koyna river is considered as the 'life line of Maharashtra'. The spillway of the dam is located at the center. It has 6 radial gates. The dam plays a vital role of flood controlling in monsoon season. The catchment area dams the Koyna river and forms the Shivajisagar Lake which is approximately 50 km in length. It is one of the largest civil engineering projects commissioned after Indian independence. The Koyna hydro-electric project is run by the Maharashtra State Electricity Board.

The Shivajisagar reservoir came into existence with the first impoundment in 1961. The reservoir was constructed at Helwak village in Patan tahsil of Satara district. Shivajisagar dam was constructed on Koyana river in a deep valley, in between two mountain ranges, running parallel to each other, over a distance of 65 km, right from Koyananagar (Helwak), the actual dam site, to Tapola and beyond, upstream. By virtue of 6182 mm/yr rainfall in the vast catchment bowl of 871.78 sq. km; huge quantity of 2797.4 mcm of water accumulates in the reservoir, just within a span of two early months of southwest monsoon, July and August.

The former seasonal stretch of the narrow and shallow Koyana river had no sizable fish production prior to the first impoundment of the Shivajisagar, which came into existence, consequent to the construction of dam. At present, about 65.7 mt of fish is harvested every year (Valsangkar, 1993). The per hectare yield in relation to maximum water spread area is 5.69 kg only. Important fish species represented in the catches are depicted in Table 16.6.

Table 16.6: Fish fauna of Shivajisagar reservoir

Sl.No.	Scientific Name	Sl.No.	Scientific Name
1.	*Tor khudree* (Sykes)	2.	*Tor mussullah* (Sykes)
3.	*Puntius sarana* (Hamilton)	4.	*Puntius kolus* (Day)
5.	*Puntius dobsoni* (Day)	6.	*Puntius amphibious* (Valenciennes)
7.	*Puntius sahyadriensis* (Silas)	8.	*Rohtee ogilbii* (Sykes)
9.	*Osteobrama vigorsii* (Sykes)	10.	*Labeo calbasu* (Hamilton)
11.	*Labeo sindensis* (Hamilton)	12.	*Labeo rohita* (Hamilton)
13.	*Catla catla* (Hamilton)	14.	*Cirrhinus mrigala* (Hamilton)
15.	*Hypophthalmichthys molitrix* (Valenciennes)	16.	*Cyprinus carpio* (Linnaeus)
17.	*Ompak bimaculatus* (Bloch)	18.	*Wallago attu* (Bloch)
19.	*Mystus vittatus* (Bloch)	20.	*Mystus seenghala* (Sykes)
21.	*Channa marulius* (Hamilton)	22.	*Salmostoma* spp. (Hamilton)

The major carp availability in commercial catches continues to be negligible. Ever since 1988, on an average 4.16 lakh of major carp fingerlings are being stocked at Tapola, 45 km upstream in the Shivagisagar reservoir, every year. So far their significance in the fish catches is not noteworthy. One remarkable feature of this reservoir is that 20 per cent of the fish catches from the reservoir consists of mahseer alone. This indicates that mahseer can adopt itself, though on a small scale, to changed habitat, if the essential factors such as running streams, rocks-gravel or lined poole or puddles and undisturbed environment are available. The mahseer is represented by two species *i.e.*, *Tor mussullah* and *Tor khudree*. They are indigenous to the Krishna river system. Mahseer in the Shivajisagar reservoir have established themselves in the lacustrine conditions, and are breeding naturally in the upstream areas, towards shallower regions of the reservoir, when the climatic conditions like low temperature (<24 °C), fresh influx of rain waters with high dissolved oxygen and fluviatile nature of the spawning grounds etc. are conducive.

The official record of fishing permits issued to the individual members of the four fishermen's co-operative societies, for undertaking fishing in Shivajisagar (Koyana) reservoir, indicated that of an average 73 fishermen are actively engaged in fishing in the reservoir.

The fishing period is restricted to 180 days/year during March to August. There is no machinery to collect daily data of fish landings. In the absence of it, estimates are made by taking into consideration the number of authentic fishermen holding fishing permits, the fishing period of 180 days, and average fish catch of 5 kg/person/day. This gives production of 65.7 mt per season.

Readymade nylon gillnets (entailing nets), below 10 cm stretched mesh size and drift nets having mesh size less than 5 cm (stretched) are being used by the local population. About 50 fishing craft made locally, by using inferior wood and technology are plying in Shivajisagar for fishing operations. These country crafts are operated by groups of 10-12 fishermen. They cannot reach distant fishing sites on their own. As such, aluminum boats discarded by the mercantile marine ships have been purchased by the co-operative societies. These powerful and stable motorized boats are employed in towing small country craft, for reaching distant fishing grounds, and returning to the landing sites. Four such motorized boats, serving as motherships, are in operation in the reservoir. Recently, about 15 outboard engines (Yamaha) have been introduced at Tapola, the distal landing centre in the reservoir.

This reform has improved the lot of fishermen considerably. They are operating their boats independently and are covering larger distant areas. Consequently the average fish catch per individual has gone to 10 kg/day, in this area (Valsangkar, 1993). Fishermen in Shivajisagar reservoir are non-traditional. They belong to higher caste, scheduled caste and nomadic tribes. This non-traditional fishermen population has adopted fishing as vocation, equal to the formation of Shivajisagar reservoir. Basically they were agriculturists, and labourers in forest areas. They lack knowledge about fish, fish behaviour, gear and craft. Fishing is done in groups of twelve fishermen.

References

Jadhav, B. V., Kharat, S.S., Raut, R.N., Paingankar, M. and Dahanukar, N., 2011. Freshwater fish fauna of Koyna river, Northern Western Ghats, India. *Journal of Threatened Taxa,* 3(1): 1449–1455.

Pathak, S. C., 1990. Harnessing reservoirs for increasing fish production. p. 9–12. In: *Reservoir Fisheries in India,* (Eds.) Jhingran Arun G. and V. K. Unnithan. Proc. of Nat. Workshop on Reservoir Fisheries, 3–4 January. Spl. Publ. 3. Asian Fisheries Society, Indian Branch, Mangalore, India.

Pawar, Sandhya and Pawar, Sourabh, 2012. Ichthyofaunal diversity of Kanher dam from Satara district of Maharashtra state of India. *International Journal of Fisheries and Aquaculture Sciences,* 2(1): 15–19.

Sakhare, V.B., 1999. Fisheries of Yeldari reservoir, Maharashtra. *Fishing Chimes.* 19(8): 45–47.

Sakhare, V. B., 2001. Reservoir fisheries in Solapur district of Maharashtra. *Fishing Chimes,* 21(5): 29–30.

Sreenivasan, A., 1991. *Reservoir Fisheries of India: An Overview Report to the IRDC.*

Sreenivasan, A., 1998. Intrgrated Development of Reservoir Fisheries of India: Production to Marketing. *Fishing Chimes,* 18(1): 60 –63.

Sugunan, V.V., 1995. *Reservoir Fisheries of India.* FAO Fisheries Tech. Report 345. Daya Publishing House, Delhi.

Valsangkar, S.V., 1980. Economic rehabilitation of fishermen in Yeldari reservoir. *India Today and Tomorrow,* 8(4): 162–163.

Valsangkar, S.V., 1993. Mahseer fisheries of Koyana river (Shivajisagar) in Maharashtra: Scrap to bonanza. *Fishing Chimes,* 12(10): 15–19.

Chapter 17

Study on Biodiversity of Ayanur Pond, Shivamogga District, Karnataka

☆ *H.M. Ashashree, H.A. Sayeswara and Nafeesa Begum*

ABSTRACT

An attempt was made to study biodiversity of Ayanur pond. The study was carried out during the period from January 2010 to December 2010. The inventory of fishes comprises 12 species of fishes belonging to 3 orders has been prepared. And year long study was conducted to measures various physico-chemical parameters, including phytoplankton and zooplankton in the pond water. The study revealed that there is indication of pollution in the pond and hence preventive measures are required to avoid further deterioration of the pond water quality.

Keywords: Biodiversity, Ayanur pond, Pollution, Phytoplankton, Zooplankton, Fishes.

Introduction

Water is considered as the source from where life began. Water bodies of every form – oceans, seas, rivers, marshes or ponds–have been the niche for thousands of species of flora and fauna. Most of the living beings require water in some form for their survival (Sayeswara *et al.*, 2011). They are also associated with an array of physical, temporal, chemical and biological characteristics. Wetlands are one of the most productive ecosystems on the earth and contains peculiar biotic components.

India has totally 67,429 wetlands, covering an area of about 4.1 million hectares (MOEF, 1990). Karnataka is one of the richest among the Indian states having inland water resources of varied types, which include 33, 612 minor tanks, 2, 762 major tanks, 18 reservoir, 3800 miles of river, innumerable irrigation channels, irrigation wells and temple ponds. Karnataka state affords nearly 10 lakh acres for fisheries development programme. In the extent of cultivable water area Karnataka ranks 1st in entire country (Fish seed committee report, Government of India, 1998).

A through study of fishes of an area will be useful for planning fisheries development as it would serve as guide for being available fish fauna and further for finding out the possibility of introducing new species which are not endemic to that area. Hence, a detailed study of fishes of an area forms the perquisite for understanding any fish culture programme on a scientific basis. Keeping the above views in mind we have undertaken the present study on diversity in Ayanur pond.

Material and Methods

Sampling of fish has been made for every fortnight days throughout study period from January 2010 to December 2010. Collection of fish was made directly from the fishermen's during the time of fishing. Two types of fish nets were used (1) Gill net (2) Cast net. The collected fishes were preserving in 4 per cent formalin. Later on in the laboratory they were identified on the basis of their morphological characters with the help of Talwar and Jhingran (1991).

The water quality parameters were analyzed by methods described by APHA (1998). Phytoplankton samples were collected on monthly basis by filtering 100 liters of water through plankton net made up of bolting silk. The zooplankton samples were preserved in 5 per cent formalin and phytoplankton and zooplankton were identified by standard literatures Sehgal (1983), Sugunan (1991), Battish (1992), Agarwal (1990), Murugan *et al.* (1998), and Dhanapathi (2000). Subsequently aquatic macrophytes such as submerged, emerged and semi aquatic plant were collected and identified. And aquatic birds also identified.

In our present study on the biodiversity of Ayanur pond has been studied. This pond is situated in 12 km for away from Shimoga-Sagara road. This pond acquires a land of 85 hectare and depth of 12 feet. The pond water is used for agricultural purposes and inland fishery development. The pond receives agricultural wastewater from the agricultural land surrounding the pond and also uses of peoples and sewage in small amount.

Table 17.1: List of Fish Species in Ayanur Pond.

Sl.No.	Common Name	Scientifc Name	Vernacular Name	Status
1.	Catla	*Catla catla*	Catla	A
2.	Common carp	*Cyprinus carpio*	Gauri menu	A
3.	Rohu	*Labeo Rohita (Ham)*	Rahu	A
4.	Mrigal	*Cirrhinus Mrigal*	Mrigal	A
5.	Olive barb	*Puntius saranana*	Gende kijan	A
6.	Black line Rasbora	*Rasbora doniconius (Ham)*	Sasalu	A
7.	Gery feather back	*Notoperus notopterus*	Chamari	A
8.	Banded snake head	*Channa striatus*	Kuchu menu	A
9.	Spotted snake head	*Channa punctatus (Bloch)*	Korava	R
10.	Magur	*Clarias batrachus (Linnaeus)*	Murugodu	R
11.	Gangetic myetus	*Mystus cavasius*	Girulu	A
12.	Tilapia	*Oreochromis massambica*	Jelabi	R

Result and Discussion

Generation of proper data based on germ plasma resources is vital to safeguard to our diversity and evolve a plan for their proper utilization and management (Dhange and Dhange, 1996). The fish fauna is an important aspect of fishery potential of a water body. Fish fauna of Indian ponds has been studied by several workers. It is observed that the distribution of fish species is quite variable because of geographical and geological conditions.

Table 17.2: List of Aquatic Plants and Birds in Ayanur Pond.

Sl.No.	Aquatic Plants	Aquatic Birds	Sl.No.	Aquatic Plants	Aquatic Birds
1.	Pistia	Little cormorant	2.	Eichornia	Little egret
3.	Hydrilla	Median egret	4.	Typha Grase	Purple heron
5.	Lemna	Grey heron	6.	Nymphaea	Pond heron
7.	Nelumbo	Duck	8.	Valisneria	King fisher
9.	Nymphoid indicum	–	10.	Ipomea	–

In our study, the result confirms that 12 species of fishes which come under 3 order of 6 families. The order cypriniformes was dominant with 7 species, order perciformes with 3 species and Siluriformes with 2 species were also reported from the study pond. We also observed the occurrence of the fish species abundant (A) and rare (R).

Table 17.3: List Phytoplankton and Zooplankton in Ayanur Pond.

Phytoplankton	Zooplankton
Closterium, Nostac caraneum, Fragilaria, Euglena acus, Euglena gracile, Euglena elongata, Euglena oxyalis, Lyngbya ceylanica Wille, Lyngbya allorgei Fremy, Lyngbya lutea, Anabaena, Microcystis, Merismopedia.	*Keratella earlinae, Brochines* sp., *Colpoda* sp., *Cyclopes* sp., *Daphnia* sp., *Elphidia* sp., *Notholca laurentiae, Notholca acuminate, Diaphanosoma brachyurum, Brochines* sp., *Colpoda* sp., *Cyclopes* sp., *Daphnia* sp., *Elphidia* sp., *Nauplius, Zoea.*

Water quality is an important criterion for the fish culture, hence the physico-chemical properties of water was analyzed. In the present study water quality in Table 17.4 reveals that the water temperature range between 23°C-25°C is within the tolerance limit of most of the cultivable fishes. The fluctuations in the water temperature had relationship with the air temperature which also showed same seasonal trend. Similar result was recorded in earlier studies by Venkateshwarlu *et al.*, 2002 and Ashashree *et al.*, 2008. Pond having water temperature more than 22°C are found to be highly productive (Jhingran and Sugunan, 1990). The average temperature recorded was 23°C.

Table 17.4: Physico-chemical Properties of Water Ayanur Pond.

Sl.No.	Parameter	Value	Sl.No.	Parameter	Value
1.	Temperature	23°C-25°C	2.	pH	6 (acidic)-7
3.	Dissolved oxygen	5.02 mg/lt	4.	Fresh CO_2	4 mg/lt
5.	Chloride	50.02 mg/lt	6.	Total hardness	164 mg/lt
7.	Total dissolved solid	103 mg/lt	8.	BOD	6.85–7.0 mg/lt

The pH is affected not only by levels of carbon dioxide but also by other organic and inorganic components of water. Further, any alteration in water pH is accompanied by changes in other physico-chemical parameters. In the present study ranged between 6-7. According to Jhingran and Sugunan

(1990) the pH range between 6.0 to 8.5 indicates medium productive, so in the Ayanur pond it was observed that the pH 6.5 deals with the medium productivity.

Oxygen is one of the most important factors in any aquatic ecosystem. The main sources of dissolved oxygen are from the atmosphere and the photosynthesis. The amount of oxygen that dissolves in water from the air depends on factors like temperature, salinity and density of phytoplankton. Aquatic macrophytes, benthic algae and phytoplankton add to the DO during photosynthesis in addition to wave action, turbulence or surface water agitation. At the same time, the zooplankton and other animal fauna may utilize this DO for respiratory activity. Hence, large variation is seen in DO levels of lakes, at different times or in different seasons. In the present investigation the dissolved oxygen was found 5.0 to 5.6 mg/L. DO less than 3 mg/L suffocates the animal (Boyd, 1982). Shanthi *et al.* (2006) recorded 4.6-6.8 mg/L of DO and 0.2-1.8 mg/L of BOD. So the present result indicate that water is unfit for drinking purpose (Delphine Rose *et al.*, 2008) but good water quality for fish survival (Sakhre and Joshi, 2002). This can be attributed to the bloom of *Microcysits* spp., photosynthetic activity of which might have released more DO. The higher level of DO is indicator of eutrophication (Arumugam and Furtado, 1979), while Salakar and Yergi (1997) stated that persistent low DO value indicate a very high degree of organic pollution. DO level can be lowered down during phytoplankton blooms, as the dead plankton are degraded by bacteria utilizing DO (Lavington, 1982).

BOD is the measure of degradable organic matter present in water and can be defined as the amount of oxygen required by the microorganisms in stabilizing the biologically degradable organic matter under aerobic condition. During present study, the BOD range between 6.85-7.0 mg/L.

The carbon dioxide content of water depends upon the temperature, depth, rate of respiration, decomposition of organic matter, chemical nature of the bottom and geographical features of the terrain surroundings the water body. The decomposition of organic matter from aquatic ecosystem also adds to CO_2 in water (Datta Munshi and Datta Munshi, 1995), but as it gets constantly released in air the amount of free carbon dioxide is kept low in water. The photosynthetic activity of aquatic flora also reduces the free CO_2 in water. During the present study free CO_2 was 4 mg/L due to it's usage by the nearby inhabitants and also fishing activity goes on this pond.

Hardness is governed by the contents of calcium and magnesium salts largely combined with bi-carbonates and carbonates giving temporary hardness. In the present study the total hardness of water 160-164 mg/L. Productive waters generally have hardness above 20 mg/L. very high hardness of water (> 300 mg/L) affects fish production because of higher pH. The optimum value of hardness ranging between 75-150 mg/L supports fish productivity (Das, 1996). The values of total hardness recorded during present study are suitable for fishery. Similar result was recorded by Sakhre and Joshi, 2002 *i.e.* 62.0–188.6 mg/L.

During present study the total dissolved solids were recorded in the range of 103 mg/L. according to Jhingran and Sugunan (1990) the TDS values up to 200 mg/L are encountered in medium productive and more than 200 in highly productive aquatic ecosystems.

The chlorides control the salinity of water and osmotic stress on biotic communities (Salaskar and Yeragi, 1997). The most important source of chlorides in the freshwater is the discharge of domestic and industrial sewage. Thus concentration of chlorides is the indicator of pollution (Kodarkar *et al.*, 1998). They also indicate presence of organic matter in the water body (Thresh *et al.*, 1994). The present study chlorides range between 50.2 – 60 mg/L, it indicates that pond is less polluted and reduction in eutrophication.

The phytoplankton constitute base line of food webs in an aquatic ecosystem. The phytoplankton of the Ayanur pond water were indicated with reference to distribution and composition. The phytoplankton species include Closterium, Nostac caraneum, Fragilaria, *Euglena acus, Euglena gracile, Euglena elongata, Euglena oxyalis, Lyngbya ceylanica* Wille, *Lyngbya allorgei* Fremy, *Lyngbya lutea, Anabaena, Microcystis* and *merismopedia* occurring in polluted water were common in this study pond. According to Khanna (1993) the zooplankton are preferred over phytoplankton by fish fry due to easy digestibility of the former. The nutritional value of phytoplankton is also less when compared with zooplankton.

The study pond also possesses varied aquatic birds and plants. Species diversity and abundance of fishes itself indicate the sufficient availability of zooplankton and phytoplankton which form the major part of food of fishes. Shallow water regions of the pond are featured with the growth of algal blooms and macro vegetations includes 10 species (Table 17.2) and this pond also includes 8 species of birds (Table 17.3). Similar work was carried out by Ashashree *et al.*, 2008 and Venkateshwarlu *et al.*, 2005 and 2007.

Conclusion

Ponds are amongst the most diverse freshwater habitats and have been recently found to support more species, as well as more uncommon, rare, and threatened species compared to lakes, rivers, and streams. Ponds may support more species than rivers (*e.g.* invertebrates and plants), and more uncommon species.

The water quality does not show much variation, only the DO and BOD have been notices which affected the fish and plants. The pond has low DO and high BOD. It needs a long term plan for reversal of eutophication and pollution control which can make pond suitable for both recreation and aquaculture purpose. The pond is dumped with agriculture field runoff. The tank required a special attention because of all the cultural fish and fingerlings are left in to pond are eaten by predatory fishes in the pond. Therefore care must be taken to increase the fish population. The entering of sewage must be prevented and washing activity of vehicles and other activity should be reduce by people. And also there is an urgent need to spread awareness and knowledge among students. Departmental officials, field guides and tourists and orient them towards inland water ecosystems. Basic taxonomic research in the universities, institutes and other research organizations in the specialized area of inland waters need to be encouraged. Also, local community and private institutions can be involved in conservation of aquatic resources and biodiversity.

References

Agarwal, S. C., 1990. *Limnology*. APH Publishing Corporation, New Delhi, 150pp.

APHA., 1988. *Standard Methods for the Examination of Water and Wastewater*. American Public Health Association, Washington D. C.

Armugam, P. T and Furtado, J. I., 1979. Eutrophication of Malasian reservoir. Effect of agroindustrial effluent. Presented at the *5th International Symposium of a Tropical Ecology*, ISTE, Kaulalumpur.

Ashashree, H. M, Venkateshwarulu, M. and Renukaswamy, H. M., 2008. Diversity of fish fauna in Nagathibelagola pond, Shimoga, Karnataka. In: *Advances in Aquatic Ecology, Vol. 2*, (Ed.) V.B. Sakhare, Daya Publishing House, New Delhi,

Battish, S. K., 1992. *Freshwater Zooplankton of India*. Oxford and IBH Publishing Co. Pvt. Ltd., New Delhi.

Boyd, C. E., 1982. *Water Quality Management of Pond Fish Culture*. Elsevier. Publication, 318.

Das, R. K., 1996. Monitoring of water quality, its importance in disease control, paper presented in Nat. Workshop on fish and prawn disease, epizootics and quarantine adoption in India. *CICFRI*, 51–55.

Datta Munshi, J and Datta Munshi, J. S., 1995. *Fundamental of Freshwater Biology*. Narendra Publishing House, New Delhi, 222 pp.

Delphine Rose, M. R, Jeyaseeli, A, Joice Mary, A and Rani, J. A., 2008. Characteristics of ground water quality of selected areas of Dindugal Distirct, Tamil Nadu. *J. Aqua. Biol.*, 23(2): 40–43.

Dhanapathit, M. V. S. S. S., 2000. *Taxonomic Notes on the Rotifers from India*. Indian Association of Aquatic Biologists, Hyderabad.

Dhange J. R and Dhange, R., 1996. Impact of habitat shrinkage on the indigenous fish genentic resources of seas drainage system. In: Proceddings of the symposium on fish genetics and fish diversity conservation for sustainable production. Sept.26–27. *NBFGR. Lucknow*, p. 9–10.

Jhingran, Arun, G. and Sugunan. V. V., 1990. General guidelines and planning criteria for small reservoir fisheries management. P 18. In: Jhingran, Arun G and Unnithan V. K (Eds), *Reservoir fisheries in India*. Proc. of the Nat. workshop on reservoir fisheries, 3–4 Jan., 1990. Special Publication. 3 : *Asian Fisheries Society*, Indian Branch, Mangalore, India.

Khanna, S. S., 1993. *An Introduction to Fishes*. Central Book Depot, Allahabad, p. 496–498.

Kodarkar, M. S., Diwan, A. D., Murugan, N., Kulkarni, K. M. and Anuradha Ramesh, 1998. *Methodology for Water Analysis: Physico-chemical, Biological and Microbiological*. Indian Association of Aquatic Biologists, Hyderabad, Publ. No. 2.

Murugan, N. P, Murgavel, P. and Kodarkar, 1998. *Cladocera: The Biology, Classification, Identification and Ecology*. Indian association of aquatic biologists (IAAB), Hyderabad. Publ. No.5.

Sakhre, V. B. and Joshi, P. K., 2002. Ecology of Palas–Nilegaon reservoir in Osmanabad district, Maharashtra. *J. Aqua. Biol.*, 18(2): 17–22.

Sayeswara, H. A, Naik, K. L, Nafessa Begum and Ashashree, H. M. 2011. Potability of water in relation to some physico-chemical parameters of Mudugodu pond, Chikkamagalur, Karnataka, India. *Environment and Ecology*, 29(1): 140–142.

Salaskar, Pramod and Yeragi, S. G., 1997. Studies of water quality characteristics of Shenala lake, Kalyan, Maharashtra, India. *J. Aquatic Biology*, 12(1 and 2): 28–31.

Sehgal, K. L., 1983. Planktonic copepods of freshwater ecosystem. *Environ. Sci. Series*. Interprint, New Delhi. 69pp

Shanthi, V, Muthumeena, S, Jeyaseeli, A and Florence Borgia., 2006. Physico-chemical status of Varaga River at Theni District, *J. Aqua. Biol.*, 21(2): 123–127.

Sugunan, V. V., 1999. Fisheries management of small water bodies in seven countries of Africa and Asia: A review. *Fishing Chimes*, 19: 7–10.

Talwar, P. K and Jhingharan, A. G., 1991. *Inland Fishes of India and Adjacent Countries in 2 Vols*. Oxford and IBH Pub. House, New Delhi, India.

Thresh, J.C., Suckling, E.V. and Beale, J.C., 1994. *Chemical and Biological Methods for Water Pollution Studies*. Environmental Publications, Karad, India: 122.

Venkateshwarulu, M., Somashekar, D.S., Jyothisri-Gowri and Ashashree, H.M., 2005. Ichthyofauna of Tunga Reservoir, Shimoga, Karnataka, *Biodiversity and its Conservation*, p. 92–93.

Venkateshwarulu, M., Jyothisri-Gowri and Ashashree, H.M., 2007. Diversity of fish fauna in Keradi Pond Sagara, Karnataka. *Diversity and Life Process from Ocean and Land*, p. 156–159.

Chapter 18

Domestic Wastewater Treatment by Electrocoagulation with Fe-Fe Electrodes

☆ *C. Sarala and P. Deepthi*

ABSTRACT

The treatment of wastewater has become an absolute necessity. An innovative cheap and effective method of purifying and cleaning wastewater before discharging into any other water systems is needed. A wide range of wastewater treatment techniques are known which include biological processes and physico-chemical processes. A host of promising techniques based on electerochemical technology are being developed and existing techniques are improved to reduce less chemical additions.

The present study was conducted to investigate the applicability of the electrocoagulation technique for the treatment of domestic wastewater at JNTU Hyderabad. Electrocoagulation is a surface reaction. Electrocoagulation is the process of passing electric current through a liquid, using anode and cathode. In this experiment iron electrodes are used and the sample is made up to run at different intervals of time *i.e.*, 5, 10, 15 and 20 minutes and different amperes of current is passed in the sample (0.12 A, 0.25A, 0.36A). The combination effects of current, pH and treatment time to the efficiency of the electrocoagulation process for the removal of Chemical Oxygen Demand, Total Dissolved Solids, pH, colour, chlorides etc., from the domestic wastewater showed that only current (C) and treatment time (t) have correlation with each other. It observed that the batch which is operated at 0.25A for 20 minutes has maximum removal efficiency of Chemical Oxygen Demand, Total Dissolved Solids, pH, colour, chlorides etc.

Keywords: Domestic wastewater, Electrocoagulation, Iron electrodes, Time intervals, Chemical oxygen demand.

Introduction

Municipal wastewater is the mixture of domestic wastewater, (the basic component), small amounts of industrial and storm water, drain water, surface infiltration, and ground water. It usually consists of a number of contaminants, such as suspended solids, biodegradable organics, pathogens, nutrients, refractory organics, heavy metals and dissolved inorganics. Direct discharge of untreated wastewater into the natural water bodies is not desirable, as the decomposition of the organic waste would seriously deteriorate the water quality (Balasubramaniam, 2001). In addition, communicable diseases can be transmitted by the pathogenic microorganisms. Nutrients such as nitrogen and phosphorous, along with organic material when discharged to the aquatic environment can also lead to excessive growth of undesirable aquatic life when discharged in excessive amounts on land can also lead to the pollution of groundwater. It was estimated that nearly half a million organic compounds have been synthesized and some 10,000 new compounds are added each year. As a result, many of these compounds are now found in the wastewater from municipalities and communities (Metcalf and Eddy, 1991). For these reasons, treatment of wastewater has become necessary for the protection of the environment keeping in view public health, economic, social and political concerns.

One of the challenging tasks faced by scientists and engineers today is to provide safe water to support healthy human life. But human activities always generate wastewaters which contain various pollutants that create problems to aquatic life and contaminate water resources. Although wastewaters may come from various sources, it mostly consists of domestic wastewaters (DWWs). Currently, domestic wastewater is normally treated by aerated biological methods. For example, the activated sludge, being the most famous biological method of wastewater treatment, produces high quality effluent, *i.e.* 90 per cent biological oxygen demand (BOD) and suspended solids (SS) removal (Metcalf and Eddy, 2003). There are some disadvantages of applying the biological method for wastewater treatment, such as requiring continuous air supply, high operating costs (skilled labour, energy, etc.), sensitivity against shock toxic loads, longer treatment time and necessary sludge disposal.

From an environmental point of view, the sewage treatment process is still far from being environmentally sustainable. There is an urgent need for the development of a more sustainable treatment process. Some of the possibilities include electrochemical treatment, improvement of the mitigation of toxic pollutants, high temperature sludge treatment processes, and membrane separation processes. Electrochemical process is a promising treatment method due to its high effectiveness. Its lower maintenance cost, less need for labour and rapid achievement of results (Feng *et al.*, 2003; Iniesta *et al.*, 2002). Other alternative solutions to wastewater treatment problems are still needed.

Using electricity to treat water was first proposed in United Kingdom in 1889 (Strokach, 1975; Mollah, 2001). The electrocoagulation of drinking water was first applied on a large scale in the United States in 1946 (Stuart, 1946; Bonilla, 1947). Because of the relatively large capital investment and the expensive electricity supply, electrochemical treatment of water or wastewater technologies did not find application worldwide (Kobya *et al.*, 2003; Pouet, 1995). However, due to extensive research in United States and USS Russia during the following half century, the process has gained large amount of knowledge. With the ever-increasing stringent environmental regulations regarding the wastewater discharge, electrochemical technologies have regained their importance worldwide during the past two decades (Chen *et al.*, 2000, Lin *et al.*, 1998). Now a days, electrochemical technologies have reached a state that they are not only comparable with other technologies in terms of cost but also potentially more efficient and for some situations electrochemical technologies may be the indispensable step in treating wastewaters containing refractory pollutants (Chen, 2003; Mattenson, 1995).

The electrocoagulation –flotation method for domestic wastewater treatment has a greater ability for the removal of chemical oxygen demand (COD) and suspended solids (SS) from effluents in comparison with treatment by conventional coagulation and so the present laboratory scale studies have been carried out to treat domestic wastewater using electrocoagulation.

Materials and Methods

Study Area

The college is located at Kukatpally, Rangareddy district, a 100 acres site, about 20 km from the heart of the city, on Bombay National Highway (NH-9). The territorial jurisdiction of the university covers the areas of Hyderabad, Ranga Reddy, Medak, Nizamabad, Adilabad, Karimnagar, Warangal, Khammam, Nalgonda and Mahboobnagar districts of Andhra Pradesh state.

Hyderabad lies between $17°$ 20' North latitude and $78°$ 30' East latitude. The climate of Hyderabad is very hot in summer and generally dry except during the southwest monsoon season. The average annual rainfall is 8500 mm. Granites, sand stones and red soils are observed in university area.

Electrocoagulation Process

All the experiments were conducted in batches. In each experimental run, a wastewater sample of 1.2 litres was collected and placed in an electrolytic cell. The sample was rigorously stirred by a stirrer. Iron electrodes were dipped into the solution upto an active surface area of 72 cm^2 and the following currents of 0.12, 0.25 and 0.36 amp (0.36 amp-15V, 0.25) were passed for a contact time of 5, 10, 15 and 20 minutes. After passing each current for each time period (*i.e.*, after each batch experiment), the sample was transferred into another beaker, and measured for pH. The measured sample was then taken to the jar test equipment, where it was rapidly mixed for 1 minute at 100 rpm. After a rapid mix for 1 minute, the sample was kept for flocculation by setting the speed of the paddles at 30 rpm for 20 minutes. Subsequently, the flocculated sample was kept undisturbed for 20 minutes, in order to allow the flocs that formed during the flocculation to settle down. After a settling time of 20 minutes, 250 ml supernatant sample was collected to perform the physical and chemical analysis according to APHA standards. Similar analysis was done with the influent raw municipal wastewater samples before starting the experiment.

After each batch experiment, samples were again filtered with 0.45 µm filter paper, to remove the electricity produced sludge, chlorides, alkalinity and a 20 ml sample was collected to perform the Total suspended solids analysis with APHA standard experiments.

Results and Discussion

The treated wastewater samples are collected from the experimental setup for the analysis. From the treatment of the wastewater by electrocoagulation process, it is observed that the colour of the water sample had changed to colourless after the treatment and for different currents the Electrical conductivity decreased with increase in the time. From the Figure 18.1 a slight increase in the pH was observed with time during the process which was within the regulatory drinking water standards. Chlorides and alkalinity also decreased with reaction time for different current amperes.

From the Figure 18.2 it is predicted that the suspended solids removal will increases with the increase of operating time and remained constant after 20 minutes of observation. With increase of current the removal efficiency of suspended solids also increased. The Figure 18.3 depicts the relationship between the percent removals of total dissolved solids and the current at various contact times. Total dissolved solids concentration is found to be decreasing with the increase in current and detention time in the settled samples. This is due to the presence of flocculent that is produced with increase in current and detention time that contributes to high removal of suspended solids. Up to 0.25 amp, total dissolved solids are found to be continuously decreasing with detention time. However, at 0.36 amp, a sharp decrease in the concentration of total suspended solids at a contact time of 5 minutes

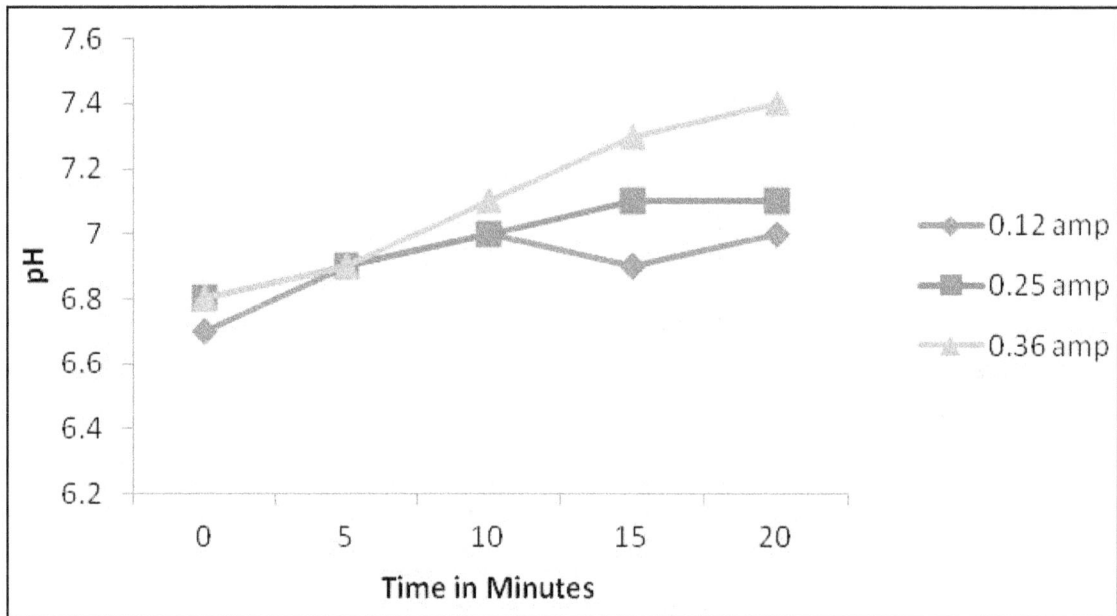

Figure 18.1: Change in pH with Time for different Currents.

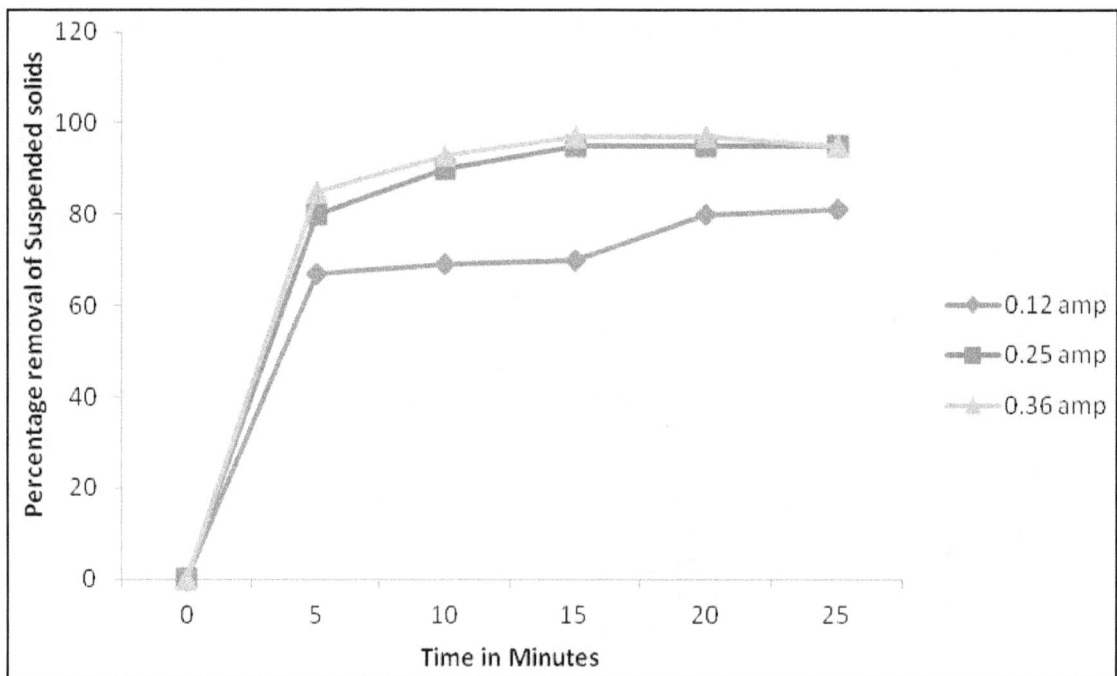

Figure 18.2: Percentage Removal of Suspended Solids with Time for different Currents.

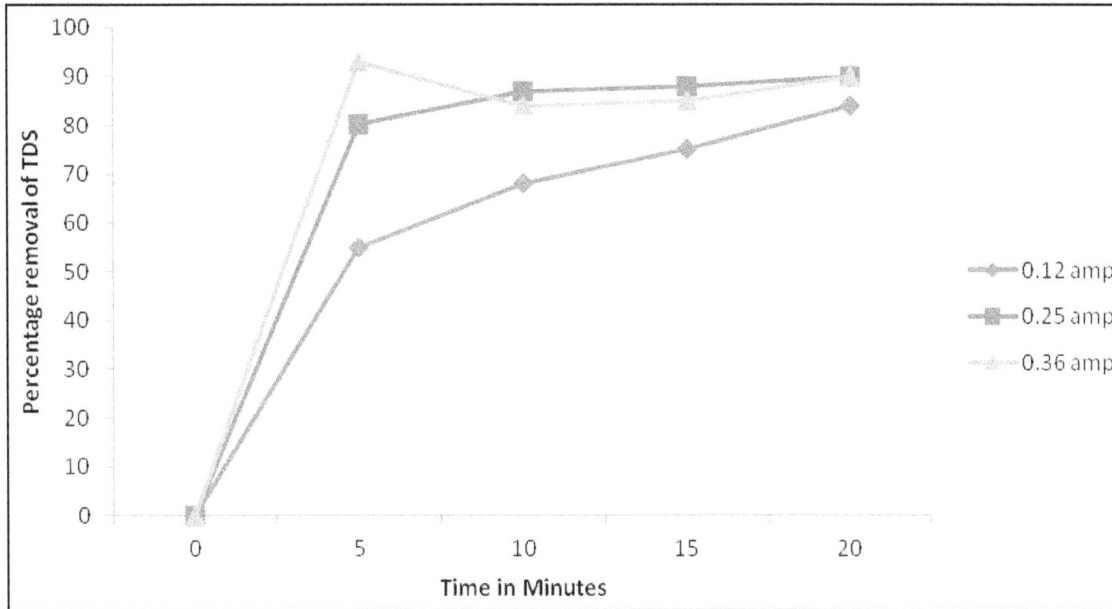

Figure 18.3: Percentage Removal of Total Dissolved solids with Time for different Currents.

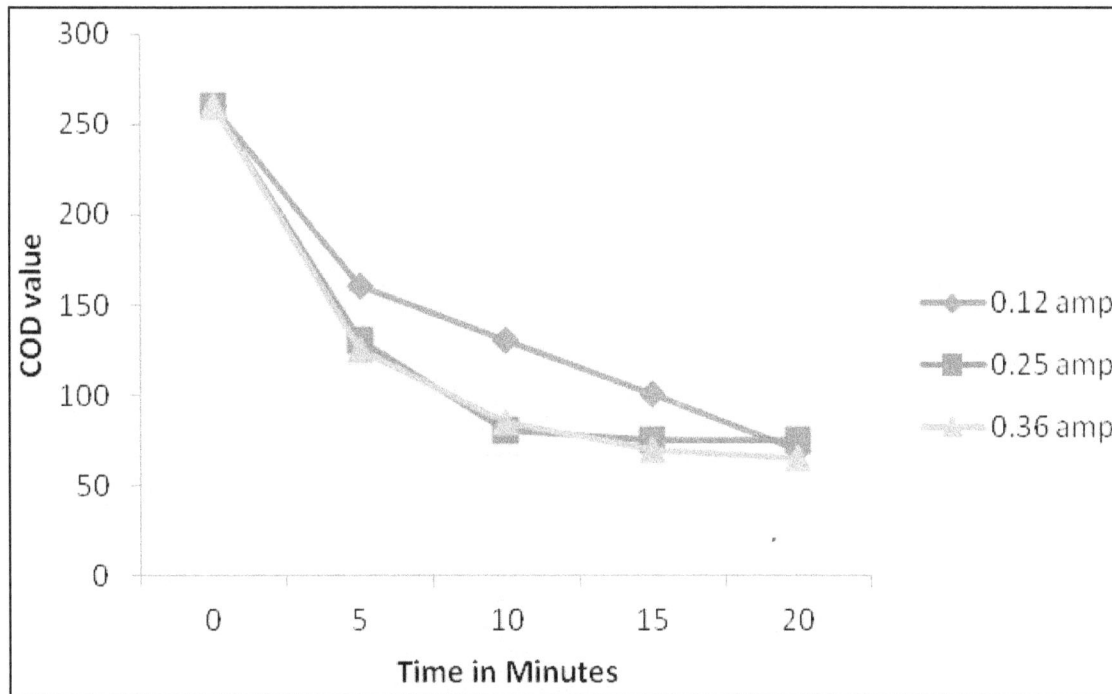

Figure 18.4: Decrease in Chemical Oxygen Demand Value with Time for different Currents.

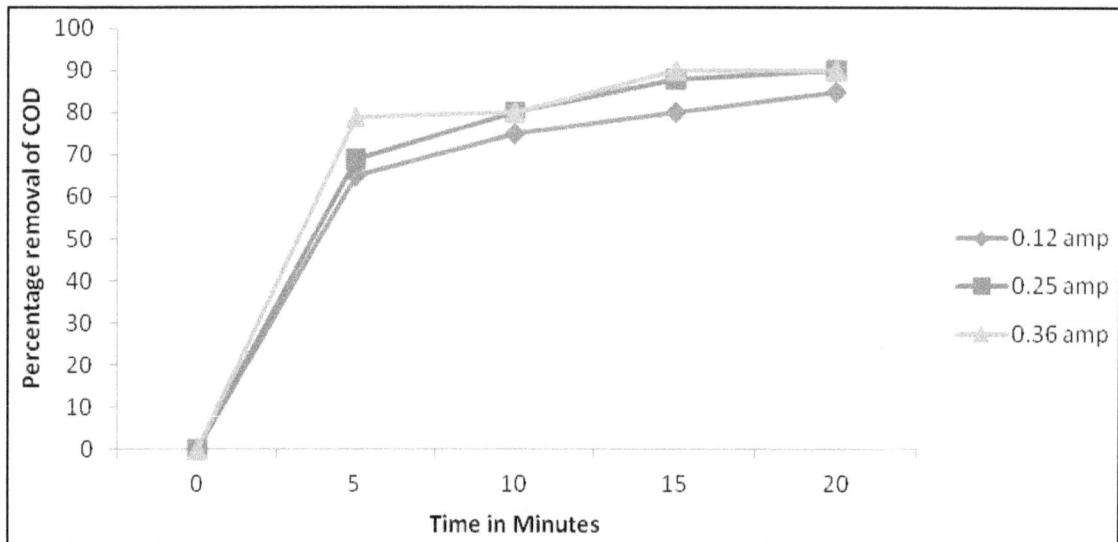

Figure 18.5: Removal Percentage of Chemical Oxygen Demand with Time for different Currents.

is observed. A gradual increase observed later. It is due to the re-stabilization phenomenon that took place because of excess coagulant dose that made particles restabilize and brought back them in suspension.

The variation of COD with electrolysis time is shown in Figure 18.4. COD decreases with increasing electrolysis time and reaches to a constant value. There is a sharp decrease in COD at the first 5 minutes of the process and after that there is a gradual decrease in the level. With the electrode combination Fe-Fe for different currents at different time intervals used and percentage removal of Chemical oxygen demand values are depicted in Figure 18.5. From the graph initially the percentage removal of chemical oxygen demand had increased and gradually the percentage removal decreases. From the results, it was obtained that the maximum reduction in the parameters is obtained at 20 minutes time interval with a variation of current ampere at 0.25 amp. The variation in parameters is presented in the Table 18.1.

Table 18.1: Reduction in Various Parameters before and after the Electrocoagulation Treatment

Sl.No.	Name of the Parameter	Units	Before Treatment	After Treatment
1.	Colour	–	Blakish brown	Clourless
2.	pH	–	6.7	7.1
3.	Chemical Oxygen Demand	Mg/l	260.4	60
4.	Total Dissolved Solids	Mg/l	672	60.48
5.	Chlorides	Mg/l	111.2	68.56
6.	Suspended Solids	Mg/l	160	20
7.	Electrical Conductivity	μs/cm	128	123.78
8.	Alkalinity	Mg/l	420	Nill

Conclusion

From the present study, it was concluded that the treatment process had shown feasible activity in removing the impurities present in the wastewater. In this process without using any chemicals and the chemicals which are present in the sample can be removed, it does not leave any additional chemicals in the sample. In this process, suspended solids formed after electrocoagulation process removed by filtration and chemical oxygen demand decreased to 90 per cent with increase of contact time for different currents, maximum reduction of chemical oxygen demand is observed at 20 minutes for 0.25 amp, 0.36 amp. The maximum reduction of total dissolved solids is 90 per cent at 20 minutes for 0.25 amp, 0.36 amp. A successful application of electrocoagulation (EC) technique for the removal of suspended solids from wastewater would address the environmental needs of reduction in the operational costs and potential saving in processing unit. A host of very promising techniques based on electrochemical technology are being developed but are not yet to the commercial stage. Among different physical and chemical methods of water and wastewater treatments, electrocoagulation method offers a special attraction due to its ecologically friendly, safety, simplicity and lower operating costs.

Based on the results, it was suggested that for successful industrial application of electrocoagulation, quantitative parameters must be identified to ensure dimensional consistency between small and large scale processes. Development of advanced materials and application of different electrode types brings a new dimension to electrocoagulation. Different electrode material can be used to assess different coagulant types for specific pollutants. For example the use of iron will produce ferric ions that are readily used in the water industry. Electrodes operation, such as periodic polarity reversal, controls passivation formation *in situ*. Development of sophisticated electrode arrangements and associated operation programs lead to significant developments for pasivation control. Further this technique will continue to make inroads into the water treatment area because of numerous advantages and the nature of the changing strategic water need in the world.

References

Balasubramaniam, N. and Madhavan, K., 2001. Arsenic removal from industrial effluent through electrocoagulation. *Chem. Eng. Technol.*, 24.

Bonilla, C.F., 1947. Possibilities of electronic coagulator for water treatment. *Water Sewage*. 85: 21–22, 44–45.

Chen Guohua, 2003. Electrochemical technologies in wastewater treatment. *Separation and Purification Technology*, 31.

Chen, G., Chen, X. and Yue, P.L., 2000. Electrocoagulation and electroflotation of restaurant wastewater. *J. Environ. Eng.*, 126: 858–863.

Feng, C., 2003. Development of a high performance electrochemical wastewater treatment system. *J. Hazard. Mater.*, B103: 65.

Iniesta, J., 2002. *Journal of Electrochemical Society*, 149: 57–62.

Lal, C.L. and Lin, S.H., 2004. Treatment of chemical mechanical polishing wastewater by electrocoagulation: system performances and sludge settling characteristics. *Chemosphere*, 54: 235–242.

Lin, S.H. and Peng, C.F., 1998. Treatment of textile wastewater by electrochemical method. *Water Res.*, 28: 277–282.

Metcalf and Eddy, 1991. *Wastewater Engineering, Treatment Disposal, Reuse,* (Eds.) G. Tchobanoglous and F.L. Burton. McGraw-Hill, New York, p. 1820.

Metcalf and Eddy *et al.,* 2003. *Wastewater Engineering, Treatment and Reuse,* Fourth Edn. Mc-Graw Hill Education, p. 1329.

Mattenson, M.J. *et al.,* 1995. Electrocoagulation and separation of aqueous suspensions of ultra fine particles. *Coll. Surf. Physico-chemical. Eng. Aspects,* 104: 101–109.

Mollah, M.Y.A. *et al.,* 2001. Electrocoagulation (EC): Science and application. *Journal of Hazardous Materials,* B84: 29–41.

Pouet, M.F. and Grasmick, A., 1995. Urban wastewater treatment electrocoagulation and flotation. *Water Sci. Technol,* 31: 275–283.

Strokach, P.P., 1975. Electrochem. *Ind. Process. Bio.,* 55: 375.

Stuart, E.E., 1946. Electronic water purification progress report on electronic coagulator. A new device which gives promise of unusually speedy and effective results. *Water Sewage,* 84: 24–26.

Chapter 19

Seasonal Changes in Bacterial Indicators in Drinking Water of Thiruvananthapuram Corporation, Kerala

☆ *A. Jayadev and V.S.G. Thanga*

ABSTRACT

Ensuring microbiological quality of potable water requires the detection of coliform bacteria as indicators of human fecal contamination. In order to verify the contamination status of water sources of Thiruvananthapuram city and their suitability for drinking purpose, a study was conducted in the Department of Environmental Sciences. The study area *i.e.* Thiruvananthapuram corporation consisting of 81 wards was classified into different zones by combining nearby wards. Samples were collected from each zone during summer and monsoon seasons of two consecutive years *viz.*, Year I and Year II. Samples were analyzed for total coliforms and fecal coliforms both by multiple tube tests for the determination of most probable number and Polymerase Chain Reaction (PCR). The results revealed varying levels of contamination in the sampling sites in different collections. The hot spots of contamination were found to be zone 15 (Karamana, Nedumcad, Kalady),7 (Nemom, Ponnumangalam, Melamcode, Pappanamcode), 12 (Thampanoor, Sreekandeswaram, Fort, Chalai) and 22 (Edavacode, Mannanthala) during summer Year I, monsoon Year I, summer Year II and monsoon Year II respectively. 5 – 25 per cent of the samples which showed nil values by multiple tube test showed to be positive in PCR based analysis.

Keywords: *Total coliforms, Fecal coliforms, Polymerase chain reaction, Multiple tube test, Most probable number, Waterborne diseases.*

Introduction

Drinking water is inevitable to sustain life and a satisfactory supply must be available to all. The source of drinking water can vary. But the most widely used source is public distribution system. Here, the water is universally used and is supplied to a large population from a few centralized sources; and hence, it is the main source of epidemics. According to WHO (1996) and BIS (1991) coliforms should not be detectable in any 100 ml of drinking water. Contaminated drinking water has its impact on human health especially in developing countries. The potential of drinking water to transport pathogens and to cause subsequent illness is well documented in countries at all levels of economies. At this juncture, as studies which ensure the bacteriological quality of drinking water in Thiruvananthapuram Corporation is rare, a study was carried out to find out the possibility and extend of contamination of public water supply system of Thiruvananthapuram Corporation so as to suggest suitable remedial measures.

Materials and Methods

Sample Collection

Water samples were collected from the whole of Thiruvananthapuram Corporation (141.74 Km^2) by dividing it into 22 zones homogenously on the basis of the availability of taps (Table 19.1). Collection of samples was made in sterile bottles according to Theroux *et al.*, 1999. Collections were done during pre-monsoon and post monsoon seasons of Year I and Year II. The number of samples varied slightly in different collections due to problems in distribution. The collected samples were maintained in controlled conditions and were subjected to analysis of TC and FC according to standard methods (Senior, 1996). On the basis of the value of TC, the samples were grouped into A (0 in 100 ml), B (1-100/100 mL), C (101-1000/100 mL) and D (>1000/100 mL). (Warrington, 1988; Al-Saimary, 2007). Those samples showing nil value in conventional technique were analyzed by PCR.

Table 19.1: Tap Water Sampling Zones

Zone No.	Ward No.	Wards
1.	2,3	Kulathoor, Attipra
2.	4,5	Kuzivila, Cheruvakkal
3.	81	Pound Kadavu
4.	7,8	Ulloor, Pongummoodu
5.	11,19,20,24	Muttada, Kuravankonam, Nalanchira, Nanthancode
6.	21,22,23,31	Peroorkada, Mannantahla, Kowdiar, Sasthamangalam
7.	52,53,54,55	Nemom, Ponnumangalam, Melamcode, Pappanamcode
8.	79,80	Karikkakom, Vettucaud
9.	78	Sangumugham
10.	70,71,72,61	Beemapalli, Valiathura, Sreevaraham, Kamaleswaram
11.	73,74,77	Perumthanni, Palkulangara, Chackai
12.	42,43,44,45,27	Thampanoor, Sreekandeswaram, Fort, Chalai, Vanchiyoor
13.	75,76	Pettah, Kadakampalli
14.	63,64,65	Ambalathara, Thiruvallam, Punchakkarai

Contd...

Table 19.1–*Contd...*

Zone No.	Ward No.	Wards
15.	47,48,56,57	Arannoor, Karamana, Nedumangadu, Kalady
16.	28,29,30,39,40,41	Secratariat, Palayam, Vazhuthakaud, Jagathy, Poojappura, Thycaud
17.	49,50,51	Mudavanmugal, Thrikkannapuram, Estate
18.	35,36,37,38	Pangode, Valiavila, Thirumalai, Chengalloor
19.	32,33,34	PTP Nagar, Kanjirampara, Vattiyoorkavu
20.	15,16,25,26	Kannammoola, Gowreesapattom, Kunnukuzhi, Rishimangalam
21.	12,13,17	Medical college, MC east, Pattom
22.	9,10	Edavacode, Mannanthala

PCR Sample Processing

PCR amplification was performed using a DNA thermal cycler (Eppendorf Personal) and gene amp kit (Geinei). PCR solution was added to tubes. The solution contained 21 µL PCR master mix, 1 µL each of forward and reverse primers and 2 µL processed sample. The total reaction mixture was 25 µL. The template DNA was initially denatured as 95°C and then a total of 30 PCR cycles were run with denaturation at 95°C for 1 minute, primer annealing and extension at 72°C for 1 minute.

A 309 bp segment of the coding region of the lamb gene of *E. coli* was amplified using 24 mer primers (Atlas *et al.*, 1995)

FP: 5'CTGATCGAATGGCTGCCAGGCTCC-3'

RP: 5'-CAACCAGACGATAGTTATCACGCA-3'

As the data of the work came from a non-normal population, non-parametric statistical procedures were applied. Kruskal-wallis test was applied to check whether there is any significant difference between various seasons and between various zones with respect to each of TC and FC (Kruskal-Wallis 1952). Pair-wise comparison was done by Mann- Whitney U test (Mann-Whitney, 1947). The level of significance of test is prefixed at 1 per cent or 5 per cent level. When the calculated P value was found to be less than 0.01 (or 0.05), it was concluded that there is a significant difference at 1 per cent (or 5 per cent) level of significance.

Results and Discussion

Total Coliforms (TC)

A wide variation in the level of contamination was observed in the drinking water of Thiruvananthapuram city, where MPN of TC in water samples ranged between 0 and 2800 during summer and between 0 and 3400 during monsoon in the Year I (Figure 19.1).

The highest value in summer was observed in samples collected from 15th zone (Arannoor, Karamana, Nedumangadu, and Kalady) and those during monsoon were observed in samples collected from the 7th zone (Nemom, Ponnumangalam, Melamcode and Pappanamcode).Based on the MPN of TC, the water samples were grouped into four categories, *viz.*, A, B, C and D. 41 per cent samples collected during summer and 55 per cent of samples collected during monsoon were grouped under A, 19 per cent of samples collected during summer and 3.1 per cent of samples collected during monsoon were grouped under B, 34.4 per cent of samples collected during summer and 20.3 per cent

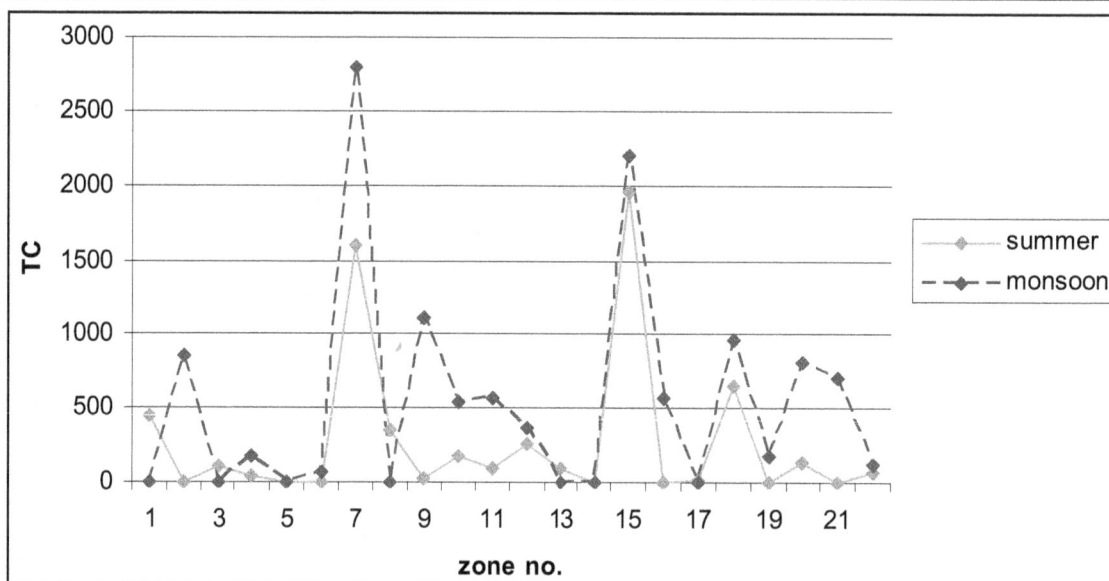

Figure 19.1: Seasonal Variation of TC (Year I).

of samples collected during monsoon were grouped under C and 6.3 per cent of samples during summer and 21.9 per cent of samples during monsoon were grouped under group D (Table 19.2).

Table 19.2: Classification of Samples Based on the Level of Contamination

Category	Level of Contamination	Year I (Per cent samples)		Year II (Per cent samples)	
		Summer	Monsoon	Summer	Monsoon
A	0	41	55	86.8	79.2
B	1 – 100	19	3.1	3.8	20.8
C	101 – 1000	34.4	20.3	7.5	0
D	>1000	6.3	21.9	1.9	0

During summer Year II, the MPN of TC in tap water samples ranged from 0 to 1600 and during monsoon, the value ranged from 0 to 16. The highest value during summer was observed in samples from zone no. 12 (Thampanoor, Sreekandeswaram, Fort, Chalai and Vanchiyoor) and that during monsoon was in zone no. 22 (Edavacode and Mannanthala). The seasonal variation in the values are shown in Figure 19.2.

Based on the MPN of TC, 86.8 per cent of samples collected during summer and 79.2 per cent of samples collected during monsoon were classified as A, 3.8 per cent of samples collected during summer and 20.8 per cent of samples collected during monsoon were classified as B, 7.5 per cent of samples collected during summer were classified as C and 1.9 per cent of samples collected during summer were classified as D. C and D levels of contamination were not noticed during monsoon (Table 19.2).

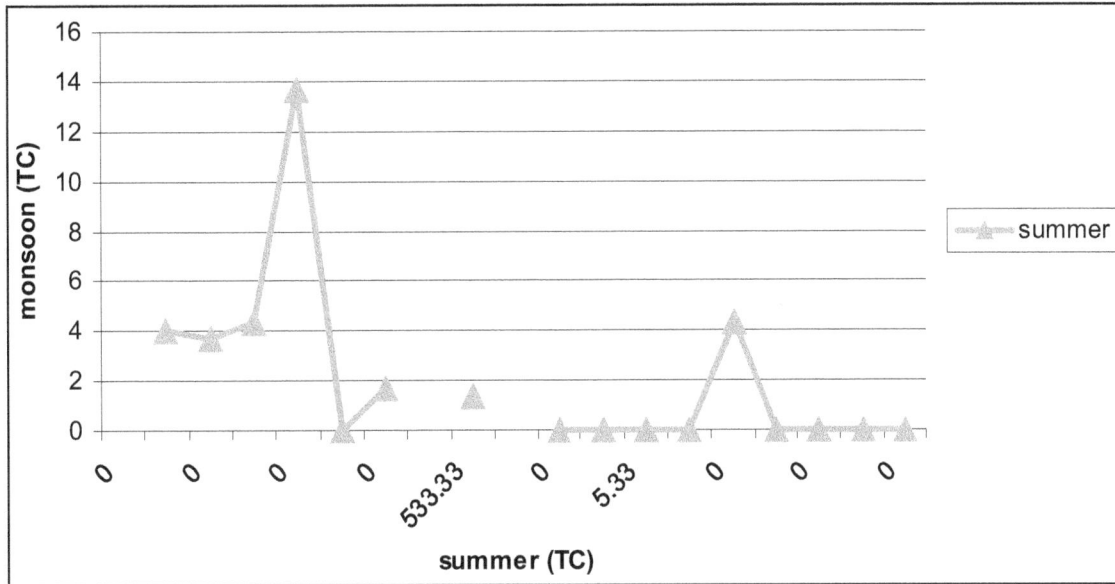

Figure 19.2: Seasonal Variation of TC (Year II).

The per cent samples in A category were 40.6, 54.7, 86.8 and 79.2 during summer Year I, monsoon Year I, summer Year II and monsoon Year II respectively. During monsoon Year II, no zones showed higher level of contamination *i.e.* C and D groups. Zone 5 (Muttada, Kuravankonam) and 14 (Ambalathara, Thiruvallam, Punchakarai) showed nil contamination for at least 3 sampling seasons, while zones 19 (PTP Nagar, Vattiyoorkavu) and 20 (Kannamoola, Gowreesapattom, Kunnukuzhi) showed nil contamination for at least 2 sampling seasons. The zones showing highest levels of contaminations differed among the sampling stations. They were zone 15 (Karamana, Nedumancad, Kalady), 7 (Nemom, Ponnumangalam, Melamcode, Pappanamcode), 12 (Thampanoor, Sreekandeswaram, Fort, Chalai) and 22 (Edavacode, Mannanthala) during summer Year I, monsoon Year I, summer Year II and monsoon Year II respectively. The higher percentage of samples in A category was observed during summer Year II; in B category during monsoon Year II; in C category during summer Year I and in D category during monsoon Year I. These results reveals the decrease in levels of contamination from Year I to Year II which was statistically significant (P < 0.01). Reasons of microbial contamination in public distribution system are various. The negative pressure in the distribution system due to intermittent supply might pave way for contamination. Of the deficiencies reported in water distribution system, loss of disinfectant residual, back siphonage and cross-contamination are the major causes of contamination (Craun, 2001). In many places, water pipes run parallel to drainage canals (Royee and Prakasam, 2004).

Some of the consequences are excessive growth and colonization of water distribution pipes by bacteria and other organisms, some of which are pathogenic; microbial re-growth in storage reservoirs and corrosion problems. Earlier reports from Haryana (Bala *et al.*, 1994) revealed high pollution levels of 1800/100 ml in public well and tap water sources. High pollution levels were also recorded by Gaur *et al.* (2000) and Dwivedi *et al.* (2000). They opined that fecal contamination and sewage contamination are the causes of high bacterial loads in water.

Fecal Coliforms (FC)

The MPN of FC collected during summer of Year I ranged from 0 to 910 (Figure 19.3).

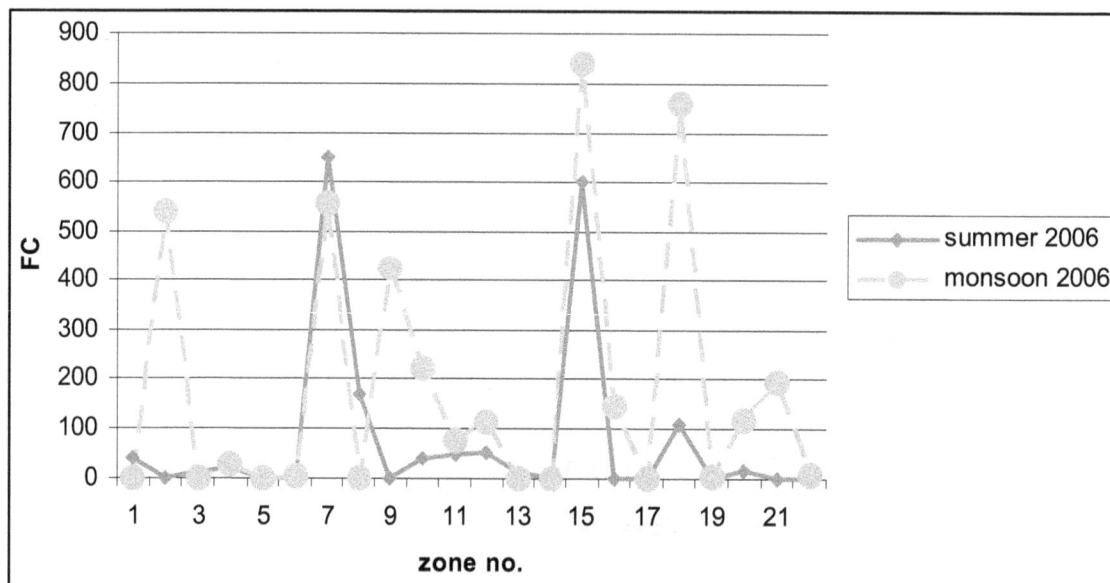

Figure 19.3: Seasonal Variation of FC (Year I).

The highest value were seen for zones 7 (Nemom, Ponnumangalam, Melamcode, Pappanamcode) and 15 (Arannoor, Karamana, Nedumangadu, Kalady). During monsoon the values ranged from 0 to 1396. The highest value observed was in zone no. 2 (Kuzhivila, Cheruvakkal). 66 per cent of samples during summer and 56 per cent of samples during monsoon showed zero contamination of FC

The results of analysis of FC in Year II showed that during summer, the value for FC ranged between 0 and 350. During monsoon, the values ranged between 0 and 16 (Figure 19.4).

The highest value during summer was seen in zone no. 12 (Thampanoor, Sreekandeswaram, Fort, Chalai, and Vanchiyoor) and during monsoon the highest value was seen in 22nd zone (Edavacode, Mannanthala). 92.7 per cent of the samples were shown to be free of contamination during summer and 81.2 per cent of the samples during monsoon were free of

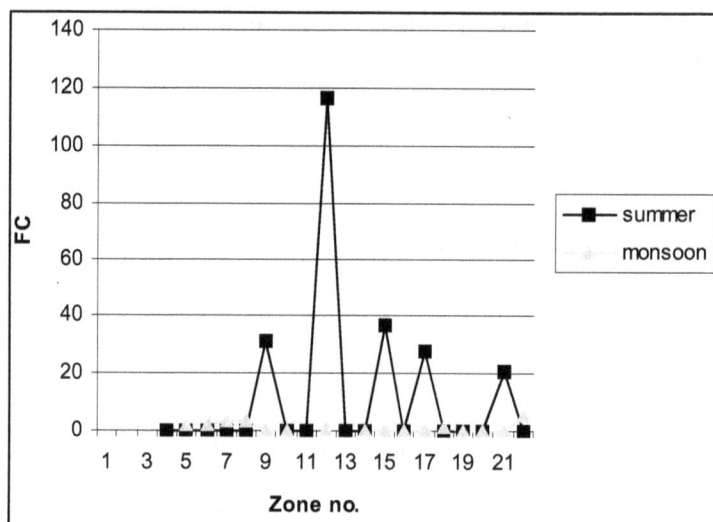

Figure 19.4: Seasonal Variation of FC (Year II).

contamination. A significant difference was observed in the values during the different sampling seasons (P < 0.01).

MPN of fecal coliforms also showed a trend similar to TC. Numbers of zones which showed nil values for FC are 10, 7, 14 and 9 during summer Year I, monsoon Year I, summer Year II and monsoon Year II respectively. Only during monsoon Year I, levels of contamination > 1000 was observed. In other seasons, such high levels were not present. Out breaks of water borne diseases occurred in Kollam municipal areas due to contamination of water supplies (Royee and Prakasam, 2004).

Drinking water quality can deteriorate during storage and transport through water distribution pipes before reaching the consumer. In the United States, 18 per cent of the outbreaks reported for public water systems are due to contamination of the water distribution system by microbial pathogens (Smith, 2002; Sobsey and Olson, 1983). Fecal contamination will not be distributed evenly throughout a piped distribution system. The more frequently the water examined for fecal indicators the more likely it is that the contamination will be detected. Frequent examination by a simple method is more valuable than less frequent examination by a complex test or series of tests.

Olowe *et al.*, Year II observed in their study that MPN of all the water sources were higher in rainy seasons as compared to winter and summer.

Polymerase Chain Reaction (PCR)

The samples which provided negative results for total coliforms are analyzed using PCR. In the summer collection of Year I, 52 samples showed nil result by the MultipleTtube Test. Of these, 13 gave positive result by PCR. During monsoon of the year, of the 46 samples which showed negative result by the conventional method of detection of total coliforms, 8 gave positive result by PCR technique (Figures 19.5 and 19.6).

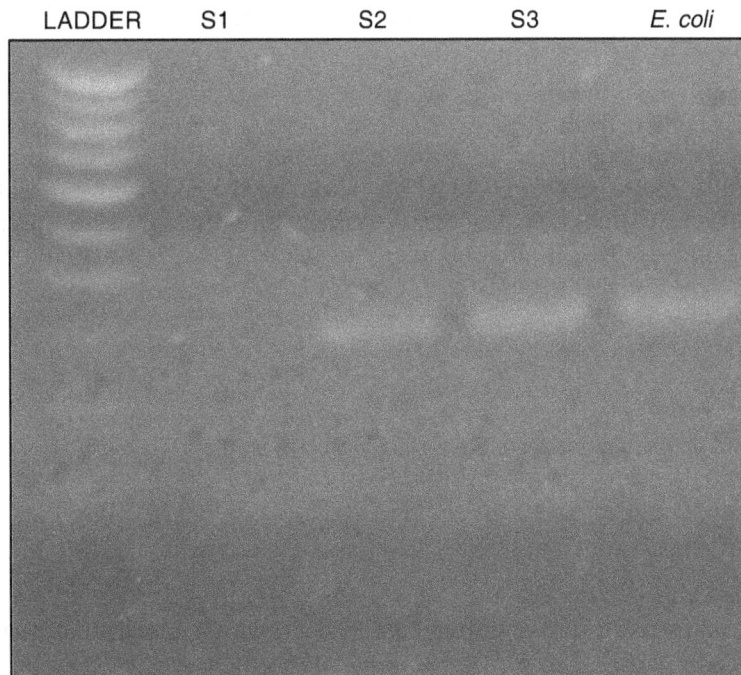

Figure 19.5: Amplified PCR Products.

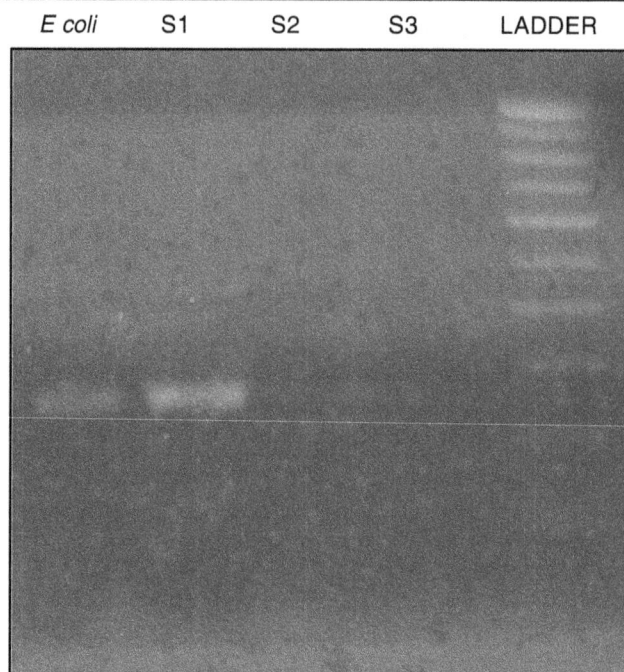

Figure 19.6: Amplified PCR Products.

Of the 54 samples which were found to be negative in the year–Year II, 10 were found to give positive result in the PCR method of detection of coliforms. Of the 56 samples analyzed by PCR in monsoon, 3 were found to be positive.

Although water is not a medium for pathogenic growth, it is a means of transmission of the pathogen to the place where the individual is able to consume and therefore starts the outbreak of disease. To reduce the incidence and prevalence of waterborne diseases, improvements in the availability, quantity and quality of water, improved sanitation, and general personal and environmental hygiene is required. The major factors that reduce the significance and impact of these diseases in public health are good sanitation, plentiful availability of good quality water, adequate disposal of human and animal excrement, and public education in hygienic practices. Frequent monitoring is required so that the precautionary measures can be taken by repair or modification of the water system to prevent outburst of waterborne diseases.

References

Al-Saimary, I.E.A.K., 2007. Evaluation of Bacterial Quality of water collected from various stations in Basrah city. *The Internet Journal of Third World Medicine*, 5(2).

Atlas, R., Bej, A.K., Dicesare, J.L., Haff, L.A. and McCarthy, S., 1995. Applications of molecular biology. In: *Environmental Chemistry*, (Eds.) R.A. Minear, A.M. Ford, L.L. Needham and N.J. Karch. Lewis Publishers, New York, pp. 3–10.

Bala, S., Gangwar, V., Yadav, L. and Kishtiwari, J., 1994. Drinking water available and Extend of its pollution in Hissar district. *Haryana Agricultural University Journal of Research*, Hissar, 25(112): 68–73.

BIS., 1991. *Indian Standard Specification of Drinking Water.*

Craun, G. F., 2001. Waterborne disease outbreaks caused by distribution system deficiencies. *J. AWWA* 93(9): 64–75.

Dwiwedi, S.I., Tiwari, C. and Bhargava, D.S., 2000. Classification of water quality of river Ganga and Varanasi. *Indian J. Environmental Protection*, 20(9): 688–697.

Gaur, R.K., Khan, A.A. and Alang, A., 2000. Bacteriological quality of river Ganga from Narora to Kannauj: a comprehensive study. *Indian J. Environmental Protection*, 20(3). 165–170.

Olowe, O.A., Ojurongbe, O. and Opaleye, 2007. Bacteriological quality of water samples in Osogbo Metro poles. *African Journal of Clinical and Experimental Microbiology*, 6(3): 219–223.

Royee, M.K.P. and Prakasam, V.R., 2004. Water supply of Kollam municipality in Kerala: Problems and solutions. In: *Water Pollution: Assessments and Management*, (Eds.) Arvind Kumar and G. Tripathi. Daya Publishing House, pp. 326–331.

Senior, B.W., 1996. Examination of water, milk, food and air. In: *Practical Medical Microbiology*, 14th edn. Mackie and McCartney. pp. 883–887.

Smith, D.B., 2002. Coliform bacteria: Control in distribution water systems. In: *Encyclopedia of Environmental Microbiology*, (Ed.) G. Bitton. Wiley-Interscience, N.Y.

Sobsey, M.D. and Olson, B., 1983. Microbial agents of water borne disease In: *Assessment of Microbiology and Turbidity Standards for Drinking Water*, (Eds.) P.S. Berger, and Y. Argaman. EPA Report No. EPA 570–9–83–001. U.S. Environmental Protection Agency, Office of Drinking Water, Washington, D.C.

Theroux, F.R., Edward, F. Eldridge and Mallmana, W.L. Roy, 1999. *Laboratory Manual of Chemical and Bacterial Analysis of Water and Sewage.*

Warrington, P.D., 1988. *Water Quality Criteria for Water Quality Indicators*. Overveiw Report. Environmental Protection Division. Updated 2001.

WHO, 1996. *Guidelines for Drinking Water Quality*, 2nd edn. Health criteria and other supporting information, Geneva.

Chapter 20

Limnology of Two Water Sheets in the Thar Desert with Special Reference to Invertebrate Diversity

☆ *Ankush Sharma, Poonam Lata, N.S. Rathore and Harbhajan Kaur*

ABSTRACT

Deserts characterized as harsh and inhospitable climatic conditions because of high temperature fluctuation and low rainfall, due to lack of water resources in desert region, village ponds acts as life line. Water quality includes physical, chemical and biological characteristics of water. The present chapter focused on physical, chemical and biological characteristics of two village ponds namely– Nal and Gajner pond respectively located in Bikaner district of North-West region of Rajasthan. A year round study (Sep. 2010 – Aug. 2011) was carried out. During this period study of physical-chemical factors included temperatures (air and water), turbidity, pH (water and soil), EC (water and soil), TDS (water and soil), dissolved oxygen, free CO_2, hardness, total alkalinity and organic matter were carried out. The invertebrate faunal studies are also carried out on these desert ponds for an annual cycle revealed 41 animal species of 10 classes belonging to 7 phyla namely–Protozoa, Coelenterata, Aschelminthes, Annelida, Mollusca and Arthropoda. It has been observed that these animals are well adapted for desert conditions and even withstand the dry period of summer.

Keywords: Physical-chemical, Desert, Village pond, Invertebrate fauna.

Introduction

Earth also known as green planet because of presence of life friendly environment such as life supporting weather conditions, life supporting gases (oxygen, carbon dioxide and nitrogen) and presence of water. Water on earth present in large amount, still usable water is in scare commodity.

According to the Biennial report of freshwater resources (Peter H. Gleick *et al.*, 2009) water spread is 1.4 billion cubic kilometer on earth surface in wide variety of forms and conditions. According Shiklomonov (1993) percentage of easy accessible water in form of lakes is 0.007 per cent, wetlands and marshes is 0.001 per cent and rivers is 0.0002 per cent. Total percentage of fresh ground water which is easy accessible to human is 0.76 per cent. Although water covers 70.9 per cent of earth surface but its distribution pattern causes scare commodity in certain areas, which gives rise to another type of ecosystem *i.e.* desert.

These areas on earth are generally inhospitable to life because of harsh and hostile climatic conditions such as high temperature fluctuations (49–18°C), low rainfall (250mm/year), low humidity, dust storms, low underground water level, intense sun radiations except certain regions like Tibet known as cold desert situated at 4000 m above sea level, scarcity of surface water, present only in the form of small villages ponds, manmade reservoirs, small tanks and few perennial lakes. These regions mostly found around Tropic of cancer and Tropic of Capricorn in northern and southern hemisphere respectively. Canopy in most desert are rear and plants are mainly ground hugging shrubs and short woody trees.

The Great Indian Desert or The Thar Desert spread in north-west of Rajasthan is one of the smallest desert in the world but most thickly populated. Most water bodies in this region are temporary and ephemeral and face dry period at least once in a year because of low rainfall, high atmospheric temperature, sandy texture of soil and high evapo-transpiration rate. Water quality directly expresses the nature and health of aquatic communities, influences the phytoplankton abundance, species composition, stability, productivity and physiological conditions of aquatic organisms and also help in flourishing other communities presents on its banks. Faunal communities of these temporary water bodies adopted strategies to face dry periods including cyst formation, high reproduction potential, and parthenogenesis. Freshwater communities contain large variety of micro, macro flora and fauna such as fishes, amphibians, insects, reptiles, nematodes, arachnids, crustaceans, sponges, molluscs, macrophytes, algae, protozoan, fungi, bacteria and viruses. Aquatic invertebrates have served as effective biological and used as a barometer of overall aquatic biodiversity.

Study Area

Rajasthan, the largest state of India has a geographical location between 23°3' to 30°12' N latitudes and 69°30' to 78°17' E longitudes. It has an area of 3, 42,239 km^2. It shares international border with Pakistan on west. The state is divided by the Aravalli range into two unequal parts- the 3/5th NW part constitutes the Indian desert and 2/5th SE part the fertile plains. The area under study is the Indian desert around Bikaner. District Bikaner of Rajasthan state having a central position in north-western region (28°N and 73°17'E, MSL 228 m).

Nal Pond

The present study was carried out at Nal pond at 15 km North-West of Bikaner city situated on the Jaisalmer highway. The maximum depth of the pond is 6 m and it covers an area of 2500 m^2. The pond is situated in low land so rain water from surroundings reaches in the pond, which is the only source of water. Nal pond rich with macrophytes like *Hydrilla* and *Vallisneria*. This village pond face dry period of two months (May and June).

Gajner Village Pond

Gajner village located at 25 km South-West of Bikaner city (27° 57' N latitude, 73° 03' E longitude and 233 meter MSL altitude). This village pond is also known as Chundasagar. Chundasagar pond was named after the late king Rao Chundaji. The pond has a water spread of about 0.2 km² and maximum depth of 6.7 m. The pond water is profusely rich in *Hydrilla, Potamogeton, Nelumbo nucifera*. A large variety of local and migratory birds inhabit the pond throughout year.

Materials and Methods

Water samples were collected from three study stations of pond. The sampling was carried out during morning hours between 06: 00 to 11: 30 hours. The water samples were collected with the help of a plastic bucket of 15 liters capacity. These were transferred to well rinse polyethylene bottles for the analysis of physical and chemical parameter. Abiotic parameter monitored included Temperatures (water and air), Turbidity, pH, Electrical Conductance (EC), TDS, Dissolved Oxygen (DO), Total alkalinity, free CO_2 and Hardness. Parameters like temperature, pH and EC were analyzed with the aid of a portable water analyzer kit (Century CK 710). For the analysis of various chemical variables, the methods as prescribed by Strickland and Parson (1972), Golterman *et al.* (1978), and APHA-AWWA-WPCF (1981) and Saxena (2001) were followed. Dissolved oxygen, alkalinity, were determined by volumetric methods. Soil samples were collected in leveled polythene bags and carried to laboratory. For soil analysis (Organic matter, pH, TDS, EC) methods prescribed by Saxena (2001) were followed.

For zooplankton study water of known volume filtered through zooplankton net made of bolting silk No. 25, transferred to plastic bottles and preserved by pouring few drops of formalin, identify with the help of microscope and suitable keys such as Edmondson (1966) and Tonapi (1980).

Macro-organisms like macro-periphyton attached on the vegetation, rocks and wall of ponds collected with the help of forceps, brush and handled net and preserved in 4 per cent formalin or 70 per cent alcohol and method prescribed by Saxena (2001). For zoo benthic study soil samples were collected from the three stations of both pond with the help of quadrate having dimensions 25 × 20 × 10 cm. Polythene bags were used to carry samples to laboratory where samples were sieved (mesh size 0.50 mm). Benthos were picked up with the help of forceps and brush and preserved in 4 per cent formalin and 70 per cent ethyl alcohol. For qualitative analysis, animals were examined using stereo and compound microscopes and standard keys (Needham and Needham, 1941; Edmondson, 1965; Subba rao, 1989; Tonapi, 1980) were followed for identification.

Results and Discussion

Air temperatures in Nal pond ranged from 9.8°C to 30.3°C, while in Gajner pond ranged from 19.6°C to 35.5°C. Wide seasonal fluctuations in air temperatures were in keeping with the thermal trend of the hot desert. Sharma (2009) reported wide fluctuation of air temperature ranged from 20.5°C to 32.5°C in three water bodies of this region. Water temperatures in of Nal pond ranged from 12.5°C to 30.7°C, while in Gajner pond ranged from 15.4 °C to 31.1 °C. Wide seasonal fluctuations in water temperatures were directly depend on air temperature. Such wide fluctuations in water temperature was also reported by Kaur (2007), Poonam Lata (2009) from the same region.

Turbidity in Nal pond ranged from 0 J. T. U. to 140 J. T. U. while in Gajner pond it ranged from 40 J. T. U. to 140 J. T. U. Desert waters are noted to be shallow and turbid, turbidity in case of ponds and canals were greater because of silt and other suspended matter. Cole (1968), one of pioneers in desert limnology, noted turbidity as one of the features of many arid zone waters. Sharma (2009) noted turbidity of three ponds of this region ranged from 40 J. T. U. to 140 J. T. U.

The pH of Nal pond water ranged from 7.5 to 8.7 and in Gajner pond water it ranged from 7.4 to 9.5, indicating that pond water investigated was alkaline throughout the period of study. Sharma (2003) reported a pH of almost same tune (7.1-10.9) in a number of waters of different nature investigated in the same area. Other study in the region also recorded the pH of water were within the range of 7 to 10 (Kaur, 2007)

The Electrical conductance of Nal and Gajner pond water ranged from 0.10 to 0.25 m mhos/cm. A still wider range of EC was documented by Sharma (2003). He reported that village ponds are little more saline as compared to large water bodies of this region. In Nal and Gajner pond water TDS ranged from 64 mg/l to 160 mg/l. Kaur (1991) reported the wide range of TDS (20 mg/l to 351 mg/l) in other waters of same region.

The dissolved oxygen value of Nal pond water ranged from 3.65 mg/l to 11.97 mg/l and Gajner pond water ranged from 8.32 mg/l to 18.06 mg/l indicates that water of Gajner pond is well oxygenated than Nal pond. In the desert waters reviewed by Sharma (2003), the dissolved oxygen ranged from 0.89 mg/l (Shivbari tank) to 12.6 mg/l (Harsolao pond). In Nal pond the free CO_2 was documented as 0 - 8 mg/l and in Gajner pond water it is nil throughout the year. In Nal pond water free CO_2 is found only two times in whole year. In most of the alkaline and hard desert waters the free CO_2 is not detected (Sharma, 2003). Sharma (2009) recorded almost same results in the other waters of this region.

As reflected from pH, water was alkaline. Total alkalinity of Nal pond ranged from 58 mg/l to 121 mg/l and of Gajner pond ranged from 53 mg/l to 122 mg/l. The hardness of these waters ranged from 85 mg/l to 157 mg/l and 58 mg/l to 116 mg/l. Sharma (2003) noted alkalinity to range from 20 to 285 mg/l and hardness from 30 to 300 mg/l in some desert waters of the same area. Kaur (2007) noted hardness of water ranged from 72 mg/l to 120 mg/l of Bikaner region. The two parameters presented almost similar trends. Values of alkalinity and hardness were comparable to those reported by Olsen and Sommerfeld (1977) in a desert reservoir in Arizona.

pH of Nal pond sediment ranged from 7.7 to 9.2 and of Gajner pond soil ranged from 7.3 to 8.8. It indicates that nature of soil was alkaline. Sharma (2009) indicates that at one occasion sediments of reservoir revealed 6.5 pH. On the rest of analysis during study period it ranged from 7.2 to 9.6. EC of Nal pond sediments ranged from 0.10 m mhos/cm to 0.25 m mhos/cm and of Gajner pond ranged from 0 to 0.20 m mhos/cm. However, in the same region considerably low EC (0.07 - 0.28 m mho/cm) were also recorded by Bugalia (1990) and Gehlot (1996). TDS of Nal pond sediments ranged from 64 mg/l to 160 mg/l while TDS of Gajner pond sediments ranged from 0 to 128 mg/l. Organic matter of Nal pond sediment ranged from 0 to 23.27 mg/g and in Gajner pond ranged from 0 to 26.92 mg/g. Bugalia, (1990) recorded organic matter ranged from 12.85 mg/g to 14.83 mg/g in Bikaner region. Sharma (2009) recorded organic matter values varied between 11.5 mg/g and 72.1 mg/g in this region.

Invertebrate faunal studies were carried out on two desert village ponds for an annual cycle revealed 41 animal species belonging to 7 phyla namely Protozoa, Porifera, Coelenterata, Aschelminthes, Annelida, Arthropoda and Mollusca (Table 20.3). 31 animal species recorded in Nal village pond while in Gajner pond 34 species of invertebrates were recorded. In Gajner village pond presence of high invertebrate diversity recorded as compared to Nal pond. Arthropoda is the dominant phylum with presence of 17 animal species in Nal pond while 16 animal species in Gajner pond. These ponds are situated in a hostile hot desert region of Bikaner, which is exposed to extremities of environmental conditions such as extreme fluctuations of temperature, greater alkalinity and salinity, low oxygen and high concentration of electrolytes due to high evapotranspiration during summer. It has been observed that Nal pond face two month of dry period but invertebrate biodiversity were not much affected by this dry period and the species are adapted for the given conditions and they even withstand the dry periods. Richness of invertebrate diversity and population is observed during monsoon season (July – October).

Table 20.1: Physico-chemical Variables of Water and Sediments at Nal village Pond, Bikaner during September 2010 to August 2011

Variables		Sep.	Oct.	Nov.	Dec.	Jan.	Feb.	Mar.	Apr.	May	Jun.	July	Aug.
											Months		
Water	Air temp.(°C)	29.9	26	20.6	9.8	13.3	15.3	19.8	25.2	–	–	29.3	30.3
	Water temp.(°C)	29.5	28.5	19.7	14.5	12.5	16	18.9	19.7	–	–	29.3	30.7
	Turbidity (JTU)	NIL	NIL	NIL	NIL	NIL	100	120	140			70	20
	pH	7.6	8.2	8.7	8.6	8.5	8.4	8.5	8.1	DRY		7.5	8.2
	EC (m mho/cm)	0.10	0.10	0.15	0.10	0.15	0.10	0.15	0.25	PERIOD		0.10	0.10
	TDS	64	64	96	64	96	64	96	160	–	–	64	64
	DO	10.75	9.33	8.92	9.53	11.97	3.65	9.74	7.91	–	–	8.72	9.74
	Free CO_2	8	NIL	NIL	NIL	NIL	NIL	NIL	6	–	–	NIL	NIL
	Hardness	101	107	89	125	116	138	155	157	–	–	109	85
	Total Alkalinity	94	62	58	69	67	79	106	121	–	–	117	95
Sediment	pH	8.1	7.7	8.1	8.8	8.8	8.8	8.7	7.7	–	–	8	9.2
	EC (m mho/cm)	0.10	0.10	0.10	0.15	0.10	0.10	0.25	0.25	–	–	0.10	0.10
	TDS	64	64	64	96	64	64	160	160	–	–	64	64
	Organic matter	NIL	NIL	NIL	1.90	3.87	23.27	1.90	1.90	–	–	NIL	NIL

Values are averages of three study stations and are expressed in mg/l in water, mg/g in sediment, except otherwise mentioned.

Table 20.2: Physico-chemical Variables of Water and Sediments at Gajner Village Pond, Bikaner during September 2010 to August 2011

Variables		Sep.	Oct.	Nov.	Dec.	Jan.	Feb.	Mar.	Apr.	May	Jun.	July	Aug.
Water	Air temp.(°C)	33.5	33.3	26.2	20.6	19.6	24.7	25.6	32.3	34.5	35.5	34.4	32.5
	Water temp. (°C)	29.9	29.5	22.2	18.4	15.4	19.9	22.3	27.5	29.5	29.6	31.1	31
	Turbidity (JTU)	100	100	40	40	40	70	100	120	120	140	140	120
	pH	7.4	7.5	7.6	8.5	8.1	8.7	9	9.5	9	8.5	8	7.9
	EC (m mho/cm)	0.10	0.10	0.10	0.10	0.25	0.20	0.15	0.10	0.10	0.10	0.20	0.10
	TDS	64	64	64	64	160	128	96	64	64	64	128	64
	DO	11.16	11.35	11.97	13.6	12.58	9.33	15.02	18.06	13.19	8.32	8.72	8.32
	Free CO_2	NIL	NIL	NIL	NIL	NIL	NIL	NIL	NIL	NIL	NIL	NIL	NIL
	Hardness	104	97	87	94	93	116	87	58	77	70	81	98
	Total Alkalinity	63	60	54	53	61	71	65	78	98	111	122	97
Sediment	pH	7.9	7.7	7.9	8.4	8	8.2	7.5	7.9	7.4	7.4	7.3	8.8
	EC (m mho/cm)	0.10	NIL	0.10	0.15	0.10	0.20	0.15	0.15	0.15	0.10	0.10	0.10
	TDS	64	NIL	64	96	64	128	96	96	96	64	64	64
	Organic matter	NIL	2.58	NIL	2.58	26.92	15.51	7.76	3.88	1.29	5.17	NIL	NIL

Values are averages of three study stations and are expressed in mg/l in water, mg/g in sediment, except otherwise mentioned.

**Table 20.3: Invertebrate Faunal Diversity in Two Water Bodies
(Nal Pond and Gajner Pond) in the Indian Desert.**

Phylum	Class	Genus and Species	Nal Pond	Gajner Pond
Protozoa	Mastigophora	*Euglena sociabilis*	+	+
	Ciliata	*Glaucoma pyriformis*	+	+
		Paramecium caudatum	+	+
		Coleps hirtus	–	+
		Stentor sp.	+	+
		Vorticella sp.	–	+
Porifera	Demospongiae	*Eunapius carteri*	+	–
Coelenterata	Hydrozoa	*Hydra vulgaris*	+	–
Aschelminthes	Rotifera	*Trichocerca longiseta*	+	+
		Keratella quadrate	+	+
		Rotaria sp.	–	+
		Brachionus havanaensis	+	+
		Filinia longiseta	–	+
Annelida	Oligochaeta	*Tubifex tubifex*	+	+
	Hirudinea	*Hemiclapsis marginata marginata*	–	+
		Hirudinaria granulosa	+	–
Arthropoda	Crustacea	*Stenocypris malcomsoni*	+	+
		Moina sp.	+	–
		Eocyzicus politus	+	–
	Insecta	*Odonata larva*	+	+
		Dragon fly larva	+	+
		Dragon fly nymph	+	+
		Mayfly larva	–	+
		Dipteran larva	–	+
		Agabus sp.	+	+
		Chironomus sp.	+	+
		Laccotrephes maculates	+	+
		Lithocerus indicum	+	–
		Corixa lima	+	+
		Crixa nymph	+	+
		Berosus sp.	+	+
		Berosus larva	+	+
		Scirtes nigropunctatus	+	+
		Cybester limbatus	+	+
		Laccophilus sp.	+	+

Contd...

Table 20.3–*Contd...*

Phylum	Class	Genus and Species	Nal Pond	Gajner Pond
Mollusca	Gastropoda	*Gyraulus rotula*	–	+
		Indoplanorbis exustus	+	–
		Gabbia orcula	+	+
		Digoniostoma pulchella	+	+
		Bellamya bengalensis	–	+
		Lymnaea acuminate	–	+

Although most of the species are also documented by other workers in the area (Bugalia, 1990; Chadha, 1999, Srivastava, 2009; Saxena, 2011), yet *Hydra vulgaris* is not found in surrounding waters in the area. Protozoan, particularly *Euglena sociabilis* found throughout whole year in both ponds. Sponge *Eunapius carteri* was recorded only in Nal pond. No sponge was recorded in Gajner pond throughout study period. Two species of class Hirudinea were found in waters of these regions. *Hirudinaria granulosa* also known as common cattle leech was recorded in Nal pond, while in Gajner pond *Hemiclapsis marginata marginata* was recorded during study period. Saxena (1996 a, b) recorded the sponges and leech fauna of Thar Desert.

References

APHA, AWWA , WPCF, 1981. *Standard Methods for the Examination of Water and Wastewater*, 15th edn, APHA, Washington DC.

Bugalia, S., 1990. Environmental monitoring of waters in and around Bikaner with special reference to macrobenthic fauna. *M. Phill. Dissertation*, Dungar college, Bikaner (India), pp. 73.

Chadha, Monika 1999. A comparative study on the quality and trophic status of some desert waters employing bioindicators and indices. *Ph.D. Thesis*, M.D.S. University, Ajmer.

Cole, G.A. and Whiteside, M.C., 1965b. Kiatuth-lanna: A limnological appraisal. I. Physical factors. *Plateau*, 38: 6–16.

Edmondson, W.T., 1965. *Freshwater Biology*. John Wiley and Sons, Inc., New York.

Gehlot, Neelam., 1996. Mollusc fauna and its ecology in some waters around Bikaner (NW Rajasthan). *M.Phil. Dissertation*, M.D.S. Univ., Ajmer, pp. 48.

Golterman, H.L. Clymo, R.S. and Ohnstand, M.A.M., 1978. *Methods for Physical and Chemical Analysis of Freshwaters*, 2nd edn. IBP Handbook No. 8, Blackwell Scientific Publications, London, pp. 213.

Kaur, H., 2007. A study on the ageing phenomenon in a reservoir created in the arid region of Rajasthan, with special reference to eutrophication. *Ph.D. Thesis*, Bikaner University, Bikaner.

Kaur, H., 1991. Environmental monitoring of Kodemdesar village pond, Bikaner. *M. Phil. Dissertation*, Bikaner University, Bikaner (India), pp. 70.

Needham, J.G. and Needham, P.R. 1941. *A Guide to Study of Freshwater Biology*. Comstock Publishing, Inc, New York, pp. 88.

Olsen, R.D. and Sommerfeld, M.R., 1977. The physico-chemical limnology of a desert reservoir. *Hydrobiologia*, 53(2): 117–129.

Peter H. Gleick, Heather, Cooly and Mari Morikawa, 2009. *The World's Water 2008–2009: The Biennial Report on Freshwater Resources.* Island press, Washington.

Poonam Lata, 2009. Comparative study of phytoplankton community of desert waters around Bikaner. *M.Phill. Dissertation,* Dungar college, Bikaner, Rajasthan.

Saxena, M.M., 2001. *Environmental Analysis: Water, Soil and Air.* Agrobotanical Publishers, India, pp. 176.

Saxena, M.M., 1996a. Freshwater sponges in the Thar Desert. In: *Faunal Diversity in the Thar Desert: Gaps in Research,* (Eds.) A.K. Ghosh, Q.H. Baqri and I. Prakash. Scientific Publ., Jodhpur, p. 37–41.

Saxena, M.M., 1996b. Leech fauna of the Thar Desert. In: *Faunal Diversity in the Thar Desert: Gaps in Research,* (Eds.) A.K. Ghosh, Q.H. Baqri and I. Prakash. Scientific Publ., Jodhpur, p. 77–80.

Saxena, S. 2011. Biodiversity and ecology of some eutrophic temple tanks in the Indian desert with special reference to bioindicators. *Ph.D. Thesis,* M.D.S. Univ., Ajmer, pp. 200.

Sharma, Ankush, 2009. Comparative physical chemical limnology of some water sheets in the desert region around Bikaner. *M.Phill. Dissertation,*. Dungar College, Bikaner (Raj.), pp. 74.

Sharma, Virendra, 2003. Physico-chemical limnology of desert waters around Bikaner: A review. *M.Phil. Dissertation,* Dungar College, Bikaner, pp. 94.

Shiklomanov, I.A., 1993: World freshwater resources. In: *Water Crisis: A Guide to the World's Freshwater Resources,* (Ed.) P.H. Gleick. Oxford University Press, New York.

Srivastava, Deepti, 2009. Faunal diversity and its ecology in some village pond ecosystems, with special reference to insect fauna, in the Indian desert. *Ph.D. Thesis,* M.G.S. Univ., Bikaner.

Strickland, J.D.H. and Parsons, T.R., 1972. *A Practical Handbook of Seawater Analysis.* Fish. Res. Bd. Canada, Ottawa, pp. 310.

Subba. Rao, N.V., 1989. *Handbook of Freshwater Molluscs of India.* ZSI, Calcutta, pp. 289.

Tonapi, G.T., 1980. *Freshwater Animals of India: An Ecological Approach.* Oxford and IBH Publishing Co., New Delhi, pp. 341.

Chapter 21

Study of Urban Sewage Water of Bindusara River Area Vicinity Beed, Maharashtra

☆ *F.I. Shaikh, Quazi Saleem and Seema Hashmi*

ABSTRACT

Bindusara is a small river situated in District Beed in Maharashtra state of India. The water of the river here has been used for the Irrigation and also for the drinking. Water pollution is due to an increase in population industries, agricultural output and concern for human welfare along with a desire to enhance the standard of living. The work is carried on the monthly basis. The study of physico-chemical properties of river water. In the present work total dissolved solids and sulphates increased. The rest parameters are in the range.

Keywords: *Bindusara river sewage water, Physico-chemical properties.*

Introduction

Bindusara is a rapid and river originates in the hill of Balaghat it is a hilly area. Various streams contributes to the River. The city of Beed is situated on the banks of Bindusara River. The river water is used for domestic, industrial and agricultural purposes. Now a days River Water sources are observed to be affected of water pollution and over exploitation. It represent as serious threat. Therefore it is essential to study the river water quality many workers such as Rao *et al.* (1984). The present work is undertaken to study the physico-chemical characteristic of sewage water Bindusara River area vicinity Beed.

Material and Methods

The water samples were collected during the investigation period January 2011 to December 2011 and analyzed the chemical properties such as pH, dissolved oxygen, total alkaline, chlorides and sulphates in laboratory using the method from standard method for examination of water and wastewater APHA (2012).

Result and Discussion

The monthly values of temperature, pH, electrical conductivity, turbidity and that dissolved solids (TDS) are represented in Table 21.1. It was observed in the investigation that the temperature is higher at Bindusara river area vicinity of Beed.

Table 21.1: Monthly Variations of Physico-chemical Parameters of Bindusara River Area Vicinity Beed

Parameters	Jan	Feb	Mar	Apr	May	Jun	Jul	Aug	Sept	Oct	Nov	Dec
Temp. (°C)	25	30	34	34.5	36	34	33	31	30	29.5	30	27
pH	8	8.2	8	7.9	8.3	8.2	8.2	8.2	8.2	7.9	8.0	8.2
TDS (mg/l)	315	340	390	401	450	410	395	380	350	390	348	330
Conductivity	190	195	202	208	240	232	215	220	228	230	219	220
Turbidity (JTU)	50	45	42	35	40	58	60	45	48	42	45	47
Alkalinity (mg/l)	125	126	128	131	135	140	141	140	138	130	128	127
DO (mg/l)	7.0	7.1	7.2	7.1	6.8	7.0	7.2	7.2	7.5	7.6	7.5	7.4
Chlorides (mg/l)	155	160	170	178	180	100	95	90	185	160	162	158
Sulphates (mg/l)	120	130	132	109	136	130	135	112	130	140	128	125
CO_2 (mg/l)	6.5	4.8	5.5	6.2	4.8	3.8	3.9	4.0	4.8	5.5	5.5	4.9

In water at high temperature solubility of oxygen and other gases decreases and water becomes tasteless temperature. It has effect on metabolic activity of aquatic biota. The maximum temperature was recorded in may. The results of present studies agreement with earlier studies of Kumar (1984) and Ramesh (1989). pH is the measurement of the free acidity or alkalinity of a water solution hence it is an important factor. For water analysis. However, pH measurement in high purity water can be extremely difficult in the present investigation. The pH maximum was recorded in summer and minimum in winter with slight increase in monsoon months similar trends has been reported by Sreenivasan (1965) studied limnology of tropical impoundments.

The lowest values of turbidity were found in post summer seasons high value in found in monsoon season. Ajmal and Raziuddin (1988) of Hiddon river and Kali nadi also have reported similar variations. It was observed that due to an increase in the turbidity the rate of photosynthesis decreases leading to decrease in the growth of phytoplankton. The latter in turn, decrease zooplankton growth the values of turbidity were found maximum at vicinity of Beed city due to the surface run of the sewage water with large amount of suspended solids etc.

In the present investigation total dissolved solids (TDS) and conductivity was found to be maximum in summer and minimum in winter season. TDS and conductivity was due to factor such as rainfalls, biota-causing changes in ionic concentration and the nature of bottom deposit. Goltermann (1975) showed inverse relationship between TDS and conductivity. The observed increase could be attributed due to the entry of sewage water. similar trends was observed by Chandrashekhar and Babu (2003) Bellandur lake, Bangalore as case study.

The total alkalinity water is said to be alkaline when the concentration of hydroxyl ion exceeds that of hydrogen ions. Chemically pure water is neutral having equal amount of hydrogen and hydroxyl ions. Oxygen is one the most important factor in an aquatic ecosystem almost all plant and animals need oxygen for respiration. Self purification of water through bio-geochemical cycling of organic matter depends on the presence of sufficient amount of oxygen dissolved in it dissolved oxygen (DO) content of Binduasara River area vicinity found maximum in monsoon while decreased in summer season. Sreenivasan (1972) and Bahura (1998) studied the relationship of D. O. with temperature, D. O. has increase relationship not only with temperature but also with free CO_2.

Water being a good solvent various salts get dissolved in natural water. In the present study the maximum chlorides recorded in summer could be due to evaporation of river water with concentration of dissolved salts and ions earlier water studies were conducted by Lohar and Patel (1998) on Manar and Amer rivers in Maharashtra.

Sulphates in the river water samples varied in the different season. The highest values was recorded during winter season. The sulphates are used as a source of oxygen by bacteria under an anaerobic condition. CO_2 is vital in the life of plant and micro organism it is produced as a result of respiration of aquatic organisms. As CO_2 is highly soluble in water it is found to be in larger amount in polluted water compared to freshwater bodies CO_2 has a great effect again on fresh growth.

References

Ajmal, M. Raziuddin and Khan, A.U., 1985. Physico-chemical aspects of pollution in kali nadi Iawde. *Tech. Ann.*, 12: 100–114.

APHA, 2012. *Standard Methods for the Examination of Water and Wastewater*, 22nd Edn. American Public Health Association 8001 street NW Washington, DC., p. 2001–3710.

Bahura, C.K., 1998. A study of physio-chemical characteristic of a highly autorophic temple tank, Bikaner. *J. Aqua. Biol.*, 13: 47–51.

Chandrashekhar, J.S. and Lenin Babu, K., 2003. Impact of urbanization on Bellandur Lake Bangalore as case study. *J. Environ. Biology*, 24(3): 223–227.

Golterman, H.L., 1975. *River Ecology*, (Eds.) B.A. Whitton. Blackwell Scientific Publication, Oxford, London, Edinburg Melbourne, pp. 39–80.

Kumar, S.R., 1984. Studies on the distribution of plankton in water of Manglore. *M.F.Sc.*, University of Agricultural Sciences, Bangalore, p. 231.

Pande, R.K. and Mishra, A., 2002. Impact of paper and pulp industry effluent on the water quality of river hindon *J. Ecophysio. Occupt.* 2(3 and 4): 173–184.

Ramesh, A.M., 1989. Distribution of copepod and copepodies in relation to Phytoplankton and hydropgraphy of the coastal water of manglore. *M.F.Sc.*, University of Agricultural Sciences, Bangalore, p. 193.

Rao, V.N.R and Valsaraj, C.P., 1984. Hydrological studies in the inshore waters of the bay of Bengal. *J. Mar Biol. Ass. India*, 26: 58–65.

Sreenivasan, A., 1965. Limnology of tropical impoundments III limnology and productivity of Amarwathy reservoir (Madras State) India. *Hydrobiologia*, 26: 501–516.

Chapter 22

Ichthyofaunal Diversity of Siddeshwar Reservoir, Hingoli District, Maharashtra

☆ *S.D. Niture and S.P. Chavan*

ABSTRACT

The present study deals with the ichthyofaunal diversity of Siddheshwar reservoir, Hingoli district Maharashtra for a period of 2 years, during 2008 and 2009. The Siddheshwar reservoir is a medium sized reservoir of about 2574 ha area, The reservoir is having catchment area of 7770 sq km. During the present investigation, 41 species of fishes belonging to 8 orders and 15 families have been obtained from the Siddheshwar reservoir.

Keywords: Ichthyofaunal, Biodiversity, River Purna, Siddheshwar reservoir.

Introduction

The Siddheshwar reservoir is a medium sized reservoir of about 2574 ha area, constructed on Purna River at Rupur camp Tq. Aundha Nagnath, Dist, Hingoli and near village Siddheshwar Tq, Aundha Nagnath, Dist, Hingoli in 1968. The reservoir lies in between north latitude 19°-0'-20" and East longitude 76°-45'-00". The reservoir is naturally situated in hilly region on both sides. Reservoir area is included in the survey of India toposheet map no 56A/10. It was the first ever-major region project in Marathwada region to initiate the process of economics development of Marathwada. Reservoirs and lakes contribute the inland fishery resource in term of production potential. Indian reservoirs has a rich variety of fish species, which supports to the commercial fisheries. The present work was mainly undertaken to investigate the fish diversity from Siddeshwar reservoir and it is first effort in this direction from this reservoir.

Material and Methods

For this new report we got the collection of these species from various fishing stations, village markets, collection from fisher folks and tribes living around the peripheral region of the reservoir. For the identification of the fish species, Day volumes and Taxonomic records from Talwar were used. Johal and Jha (2007) suggested that periodic fish faunal surveys must be undertaken, so that the loss or gain of fish diversity can be evaluated. During present investigation Ichthyofaunal diversity, their economic importance and abundance was studied.

Results

During the present investigation, 41 species of fishes belonging to 8 orders and 15 families have been obtained from the Siddheshwar reservoir. Earlier Sakhare (2002) reported the occurrence of 29 fish species from Yeldari reservoir which was constructed on same river in 1968, however we found the occurrence of 13 different fish species as new report as not identified in earlier study from this reservoir. The species like *Labeo calbasu, Labeo boggut, Cirrhina reba, Rohtee cotio, Rohtee vigorsi, Chela phulo, Channa punctatus, Amblypharyngodon mola, Belone cancila, Mastacembelus aculeatus, Nemacheilus botia, Rasbora daniconius* and *Discognathus modestus* were newly reported during this investigation.

Table 22.1: Check List of Ichthyofauna.

Phylum: Chordata

Sub-phylum: Vertebrata

Class: Osteichthyes

Sub-class: Teleostei

Order: Cypriniformes

Family: Cyprinidae

1)	*Labeo rohita* (Gunt)	2)	*Labeo calbasu* (Gunt)
3)	*Labeo boggut* (Cuv)	4)	*Cirrhina reba* (Cuv-Val)
5)	*Cirrhina mrigal* (Cuv-val)	6)	*Cyprinus carpio communis* (Linn)
7)	*Catla-catla* (Cuv-val)	8)	*Rohtee cotio* (Sykes)
9)	*Rohtee vigorsi* (Sykes)	10)	*Rohtee ogilbi* (Sykes)
11)	*Chela phulo* (Ham-Buchanan)	12)	*Chela bacila* (Ham.-Buchanan)
13)	*Puntius sarana* (Ham)	14)	*Puntius ticto* (Ham)
15)	*Puntius hexacticus kolus* (Sykes)	16)	*Puntius sophore* (Ham)
17)	*Ctenopharyngodon idella*	18)	*Hypophthalmichthys molitrix* (Val)

Family: Poeciidae

19)	*Gambusia affinis* (Bird and Gira)		

Order: Channiforms

Family: Channidai

20)	*Channa morulius* (Ham)	21)	*Channa straitus* (Ham)
22)	*Channa punctatus* (Ham)	23)	*Channa gachua* (Ham)

Family: Cobitida

24)	*Amblypharyngodon mola* (Ham-Buch)		

Contd...

Table 22.1–*Contd...*

Order: Siluriformes

Family: Siluridae

  25) *Wallago attu* (Schne)   26) *Ompok bimaculatus* (Bloch)

Family: Bagridae

  27) *Mystus seenghala* (Sykes)   28) *Mystus cavassius* (Ham)

Family: Heteropneustidae

  29) *Heteropneustes fossilis* (Bloch)

Family: Clairidae

  30) *Clarius batrachus* (Cuv-val)

Family: Scombersocidae

  31) *Belone concila* (Cuv-val)

Order: Clupeiformes

Family: Notopteridae

  32) *Notopterus kapirat* (Lacep)   33) *Notopterus chitala* (Gunt)

Order: Perciformes

Family: Percidae

  34) *Ambasis nama* (Cuv-val)   35) *Ambassis ranga* (Cuv-val)

Order: Mastacembeliformes

Family: Mastacembelidae

  36) *Mastacembelus armatus* (Cuv-val)   37) *Mastacembelus aculeatus* (Cuv-val)

Order: Cypriformes

Family: Cobitidae

  38) *Nemacheilus botia* (Gunt)   39) *Rasbora daniconius* (Bleek)

Order: Discognathiformes

Family: Discognathidae

  40) *Discognathus modestus* (Hackel)

Order: Gobiformes

Family: Gobiidae

  41) *Gobius guiris* (Ham-Buch)

Discussion

Kumar (1990) reported 51 ichthyofaunal species of 09 families from Govindsagar reservoir of Himachal Pradesh, out of which 12 fish species are commercially important as in this reservoir we got commercially important 14 fish species.

Devi (1997) studied the ichthyofauna of Ibrahimbagh and Shatamrai reservoirs of Hyderabad and reported order cypriniformes fishes dominated and followed by order siluriformis, channiformis and perciformis. Jain (1998) reported of 53 fish fauna and was grouped into seven categories in Rajasthan state. Sukumaran and Rahaman (1998) stated that majority of reservoirs of Karnataka state

Table 22.2: Economic Importance of Ichthyofauna of Siddheshwar Reservoir

Sl.No.	Name of Fishes	Abun-dance	Commercial Food	Fine Food	Corse food	Aquarium Fish	Other
1.	*Ambassis nama* (Cuv-val)	−	−	−	−	−	LV
2.	*Ambassis ranga* (Cuv-val)	−	−	−	−	−	LV
3.	*Amblypharyngodon mola* (Sykes)	++	−	−	√	√	LV
4.	*Belone concila* (Cuv)	−	−	−	√	−	MD
5.	*Catla-catla* (Cuv-val)	+++	√	√	−	−	−
6.	*Channa punctatus* (Ham)	++	√	√	−	−	MD
7.	*Channa gachua* (Bloch)	+	√	√	−	−	MD
8.	*Channa morulius* (Bloch)	+	√	√	−	−	MD
9.	*Channa straitus* (Bloch)	++	√	√	−	−	MD
10.	*Chela phulo* (Ham-Buch)	−	−	−	√	−	−
11.	*Chela bacila* (Ham-Buch)	−	−	−	√	−	−
12.	*Cirrhina mrigal* (Cuv-val)	+++	√	√	−	−	−
13.	*Cirrhina reba* (Cuv-val)	+	√	√	−	−	−
14.	*Clarius batrachus* (Cuv-val)	++	√	√	−	−	−
15.	*Ctenopharyngedon idella* (Val)	+++	√	√	−	−	−
16.	*Cyprinus carpio communis* (Linn)	+++	√	√	−	−	−
17.	*Discognathus modestus* (Hackel)	−	−	−	√	−	−
18.	*Gambusia affinis* (Bird and Gira)	−	−	−	√	−	LV
19.	*Gobius guiris* (Ham-Buch)		−	−	−	−	−
20.	*Heteropneustes fossilis* (Bleck)	++	√	√	−	−	MD
21.	*Hypophthalmichthys molitrix* (Val)	++	√	√	−	−	−
22.	*Labeo boggut* (Cuv)	+		√	−	−	−
23.	*Labeo calbasu* (Gunt)	+	√	√	−	−	−
24.	*Labeo rohita* (Gunt)	+++	√	√	−	−	−
25.	*Mastacembelus aculeatus*(Cuv-val)	+	−	√	−	−	MD
26.	*Mastacembelus armatus* (Cuv-val)	++	−	√	−	−	MD
27.	*Mystus cavassius* (Ham)	+	−	√	−	−	−
28.	*Mystus seenghala* (Ham)	++	−	√	−	−	−
29.	*Nemacheilus botia* (Gunt)	++	−	−	√	√	−
30.	*Notopterus chitala* (Gunt)	+	−	−	√	−	MD
31.	*Notopterus kapirat* (Gunt)	++	−	−	√	−	MD
32.	*Ompok bimaculatus* (Bloch)	−	−	−	√	−	MD
33.	*Puntius hexacticus* (Sykes)	−	−	−	√	√	BT, LVMD
34.	*Puntius sarana* (Ham)	+	−	−	√	√	BT, LV
35.	*Puntius sophore* (Ham)	+	−	−	√	−	BT
36.	*Puntius ticto* (Ham)	++	− −	−	√	√	BT, LV
37.	*Rasbora daniconius* (Bleek)	+	−	−	√	− −	−
38.	*Rohtee cotio* (Sykes)	−	−	−	√	−	−
39.	*Rohtee vigorsi* (Schne)	−	−	−	√	−	−
40.	*Rohtee ogilbii* (Schne)	−	−	−	√	−	MD
41.	*Wallago attu* (Schne)	++	−	√	−	−	−

+++: Most abundant; ++: Abundant; +: Less abundant; −: Rare.

LV: Larvivorus fish; BT: Bait; MD: Medicinal value.

Figure 22.1: View of Sluice Gate of Siddheshwar Reservoir.

Figure 22.2: Collection of Fish Sample by Netting in Siddheshwar Reservoir.

Figure 22.3: Collection of Fish Sample after Netting from Siddheshwar Reservoir.

have a large population of predatory fish. A total of 27 species belonging to six families have been encountered in pong reservoir (Singh, 2001).

Sakhare and Joshi (2003) reported the icthyofauna of Bori reservoir in Maharashtra, total 20 species of fishes belonging to 14 genera falling under 4 orders (cypriniformes, perciformis, siluriformis and osteoglossiformis). Cypriniformes order dominated with seven species with Puntius was abundant. Suresh (2003) reported 54 fish species in Loktak Lake, Manipur and 15 species were commercially important. Mahapatra (2003) recorded abundance of cat fishes in Hirakund reservoir and reported 43 species, out of which 18 were economically important. Sakhare and Joshi (2003) reported 34 species of fishes in reservoirs of Parbhani District of Maharashtra.

Acknowledgements

We are thankful to Dr. V. B. Garad, Head, Department of Zoology and Fishery Science, D. S. M. College Parbhani for providing Laboratory facilities.

References

Ahirao, S.D. and Mane, A.S., 2000. The diversity of ichthyofauna, taxonomy and fisheries from freshwater of Parbhani district, Maharashtra State. *J. Aqua. Biol.*, 15(1 and 2): 40–43.

Biswas, K.P., 1990. *A Textbook of Fish, Fisheries and Technology*. Narendra Publishing House, Delhi, p. 143–164.

Day, Francis, 1889. *The Fauna of British India including Ceylon and Burma Fishes*.

Devi, B.S., 1997. Present status Potentialities, management and economics of fisheries of two minor reservoirs of Hyderabad.

Dhimdhime, S.D., 2004. Hydrobiology of Siddeshwar dam. *Ph.D. Thesis*, Swami Ramanand Teerth Marathwada University, Nanded.

Gopinath, P. and Jayakrishnan, T.N., 1984. A study on the pisicifauna of the Idukki Reservoir and catchment area. *Fish. Technol.*, pp. 131–136.

Jain, A.K., 1998. Fisheries resource management in Rajasthan: An overview of present status and future scope. *Fishing Chimes*, 17(11) 9–15.

Jayaram, K.C., 1981. *The Freshwater Fishes of India, Pakistan, Bangladesh, Burma and Sri Lanka: A Handbook*. ZSI, Calcutta, India.

Jayaram, K.C., 2002. *Fundamental of Fish Taxonomy*. Narendra Publication, Delhi.

Jhingran, A.G., 1980. Riverine fishery resources of India and their socio-cultural impact. *Tropical Ecology and Development*, p. 747–756.

Jhingran, Arun G., 1988. Reservoir fisheries in India. *J. Indian Fisheries Assoc.*, 18: 261–273.

Jhingran, V.G., 1985. *Fish and Fisheries of India*. Hindustan Publishing Corporation, India, New Delhi. p. 106, 171–191.

Johal, M.S. and Jha, S.K., 2007. Fish diversity of Haryana state and its conservation status. *Fishing Chimes*, 27(1): 107–108.

Kumar, K., 1990. Management and development of Govindsagar reservoir: A case study. *Proc. Nat. Workshop Reservoir Fish*, p. 13–20.

Pandey, K.C., 1998. *Concepts of Indian Fisheries*. Shree Publishing House, New Delhi.

Parihar, R. P., 1999. *A Textbook of Fish Biology and Indian Fisheries*. Central Publishing House, Allahabad. p. 310–313.

Sakhare, V.B. and Joshi, P.K., 2002. Ecology and ichthyofauna of Bori reservoir in Maharashtra. *Fishing Chimes*, 22(4): 40–41.

Sakhare, V.B. and Joshi, P.K., 2002. Ecology of Palas-Nilegaon reservoir in Osmanabad district, Maharashtra. *J. Aqua Bio.*, 18(2): 17–22.

Sakhare, V.B., 2002. Studies on some aspects of fisheries management of Yeldari reservoir. *Ph.D. Thesis*, Swami Ramanand Teerth Marathwada University, Nanded.

Sakhare, V.B. and Joshi, P.K., 2003. Reservoir fishery potential of Parbhani district of Maharashtra. *Fishing Chimes*, 23(5): 13–16.

Sakhare, V.B. and Joshi, P.K., 2003. Water quality of Hingni (Pangaon) Reservoir and its significance to fisheries ABN–008. *Nat. Conf. Resent Trends Aquat. Biol.*, p. 56.

Srinivas, 2007. Fisheries of Edulabad reservoir in A. P. *Fishing Chimes*, 26(10): 105–107.

Talwar, P.K. and Jhingran, A.G., 1991. *Inland Fishes of India and Adjacent Countries*. Oxford and IBH Publishers, New Delhi.

Chapter 23

Plankton Diversity in Highly Human Interfered Temple Pond

☆ *Poonam Lata, Ankush Sharma,*
Harbhajan Kaur and N.S. Rathore

ABSTRACT

In earlier days water bodies like lakes, ponds and streams, water was used only for drinking purpose but now a day's water bodies are being utilized for various purpose. Anthropogenic activities are disturbing the water quality due to which many water bodies are undergoing to accelerated aging or eutrophication. Certain species of plankton are very sensitive to water pollution and hence referred as biological indicators. The present research work was carried out in an ancient temple pond at Kodamdesar in Bikaner District from September 2010 to August 2011. During the study period 18 species of three algal groups namely chlorophyceae, cynophyceae and bacillariophyceae were recorded. In case of zooplankton 16 species related to three groups namely protozoa, rotifera and crustacea were recorded.

Keywords: Plankton diversity, Temple pond, Bikaner district.

Introduction

The word 'Plankton' is derived from the Greek word 'Planktons' which means 'drifter'. Plankton refers to microscopic aquatic plants (phytoplankton) or animals (zooplankton) having little or no resistance to water current and living free floating and suspended in open or 'pelagic waters'. These are the principal component of aquatic ecosystem. Plankton formulates the food chains and food webs in all aquatic ecosystems. Plankton communities are very sensitive first targeted by water pollution so any undesirable change in water bodies, affects the diversity and biomass of plankton. In aquatic ecosystems, plankton also play important role in the biogeochemical cycles of many chemical elements particularly in the carbon cycle.

Rajasthan is the largest state of India in area wise. Rajasthan has India's largest desert known as Great Indian Desert (Thar Desert). A scorching summer, chilly winter, dry monsoon and dust storms are the main characteristics of Rajasthan climate. The city Bikaner lies in the western region of the state. Bikaner witnesses for extreme conditions of temperature. In summer temperature rises up to 48°C and in winter comes down to 4°C to 5°C. During the months of May and June thunderstorms are regular in the afternoon. The region receives less than 400 mm of rain in a year. The soil type of the area is basically alkaline. In this region water bodies are present in the form of temple tanks, village pond, lakes and man-made reservoirs.

Study Area

Kodamdesar is situated 28°3' North, 73° 4' East in Bikaner region of Rajasthan. It is situated 24 kms away from Bikaner city on national highway leading to Jaisalmer. Temple of Bhainru Ji was installed by Rao Bikaji on the bank of pond. It is another place of worship and picnic. In the month of *Bhadrapad*, every year a fair is celebrated by people in this temple. The pond is utilized by village people for the purpose of washing and bathing. The benthic fauna are mainly chironomids, red tail maggot, beetle and bugs. This pond acts as habitat for various local and migratory birds.

Materials and Methods

Collection and Examination of Phytoplankton

For analysis of phytoplankton water samples were taken in wide mouth polythene bottles of 500 ml. To this Lugol's iodine solution and 4 per cent formaldehyde were added. The solutions provide them stain and preserve them. Samples of phytoplankton were brought to the laboratory and were allowed to undergo sedimentation. The supernatant from phytoplankton sample was removed after sufficient sedimentation and sediment was made to a fixed volume. Phytoplankton were identified and counted using Sedgwick-rafter slide under research microscope. They were identified using Needham and Needham (1978), Edmondson (1966), Chapman and Chapman (1975).

Collection and Examination of Zooplankton

Zooplankton samples were collected by filtering 50 l of water through a plankton net made of bolting silk (No. 25, 0.3 mm mesh). The samples so collected were transferred to the narrow mouthed bottles of 100 ml and preserved with 4 per cent formaldehyde. Zooplankton sample were shaken and known volume was transferred in zooplankton counting chamber. These were identified and counted under binocular microscope. The population were expressed in No./l. The identification of zooplanktons were made following Edmondson (1996), Michael (1973), Needham and Needham (1978) and Tonapi (1980).

Results and Discussion

Phytoplankton in the Kodamdesar pond have been noted and presented in the Table 23.1. Totally 18 species of phytoplankton belonging to three different algal groups namely Chlorophyceae, Cyanophyceae and Bacillariophyceae were noticed. In Chlorophyceae six species such as *Protococcus*, *Coelastrum*, *Closterium*, *Selenastrum*, *Scenedesmus* and *Netrium* were found. Chlorophyceae constitutes one of the major groups of algae occurring in aquatic ecosystems, the cells are typically green in colour due to the presence of chlorophyll *a* and *b*. Khatri (1980), Bahura (1989), Sharma and Saxena (1995),

Table 23.1: Phytoplankton Diversity of a Temple Pond

Phytoplankton	Months											
	Sep.	Oct.	Nov.	Dec.	Jan.	Feb.	Mar.	Apr.	May	Jun.	Jul.	Aug.
Chlorophyceae												
Protococcus	+	+	+	+	+	+	+	+	+	+	+	+
Coelastrum	+	+	+	+	+	+	+	+	+	+	+	+
Closterium	+	+	+	+	+	+	+	+	+	+	+	+
Selenastrum	+	+	+	+	+	+	+	+	+	+	+	+
Scenedesmus	+	+	+	+	+	+	+	+	+	+	+	+
Netrium	+	+	+	+	+	+	+	+	+	+	+	+
Cynophyceae												
Synechocystis	+	+	+	+	+	+	+	+	+	+	+	+
Coelosphaerium	+	+	+	+	+	+	+	+	+	+	+	+
Anabaena	−	−	−	−	−	−	+	+	+	+	+	+
Aphanocapsa	+	+	+	+	+	+	+	+	+	+	+	+
Bacillariophyceae												
Navicula	+	+	+	+	+	+	+	+	+	+	+	+
Nitzschia	+	+	+	+	+	+	+	+	+	+	+	+
Diatoma	+	+	+	+	+	+	+	+	+	+	+	+
Melosira	+	+	+	+	+	+	+	+	+	+	+	+
Coscinodiscus	+	+	+	+	+	+	+	+	+	+	+	+
Stephanodiscus	+	+	+	+	+	+	+	+	+	+	+	+
Gyrosigma	+	+	+	+	+	+	+	+	+	+	+	+
Frustulia	−	−	+	+	+	+	+	+	+	+	+	+

Kaur (2007) observed that the chlorophyceae contributed the maximum number of species to the phytoplanktons. The algal group cyanophyceae comprises prokaryotic organisms popularly known as blue-green algae. They possess the oxygen evolving photosynthetic system. These are truly cosmopolitan organisms occurring in habitats of extreme conditions of light, pH and nutritional resources. In cyanophyceae four species namely *Synechocystis, Coelosphaerium, Anabaena* and *Aphanocapsa* were observed. *Anabaena* was recorded only in March to July. Eutrophic waters are known to be characterized by cyanophyceae (Wetzel, 1975). Bohra, 1976; Misra *et al.,* 1978; Saxena, 1982; Arora, 1994 noted a general dominance of blue-greens. Bacillariophyceae group are popularly known as diatoms. Diatomas are basically unicellular, in some cases become pseudo filamentous or aggregated into colonies and cell wall of diatoms is impregnated with silica. Eight species of bacillariophyceae namely *Navicula, Nitzschia, Diatoma, Melosira, Coscinodiscus, Stephanodiscus, Gyrosigma,* and *Frustulia* were noticed. *Frustulia* was not reported in September and October months. Chadha (1999) recorded the dominance of diatoms in respect of diversity and density. Bahura, 1989; Saigal, 1998; are also recorded diatoms in the Bikaner region.

In an aquatic ecosystem Zooplankton form an important link in the food chain from primary to tertiary level leading to the production of fishery. According to Gajbhiye (2002) due to their large density and different tolerance to stress, these organisms are used as the indicator for the various physical, chemical and biological processes in the aquatic ecosystem. Organisms of three classes namely Protozoa, Rotifera and Crustacea were recorded (Table 23.2). In Protozoa four species namely *Vorticella* sp., *Stentor* sp., *Paramecium caudatum* and *Euglena sociabilis* were observed. Bahura (1989), Bugalia (1990) and Bahura, R. (1990) are also reported these species in this region. Totally six species of Rotifers namely *Philodina roseola, Roteria* sp., *Esplenchna* sp., *Brachionus calyciflorus, Brachionus havanaensis* and *Keratella quadrata* were found. Arora and Saxena (1997), Saxena (1998) and Kaur (2007) reviewed the faunal diversity in the waters of Indian desert. Crustacea was represented by the six species namely *Diaptomus glacialis, Nauplius larva, Bosmina longisita, Cyclop* sp., *Daphnia carineta*, and *Stenocypris* sp. *Bosmina longisita* was not observed in September, October and April months. Bugalia (1990), Kaur (2007), Srivastava (2009) were also observed these species.

Table 23.2: Zooplankton Diversity of a Temple Pond

Zooplankton	Months											
	Sep.	Oct.	Nov.	Dec.	Jan.	Feb.	Mar.	Apr.	May	Jun.	Jul.	Aug.
Protozoa												
Stentor sp.	+	+	+	+	+	+	+	+	+	+	+	+
Paramecium caudatum	+	+	+	+	+	+	+	+	+	+	+	+
Euglena sociabilis	+	+	+	+	+	+	+	+	+	+	+	+
Vorticella sp.	+	+	+	+	+	+	+	+	+	+	+	+
Rotifera												
Philodina roseola	+	+	+	+	+	+	+	+	+	+	+	+
Roteria sp.	+	+	+	+	+	+	+	+	+	+	+	+
Esplenchna sp.	+	+	+	+	+	+	+	+	+	+	+	+
Brachionus calyciflorus	+	+	+	+	+	+	+	+	+	+	+	+
Brachionus havanaensis	+	+	+	+	+	+	+	+	+	+	+	+
Keratella quadrata	+	+	+	+	+	+	+	+	+	+	+	+
Crustacea												
Diaptomus glacialis	+	+	+	+	+	+	+	+	+	+	+	+
Nauplius larva	+	+	+	+	+	+	+	+	+	+	+	+
Bosmina longisita	–	–	+	+	+	+	+	–	+	+	+	+
Daphnia carineta	+	+	+	+	+	+	+	+	+	+	+	+
Cyclop sp.	+	+	+	+	+	+	+	+	+	+	+	+
Stenocypris sp.	+	+	+	+	+	+	+	+	+	+	+	+

References

Arora, Meeta and Saxena, M. M., 1997. Planktonic fauna of a desert village pond in relation to certain abiotic factors. *Env. and Ecol.*, 15 (2): 367–369.

Bahura, C. K., 1989. Limnoligcal studies on Shivbari Temple Tank, Bikaner: Phytoplanktonic community and productivity. *M. Phil Dissertation*, Dungar College, Bikaner, pp. 64.

Bahura, R., 1990. A planktonological survey of some water bodies in and around Bikaner (Rajasthan). *M.Phil Dissertation*, Dungar College, Bikaner, pp. 51.

Bohra, O. P., 1976. Some aspects of limnology of Padamsagar and Ranisagar, Jodhpur. *Ph. D. Thesis*, University of Jodhpur, Jodhpur.

Bugalia, S., 1990. Environmental monitoring of waters in and around Bikaner with special reference to macrobenthic fauna. *M. Phil. Dissertation*, Dungar College, Bikaner (India), pp. 73.

Chadha, Monika 1999. A comparative study on the quality and trophic status of some desert waters employing bioindicators and Indices. *Ph.D. Thesis*, M.D.S. Univ., Ajmer, pp. 127.

Edmondson, W.T. (Ed.), 1996. *Freshwater Biology*, 2nd edn. John Wiley and Sons, Inc., New York, USA.

Gajbhiye, S.N., 2002. Zooplankton: Study, methods, importance and significant observations. *Proceedings of the National Seminar on Creeks, Estuaries and Mangroves: Pollution and conservation*, p. 21–27.

Kaur, H., 2007. A study on the ageing phenomenon in a reservoir created in the arid region of Rajasthan, with special reference to eutrophication. *Ph.D. Thesis*, Bikaner University.

Khatri, T.C., 1980. Limnological studies of Lakhotia lake, Pali (Rajasthan). *Ph.D. Thesis*, University of Jodhpur, Jodhpur (India), pp. 134.

Michael, R. G., 1973. *A Guide to the Study of Freshwater Organisms*. J. Madurai Univ., Suppl. I, 23–36.

Misra, S.D., Bhargava, S.C., Jhaker, G.R. and Dey, T., 1978. Hydrobiology and productivity of some freshwater reservoirs and lakes of semi-arid zone, Jodhpur (Rajasthan), U.G.C. Project report, Univ. of Jodhpur, Jodhpur, India.

Needham, J.G. and Needhan, P.R., 1978. *A Giude to the Study of Freshwater Biology*. Halden Day. Inc. Publ., San Francisco.

Saigal, Deepak, 1998. Senescence in a manmade water sheet in the India desert. *M. Phil Dissertation*, M. D. S. Univ., Ajmer.

Saxena, M.M., 1982. Limnological study of freshwater reservoirs: Sardarsamand. *Ph.D. Thesis*, University of Jodhpur, Jodhpur (India), pp. 122.

Saxena, M.M., 1998. A survey of Cyclops (Crustacea: Copepoda) popultion in some bodies of water in Rajasthan in relation to the disease dracotiasis. *Abst. Nat. Sem. Adv. Econ. Zool.*, pp. 93.

Sharma, L.K. and Saxena, M.M., 1995. The biotope and community of newly constructed reservoir in the Indian desert. *Acta Ecol.*, 17(2): 106–108.

Srivastava, Deepti, 2009. Faunal diversity and its ecology in some village pond ecosystems, with special reference to insect fauna, in the Indian desert. *Ph.D. Thesis*, M.G.S. Univ., Bikaner.

Tonapi, G.T., 1980. *Freshwater Animals of India*. Oxford and IBH Publ. Co., New Delhi, India.

Wetzel, R.G., 1975. *Limnology*. W. B. Saunders Co., Philadelphia, 743 pp.

Chapter 24

Plankton Diversity of some Freshwater Lakes of Jath Tahsil of Sangli District, Maharashtra

☆ *M.H. Karennawar and S.A. Khabade*

ABSTRACT

In present investigation during monsoon and post-monsoon season, the bacillariophyceae is dominantly present followed by chlorophyceae, cyanophyceae, protozoa and crustacea. The phytoplanktons studied indicate that 4 species of cyanophyceae, 4 species belonging to chlorophyceae, 7 species bacillariophyceae were reported in the investigation. The occurrence of zooplanktons was also observed in the study which indicated that 3 species of protozoa and 2 species of crustacea were reported in the investigation.

Keywords: Planktonic diversity, Freshwater lakes, Jath Tahsil.

Introduction

In India, Maharashtra is one of the agriculturally important state which receives rainfall in moderate amount. In Western Maharashtra, average rainfall is comparatively high. The Sangli district lies on the border of western and central Maharashtra. The western region of Sangli district receives heavy rainfall whereas eastern region receives comparatively low rainfall and region is drought prone region.

Jath tahsil of Sangli district is one of the drought prone tahsil, which receives average rainfall of about 449.5 mm. It is considered that the rainfall is below average rainfall. Total area of Jath tahsil is about 225828 hectares and the population is about 238862.

Generally, the water resources from heavy rainfall regions are assessed for their hydro biological study and the drought prone regions are neglected. A drought is the condition in which region suffers a severe deficiency in its water supply. Generally, this condition arises when a region receives constantly below average rainfall. Drought leads to adverse effects on the ecosystem and agriculture of the region. (Dave Deeksha and Katewa, 2008).

Many workers have studied the macrofauna and flora of the draught prone regions but still today nobody has studied the aquatic macrofauna, macroflora, microfauna and microflora of the water tanks of Jath tahsil, so that the present study was carried out to investigate the diversity of planktons of the Jath tahsil.

Material and Methods

The lakes studied are situated western part of the tahsil. These are situated in between Lat. 16°46' to 17°1' N and Long. 70°42' to 75°4' E. All the lakes were constructed by the Government of Maharashtra under Employment Guarantee Scheme (EGS). The study of planktons was conducted from June 2012 to November 2012, during monsoon and post monsoon season. Four sampling sites were selected namely SI, SII, SIII and SIV. The water samples were collected by using plankton net of bolting silk of mesh size 125μ. The surface and subsurface water samples from the tanks were collected during morning and evening. The collected samples were fixed in 4 per cent formalin on the spot and then carried to the laboratory for qualitative analysis.

The planktons were identified by using standard literature of Fritsch (1965), Adoni *et al.* (1985), Biswas (1980), Tonapi (1980), Sarode and Kamat (1984), Cox(1996), Ward and Whipple (1959) and Sreenivas and Duthie (1973) and other standard published literature.

Results and Discussion

Plankters, according to their quality may be classified as phytoplankton and zooplankton. Phytoplankton consists of chlorophyll bearing organisms *e.g.* Microcystis, Volvox etc. and the non-photosynthetic plants or saproplanktons *e.g.* Bacteria and Fungi. Zooplankton consists of plankters of animal origin (Jhingran 1983).

Phytoplankton mainly pertains to the groups chlorophyceae, bacillariophyceae, euglenineae and myxophyceae and rarely to a few dinophyceae. Zooplankton are mainly protozoans, rotifiers and planktonic forms of crustacea. Phytoplanktons are small plants, mostly microscopic either are weakly motile or drifted in water subject to the action of the waves and currents. Phytoplanktons forms the base of the food for many animals and therefore is extremely important to the ecosystem. Phytoplanktons are the product and belong to first tropical level (autotrophs). (Dholakia, 2004).

Chakraborty and Konar (2003), reported 37 genera of phytoplanktons at different sites of river Damodar from Raniganj region (West Bengal).

Diversity of zooplanktons in the Kamalapur lake was studied by Thirupathaiah *et al.* (2011) and reported about 18 species of zooplanktons which belongs to rotifera, cladocera, copepod and protozoa.

In present study, Birnal, Daphalapur and Pratappur lakes of Jath tahsil were investigated for the planktonic diversity study and the results obtained were summarized in the Table 24.1.

Phytoplankton being the primary producers, which play a significant role in primary production of freshwater ecosystem. Phytoplankton forms a base food for zooplanktons and other higher organisms like molluscans. Bhosale *et al.* (1994) reported about 18 species of Phytoplanktons from water bodies of Sangli district.

Table 24.1: Occurrence of Planktons in Birnal, Daphalapur and Pratappur Tanks from June 2012 to November 2012

Sl.No.	Name of the Planktons	Birnal						Daphalapur						Pratappur					
	Season →	Monsoon			Post-monsoon			Monsoon			Post-monsoon			Monsoon			Post-monsoon		
	Months →	Jun	Jul	Aug	Sept	Oct	Nov	Jun	Jul	Aug	Sep	Oct	Nov	Jun	Jul	Aug	Sep	Oct	Nov
A	**CYANOPHYCEAE (Blue Green Algae)**																		
1.	Nostoc sp.	+	–	+	–	+	–	–	+	+	–	+	–	+	–	–	+	–	+
2.	Spirulina princeps	–	+	–	+	–	+	+	–	–	+	+	–	+	–	–	–	–	+
3.	Microcystis aeruginosa	+	–	+	–	+	–	–	+	–	+	+	–	–	–	–	+	–	–
4.	Coelosphaerium sp.	–	–	+	–	–	–	–	+	–	–	–	–	–	–	+	–	–	–
B	**CHLOROPHYCEAE (Green Algae)**																		
1.	Spirogyra sp.	+	–	–	–	–	+	+	–	–	–	–	–	+	–	–	–	–	–
2.	Pediastrum simplex (4 celled stage)	–	–	–	–	–	–	–	–	–	–	–	–	+	–	+	–	+	+
3.	Pediastrum simplex (6 celled stage)	–	–	+	–	–	–	–	+	–	–	–	–	–	+	–	+	–	+
4.	Oedogonium sp.	–	–	+	–	–	–	–	+	+	–	–	–	–	+	+	–	–	–
C	**BACILLARIOPHYCEAE**																		
1.	Cyclotella striata	–	+	–	+	–	–	–	+	–	–	–	–	+	–	–	–	–	–
2.	Navicula sp.	+	+	–	–	–	–	+	–	+	–	–	–	–	–	+	–	–	–
3.	Achnanthes hungarica	–	–	–	–	–	–	–	+	–	–	–	–	–	–	–	–	–	–
4.	Cyclotella catenata	–	+	–	+	–	–	–	–	–	–	–	–	–	–	–	–	+	–
5.	Nitzschia closterium	+	+	–	–	–	–	–	–	+	–	–	–	+	–	–	–	–	–
6.	Nitzschia palea	–	+	–	–	–	–	–	–	–	–	–	–	+	–	–	–	–	–
7.	Pinnularia sp.	–	–	–	–	–	–	–	+	–	–	–	–	–	–	–	–	–	–
D	**PROTOZOA**																		
1.	Oxytricha oblongngatus	+	–	–	–	–	–	–	–	–	–	–	–	–	–	–	–	–	–
2.	Vorticella	–	+	–	–	–	–	–	+	–	–	–	–	–	+	–	+	–	–
3.	Lacrymaria sp.	–	–	–	–	–	–	–	–	–	–	–	–	–	–	+	–	–	–
E	**CRUSTACEA**																		
1.	Cyclops sp.	–	+	–	–	–	–	–	+	+	–	–	–	+	–	–	–	–	–
2.	Bosmina sp.	+	–	–	–	–	–	+	–	–	–	–	–	–	–	–	–	–	+

+: Presents; –: Absent.

Goel *et al.* (1985) listed about 15 phytoplanktons from freshwater bodies of South-western Maharashtra where no species was common to all water bodies.

The present investigation and the prepared check-list reflects that the bacillariophyceae is dominantly present followed by chlorophyceae, cyanophyceae, protozoa and crustacea.

The phytoplanktons studied indicate that 4 species of cyanophyceae, 4 species belonging to chlorophyceae, 7 species bacillariophyceae were reported in the investigation.

The occurrence of Zooplanktons was also observed in the study which indicated that 3 species of Protozoa and 2 species of Crustacea were reported in the investigation.

References

Adoni, A.D., Joshi, Gunwant, Ghosh, Kartik, Chourasia, S.K., Vaishya, A.K., Yadav, Manoj and Verma, H.G., 1985. *Workbook on Limnology*, Pratibha Publisheres, Sagar, India, p. 48.

Bhosale, L.J., Sabale, A.B. and Mulik, N.G., 1994. Survey and status report on some wetlands of Maharashtra. Final report submitted to Shivaji University, Kolhapur. India p 60.

Biswas, Kalipada, 1980. Common fresh and brackish water algal flora of India and Burma. *Records of the Botanical Survey of India*, Vol. XV Part I. International Book Distributors.

Chakraborty Dibyendu and Konar, S.K., 2003. Studies on the hydrobiology of river Damodar from Raniganj region (West Bengal) with reference to phyto plankton. In: *Aquatic Ecosystems*, (Ed.) Arvind Kumar. A.P.H. Publishing Corporation, Darya Ganj, New Delhi, 7: 63–70.

Cox Eileen, J., 1996. *Identification of Freshwater Diatoms from Live Material London*.

Dave Deksha and Katewa, S.S., 2008. *Textbook of Environmental Studies*. Cengage Learning India Pvt. Ltd., Patparganj, Delhi, p. 32.

Dholakia, A.D., 2004. *Fisheries and Aquatic Resources of India*. Daya Publishing House, Delhi, pp. 171–192.

Fritsch, F.E. 1965. *Structure and Reproduction of Algae Vols. 1 and 2*, Cambridge.

Goel, P.K., Trivedy, R.K. and Bhave, S.V., 1985. Studies on the limnology of a few freshwater bodies in south-western Maharashtra. *Indian J. Environ. Prot.*, 5(1): 19–25.

Jhingran, V.G., 1983. *Fish and Fisheries of India*. Hindustan Publishing Corporation, New Delhi, India. pp. 9–292.

Sarode, P.T. and Kamat, N.D., 1984. *Freshwater Diatoms of Maharashtra*. Saikripa Prakashan, Aurangabad.

Sreenivas, M.R. and Duthie, H.C., 1973. Diatom flora of the Grand river, Ontario, Canada. *Hydrobiol.* 42: 161–224.

Thirupathaiah, M., Somatha, Ch and Simmaiah, Ch., 2011. Diversity of Zooplankton in freshwater lake of Kamalapur, Karimnagar Dist. (A.P.) India. *The Ecoscan.*, 5(1 and 2): 85–87.

Tonapi, G.T., 1980. *Freshwater Animals of India: An Ecological Approach*. Oxford and IBH Publ. Co., New Delhi.

Ward, H.B. and Whipple, G.C., 1959. *Freshwater Biology*, 2nd Edn. John Wiley and Sons, New York.

Chapter 25

Drinking Water Quality Assessment of Aurangabad Municipal Corporation, Aurangabad, Maharashtra

☆ *Bhagwaan Maknikar, A.M. Late, B.J. Bhosale,*
A.S. Dhapate and M.B. Mule

ABSTRACT

The present chapter deals with the studies on drinking water quality of municipal water supply of Aurangabad city with respect to physico-chemical parameters. The water samples were collected from 25 different sampling sites from residential area of Aurangabad city. During the present investigation some physico-chemical and bacteriological parameters such as colour, taste, pH, temperature, turbidity, total alkalinity, total dissolved solids, chlorides, total hardness, residual chlorine, chloride, dissolved oxygen, fluoride and nitrate were studied. The obtained results were compared with ISO standard.

The results obtained during the present study reveals that the collected water samples were free from pathogenic organisms such as total coli form. Whereas, the parameters such as total hardness and residual chlorine were found out of desirable limit. However, during the present investigation the drinking water quality of municipal water supply for Aurangabad were found potable.

Keywords: Water quality, Drinking water, Aurangabad city, Physico-chemical parameters.

Introduction

Water is a precious gift of nature to all kinds of life on this planet. Water is one of the most abundant compounds found in value, covering approximately three fourths of the surface of the earth.

In spite of this apparent abundance, several factors sever to limit the amount of water available for human use. About 97 per cent of the total water supply in contained in the oceans and other saline bodies of water and in not readily usable for most purposes of the remaining 3 per cent a little over 2 per cent is tied up in ice caps and glaciers and only 1 per cent is available as freshwaters in rivers, lakes, streams etc.

Rapid population growth, urbanization and industrialization have led to a greater demand for an increasingly smaller supply of water resources in the country. Of the present water usage in the country, majority is consumed in agriculture (70–90 per cent), and the remaining is consumed in industrial activities and for domestic purposes such as drinking water and sanitation.

Aurangabad is situated at latitude 19° 53′ 59″ North and longitude 75° 20′ East. The city established on the bank of the Kham river. Topographically it is located in the valley region between the Chauka hills on the North and Satara hills on the South. Ajantha and Ellora caves have put the city on the tourist map of the world. It is the cultural, religious, educational and industrial center. The average altitude of the city is about 581 m above mean sea level. The total area of the city is about 138.5 sq. km containing about 12 to 13 lakh floating population.

Aurangabad is a fastly growing township in Asia. The total number of small, medium and large scale industrial units are about 1020 and about 35000 workers are employed in these units. The industrial development has positive impact on the growth of city.

Initially the drinking water supply was provided from three different sources to Aurangabad city *i.e.* Harsul Dam, Kham river and Jaikwadi dam. Prior to construction of Jaikwadi dam the water supplies to Aurangabad city from Harsul dam which was built in 1954 on the Kham River.

However, with the progress of urbanization and increasing population the demand of water supply is increased. In 1975 the water was supplied to Aurangabad city from Jaikwadi dam (Paithan) at a rate of about 28 MLD. After 1984 the capacity of water supply increased by booster system and 50 MLD water was supplied every day (AMC, 2006).

Presently water is supplied to Aurangabad city at a rate of about 135 MLD from Jaikwadi dam. This water is treated in Pharola water treatment plant.

The quality of drinking water is measured in term its physico-chemical parameters. Specific range of physico-chemical parameters decides its quality and utility.

Hence, present investigation aims to asses the drinking water quality supplied by Municipal Corporation of Aurangabad city.

Materials and Methods

The present investigation was carried out to asses the drinking water quality which is supplied by Aurangabad Municipal Corporation, Aurangabad. During the present study, water samples were collected from 25 different sampling stations selected from residential areas of Aurangabad city. The collected water samples from selected stations were analyzed for its physico-chemical and bacteriological parameters.

The samples were collected by using plastic cans. The parameters such as pH and temperature were measured at sampling sites. However, the parameters such as pH, temperature, turbidity, total alkalinity, total dissolved solids, chlorides, total hardness, residual chlorine, chloride, dissolved oxygen, fluoride and nitrate were analyzed in the laboratory by following the widely accepted methods as described by APHA (1998), Goel and Trivedy (1986).

Results and Discussion

During the present study the parameters such as colour, taste, temperature and pH were observed at sampling sites. Temperature of water samples was observed between 21° C to 32° C whereas colour and taste of water samples were found non-objectionable. However, the results of other remaining parameters were summarized in Table 25.1.

In the present investigation the maximum pH values 8.2 were observed at N – 1 CIDCO and minimum 7.0 at Samarth Nagar. The lower concentration of pH values may responsible for the corrosion of metallic pipes used in distribution system.

During the present study, the maximum value of turbidity 3.7 NTU recorded at Shahagunj and minimum 1.1 NTU at Padampura. Turbidity in water may causes due to fine particles of mud, sand, clay, loam and organic matter that may present (Gupta *et al.*, 2003). Turbidity values ranges from 1.8 – 2.2 NTU in intermittent water supply and 0.9 – 1.6 NTU in a continuous water supply (Kelakar *et al.*, 2001).

The total alkalinity values varied from 74 mg/lit to 371 mg/lit at Mill Corner and Jawahar colony respectively during the study. Similar results were recorded by Gajghate *et al.* (1990) studied on water quality, evaluation studies in rural drinking water supply in Latur district and reported total alkalinity in between 64 – 508 mg/lit at Latur Taluka, 60 – 548 mg/lit, 48 – 368 mg/lit, 44 – 692 mg/lit and 112 – 644 mg/lit at Ausa, Nilanga, Udgir and Ahmedpur Taluka respectively.

In the present investigation the maximum concentration of total dissolved solids 668 mg/lit at Mill corner and minimum 115 mg/lit at labour colony. The similar results were reported Kelkar *et al.*, (2001) the value of total dissolved solids ranging from 620 – 680 mg/lit.

During the present study, the chloride values ranged from 181.7 to 25.56 mg/lit at Mill corner and Khadkeswar respectively. Similar observations were noted by Sravanthi and Siddharth (1998).

In the present investigation, maximum total hardness values 482 mg/lit were recorded at sampling site N – 5 ESR and minimum were 96 mg/lit at Jay Tower. Sharma (2003) reported the total hardness values ranging from 80 – 340 mg/lit.

The maximum concentration of residual chlorine were 2.1 mg/lit at Ambedkar Nagar and minimum were 0.27 mg/lit at Chikalthana. The residual chlorine concentration ranged from 0.1 mg/lit to 1.0 mg/lit (Govindan *et al.*, 1993). The analysis of residual chlorine is usually done for verify its level, which inhibits the growth of pathogenic organism.

Dissolved oxygen is key parameter which reflecting the quality of water and its utility for drinking. In the present study the dissolve maximum oxygen concentration 7.8 mg/lit at labour colony and Chikalthana, whereas minimum were 5.5 mg/lit at Shahagunj site.

During the present investigation fluoride concentration of water samples were ranges from 0.1 mg/lit to 0.4 mg/lit at selected sampling stations. In India 20 million people are severely affected due to flourosis and about 40 million people are exposed to risk to endemic flourosis (Chinoy, 1991). According to WHO (1984) and Indian Drinking Water Specification (1991) the maximum permissible limit of fluoride in drinking water is 105 mg/lit and highest desirable limit is 1.0 mg/lit.

The maximum nitrate concentration 3.76 mg/lit was observed at Hudco Corner and minimum 1.09 mg/lit at the site S. B. Colony. Increase in concentration of Nitrate due to excessive use of chemical fertilizers in agriculture, decayed organic matter, domestic sewage etc becomes a serious problem

Table 25.1: Physico-Chemical Parameter of Drinking Water Supply by Aurangabad Municipal Corporation

Sample No.	Name of the Sampling Station	pH	Turbidity (NTU)	Alkalinity	Total Dissolved Solids	Chlorides	Total Hardness	Residual Chlorine	Dissolved Oxygen	Fluoride	Nitrate
1.	Padampura	7.91	1.1	312	230	58.22	312	1.18	5.97	0.3	2.42
2.	Aurangpura	7.52	1.2	112	207	46	160	1.9	6.4	0.1	1.98
3.	Chikalthana	8.01	1.4	98	486	58.4	179	0.27	7.87	0.2	3.03
4.	Shanoorwadi	7.91	1.9	104	327	123.5	269	1.19	6.39	0.2	1.34
5.	Nandanvan Colony	7.47	2.3	331	291	93.8	407	1.76	7.25	0.1	1.46
6.	Swantatraya Sainik Colony	7.62	1.5	242	314	55	311	1.32	6.97	0.4	1.49
7.	Padegaon	7.25	1.8	279	147	36.9	392	1.18	6.83	0.3	2.12
8.	S. B. Colony	7.06	1.6	193	294	58.2	108	1.27	7.1	0.4	1.09
9.	Raman agar	7.78	2.1	224	221	124.9	276	1.13	6.92	0.3	1.34
10.	Jawahar colony	8.04	2.4	371	213	107	238	0.14	7.13	0.3	2.63
11.	Bansilal Nagar	7.41	1.3	117	321	127.8	180	1.18	7.42	0.2	1.12
12.	Jadhavwadi	7.61	1.3	301	532	66.2	244	0.72	6.91	0.3	2.76
13.	Ambedkar Nagar	7.8	1.8	277	324	31.24	194	2.1	5.98	0.2	1.56
14.	N – 5 ESR	7.98	1.3	243	141	76.8	482	1.78	7.43	0.1	1.95
15.	Khokadpura	7.68	1.7	111	238	65.28	378	1.17	7.64	0.3	2.85
16.	Mill Corner	7.53	2.2	74	668	181.7	114	1.87	7.04	0.4	2.34
17.	Khadkeswar	7.35	1.4	219	274	25.56	342	1.29	6.48	0.3	2.49
18.	University Gate	7.1	2.5	369	313	76.68	210	1.16	6.84	0.4	2.13
19.	Begumpura	7.23	1.9	283	294	63.9	467	1.28	6.32	0.1	1.42
20.	N – 1	8.2	2.7	203	238	58.22	360	1.13	6.78	0.1	1.78
21.	HUDCO Corner	7.14	2.6	364	230	71	194	1.98	5.92	0.3	3.76
22.	Vivekananda Colony	7.02	1.3	241	371	75.26	332	1.76	6.83	0.3	2.12
23.	Labour Colony	7.24	2.4	237	115	46.86	316	2.03	7.87	0.4	1.86
24.	Shahagunj	7.74	3.7	223	204	69.58	138	1.13	5.55	0.2	2.82
25.	Near Jay Tower	7.62	1.4	273	185	63.7	96	1.29	7.54	0.3	3.23
	Average	7.56	1.872	232.04	287.12	74.46	267.96	1.32	6.85	0.26	2.12
	Standard Deviation (S. D)	**0.33**	**0.61**	**88.87**	**124.6**	**35.2**	**110.9**	**0.49**	**0.61**	**0.10**	**0.70**

All values are expressed in mg/lit except pH and turbidity.

(Makhijani *et al.*, 1999). Excessive concentration of nitrate in drinking water is hazardous for infant tract; nitrates are reduced to nitrites which may cause methemoglobinemia.

Table No. 2: Drinking Water Quality Standards

Sl.No.	Parameters	ISI	WHO
1.	Colour (Hazen)	10	5.0
2.	Taste	–	–
3.	pH	7.0 – 8.5	7.0 – 8.5
4.	Turbidity (NTU)	5.0	5.0
5.	Total Alkalinity	300	300
6.	Total Dissolved Solids	500	500
7.	Chlorides	250	200
8.	Total Hardness	300	100
9.	Residual Chlorine	0.2 – 0.5	
10.	Dissolved Oxygen	4 – 6	
11.	Fluoride	0.6 – 1.2	1.0
12.	Nitrate	45	45

*: All values are expressed in mg/lit except pH and temperature.

Conclusion

The drinking water quality of Aurangabad Municipal Corporation was studied for its physico-chemical parameters. The water samples were collected from 25 different sampling stations from residential area of Aurangabad city.

The drinking water quality parameters of Aurangabad city such as total hardness and residual chlorine were found out of desirable limit. However, during the present investigation the drinking water quality of municipal water supply for Aurangabad were found potable.

However, some important findings at the time of survey are that in some areas, the pipelines for drinking water supply were very close to sewers. So, there are chances of contamination during the leakage of metallic pipeline. Therefore, present study suggests that there are some primary treatments for water to reduce the total hardness and residual chlorine concentration were needed. Whereas some precautions should be taken during the layout procedure of pipeline allotment in the residential areas.

References

AMC, 2006–2007. Aurangabad Municipal Corporation; Environment Status Report, 2006 – 2007; Govt. Engg. College, Aurangabad pp. 18–19.

APHA, 1998. *Standard Methods for the Examination of Water, Sewage and Industrial Waste*. APHA, AWWA, Wahington.

Chinoy, J.N., 1991. Effects of fluoride on physiology of animal and human being. *Ind. J. Env. Toxicol.*, 1: 17–32.

Gajghate, D.G., Murthy, M. R. K., Rao, D. V. and Rao, M. Vittal, 1990. Water quality evaluation studies – Assessment of problems in Rural Drinking Water Supply in Latur District, *Journal of Indian Water Works Association*, 12(1): 117–120.

Govindan, V.S. and Thansasekaran, K., 1993. Quality control in rural and urban water supplies, *Journal of Indian Water Works Association*, 25(1): 123 – 128.

Goel, P.K. and Trivedy, R.K., 1986. *Chemical and Biological Methods for the Water Pollution Studies*. Environmental Publication, Karad, India.

Gupta, M.C. and Mahajan, 2003. *Textbook of Preventive Social Medicine*. Jaypee Brothers Medical Publishers (P) Ltd., EMCA House, New Delhi, pp. 478.

Indian Standard Drinking Water, 1991. *Specifications IS 10500.*

Kelkar, P.S., Talkhande, A.V., Joseph, M.W. and Pandey, S.P., 2001. Water quality assessment in distribution system under intermittent and continuous modes of water supply. *Journal of Indian Water Works Association*, 33(1): 39–43.

Makhijani, S.D. and Manoharan, 1999. Nitrate pollution problem in drinking water sources monitoring and surveillance. In: Paper Presented in the *Workshop of water Quality Field Taste Kits for Arsenic, Fluoride and Nitrate*, held form 8–9 Sept. at IRTC, Lucknow, U.P.

Sharma, 2003. Water quality of traditional drinking water resources in Haminpur district (H.P.). *Journal of Indian Water Works Association*, 25(1): 57–59.

Sravanti, K. and Sudharshan, 1998. Geochemistry of groundwater, Nacharam industrial area, Ranga Reddy District, A.P., India. *Environmental Geo. Chemistry*, 1(2): 81–88.

WHO, 1984. *Guidelines for Drinking Water Quality*, World Health Organization, 2: 307.

Chapter 26

Plankton Diversity in Panchganga River Water Near Ichalkaranji City, Maharashtra

☆ *S.A. Khabade, Archana Patil and Rekha Murgunde*

ABSTRACT

Plankters, according to their quality may be classified as phytoplanktons and zooplanktons. The study was carried out during winter season 2012. The phytoplanktons reported were 3 species of cyanophyceae, 20 species of chlorophyceae and 6 species of bacillariophyceae. The zooplanktons reported were 4 species of rotifera, 4 species of crustacea and 1 species of arthropoda.

Keywords: Plankters, Panchganga river, Ichalkaranji city, Winter season.

Introduction

The planktons or plankters are those organisms that, because of their size or immobility or both are at the mercy of water movements. The limnologists generally consider these to be tiny forms of life. Plankton includes forms of aquatic bacteria and ultra algae only a few microns in diameter and macroscopic forms of crustacean several millimeters long. Planktons are important components of aquatic systems. They lack both motility and attachment devices. They are not commonly found in rivers. The plankton often serves as indicators of water quality.

The study of phytoplanktons from water has got applied significance as it reflects the potential of aquaculture. The phytoplankton mark the lowest trophic level and estimation of trophic status is essential to evaluate the feasibility of pisiculture in any freshwater body.

Zooplanktons are those organisms within the aquatic ecosystem they are intermediators in the food chain. The zooplanktons community is a major link in the energy transfer at secondary level of food chain of aquatic ecosystem. They are first consumers and form an important link of food chain. The study of zooplankton in view of their composition abundance and seasonal variations help in planning and successful fishery management. Keeping in view above the present study was carried out during winter season 2012.

Material and Methods

Ichalkaranji city is located at 16°42' N, 74°28'E/16.7° N, 74.47°E. The water samples were collected in small plastic bottles during morning 8.00 O'clock by using plankton net of bolting silk of mesh size 125 micron. The collected samples were fixed in 4 per cent formalin on the spot and then observed and studied under microscope in the laboratory for qualitative purpose. The planktons were identified by using standard literature of Fritsch (1965), Adoni *et al.* (1985), Biswas (1980), Tonapic (1980), Sarode and Kamat (1984), Cox (1996), Ward and Whipple (1959) and Sreenivas and Duthie (1973) and other standard published literature.

Results and Discussion

The present investigation the phytoplanktons reported were 3 species of cyanophyceae, 20 species of chlorophyceae and 6 species of bacillariophyceae.

Table 26.1: Occurrence of Phytoplanktons of Panchganga River Water

Sl.No.	Name of the Phytoplanktons	Month			Nature of Occurrence
		October 2012	November 2012	December 2012	
A.	**CYANOPHYCEAE (Blue Green Algae)**				
1.	*Microcystis incerta*	++	++	++	A
2.	*Microcystis aeruginosa*	++	++	++	A
3.	*Spirulina* sp.	—	—	++	A
B.	**CHLOROPHYCEAE (Green Algae)**				
1.	*Zygnema* sp.	++	—	++	R
2.	*Pediastrum*-5 celled	++	++	++	A
3.	*Pediastrum*-6 celled	++	++	++	A
4.	*Pediastrum*-7 celled	++	++	++	A
5.	*Pediastrum*-15 celled	++	++	—	A
6.	*Pediastrum*-16 celled	—	—	++	A
7.	*Pediastrum*-18 celled	++	—	—	R
8.	*Tribonema* sp.	—	—	++	R
9.	*Scenedesmus quadricauda* (4-celled)	++	—	++	R
10.	*Closterium parvulum*	—	—	++	R
11.	*Closterium Ralfsii* var. *hybridum*	—	—	++	R
12.	*Characium* sp.	—	—	++	R
13.	*Mugeotia* sp.	++	—	++	R

Contd...

Table 26.1–*Contd...*

Sl.No.	Name of the Phytoplanktons	Month			Nature of Occurrence
		October 2012	November 2012	December 2012	
14.	*Ankistrodesmus* sp.	—	—	++	R
15.	*Protococcus* sp.	++	—	++	R
16.	*Cladophora* sp.	—	—	++	R
17.	*Mesotaenium* sp.	—	++	++	R
18.	*Chlorella vulgaris*	++	++	++	R
19.	*Selenastrum* sp.	—	—	++	R
20.	*Characium* sp.	—	—	++	R
C.	**BACILLARIOPHYCEAE (Diatoms)**				
1.	*Eunotia* sp.	++	++	—	R
2.	*Tryblionella* sp.	—	—	++	A
3.	*Synedra* sp.	++	++	++	A
4.	*Tabellaria* sp.	++	—	++	A
5.	*Cymbella ventricosa*	—	++	++	R
6.	*Fragilaria* sp.	—	—	++	A

++: Present; —: Absent; R: Rare; A: Abundant.

Table 26.2: Occurrence of Zooplanktons of Panchganga River Water

Sl.No.	Name of the Zooplanktons	Month			Nature of Occurrence
		October 2012	November 2012	December 2012	
A.	**ROTIFERA**				
1.	*Kellicottia* sp.	—	++	++	R
2.	*Notholca* sp.	++	—	++	R
3.	*Brachionus forficula*	—	++	++	A
4.	*Keratella tropica*	—	++	++	A
B.	**CRUSTACEA (Cladocera, Copepoda and Ostracoda)**				
1.	*Eurycerus* sp.	—	++	++	R
2.	*Nauplius larva*	++	—	++	R
3.	*Cyclops* sp.	++	++	++	A
4.	*Chydorus sphaericus*	—	—	++	R
C.	**ARTHROPODA**				
1.	*Mosquito larva*	—	—	++	R

++: Present; —: Absent; R: Rare; A: Abundant.

In India, little work has been done on the phytoplankton of different waters. According to Gaikwad *et al.* (2012) the planktonic population may vary qualitatively and quantitatively depending on the depth of water bodies, site, time and season and also on the source of water. Further, it may depend on organic and inorganic contents, geological, biological and climatic factors.

In present investigation, the checklist shows that chlorophyceae is dominantly present followed by bacillariophyceae and cyanophyceae.

Freshwater zooplankton is an important component in aquatic ecosystems, whose main function is to act as primary and secondary links in the food chain (Hutchinson, 1957). Zooplankton have long been used as indicators of the eutrophication (Vandysh, 2004).

The zooplanktons investigated in the present study shows that there are about 4 species of rotifera, 4 species of crustacea and 1 species of arthropoda. The rotifera and crustacea were dominantly present over arthropoda.

The seasonal changes in plankton population of freshwater bodies have been studied by many workers. The important contributors are Ramakrishna (1993), Siddiqui and Chandrasekhar (1993) and Chandrasekhar and Kodarkar (1994).

Acknowledgements

The authors are thankful to Principal Dr. Milind S. Hujare for providing necessary laboratory facilities during this work. The authors are also thankful to Rane R. G., Yadav B. T. and Mane L. G. for their help during the work.

References

Adoni, A.D., Joshi, Gunwant, Ghosh, Kartik, Chourasia, S.K., Vaishya, A.K., Yadav, Manoj and Verma, H.G., 1985. *Workbook on Limnology*, Pratibha Publisheres, Sagar, India, p. 48.

Biswas, Kalipada, 1980. Common fresh and brackish water algal flora of India and Burma. *Records of the Botanical Survey of India*, Vol. XV Part I. International Book Distributors.

Chandrasekhar, S.V.A. and Kodarkar, M.S., 1994. Biodiversity of Zooplanktons in Saroornagar lake, Hyderabad. *J. Aqua. Biol.*, 10(1): 44–47.

Cox Eileen, J., 1996. *Identification of Freshwater Diatoms from Live Material London*.

Fritsch, F.E., 1965. *Structure and Reproduction of Algae Vols. 1 and 2*, Cambridge.

Gaikwad, M.M., Kulkarni, D.A. and Kumbhar, A.C., 2012. Qualitative and quanitative study of phytoplanktons in Seena river water at Mohol district, Solapur. In: *Aquatic Ecology and Toxicology*, Jotichandra Publication Pvt. Ltd., Latur (M.S.), India, pp. 137–145.

Hutchinson, G.E., 1957. *A Practical on Limnology and Primary Production on a Tropical Lake*.

Ramakrishna, 1993. Biomonitoring of inland water: Physical, chemical and biological parameters of river Moosi, Hyderabad, Andhra Pradesh, India. *Rec. Zool. Survey India*, 93(3–4): 367–392.

Sarode, P.T. and Kamat, N.D., 1984. *Freshwater Diatoms of Maharashtra*. Saikripa Prakashan, Aurangabad.

Siddiqui, S.Z. and Chandrasekhar, S.V.A., 1993. New distributional records of freshwater Cladocera from Hyderabad: A taxo-ecological profile. *Geobios New Report*, 12: 105–110.

Sreenivas, M.R. and Duthie, H.C., 1973. Diatom flora of the Grand river, Ontario, Canada. *Hydrobiol.* 42: 161–224.

Tonapi, G.T., 1980. *Freshwater Animals of India: An Ecological Approach*. Oxford and IBH Publ. Co., New Delhi.

Vandysh, O.I., 2004. Zooplankton as an indicator of state of lake of ecosystems polluted with mining wastewater in the Kolapeninsula. *Russian J. Ecol.*, 35(2): 110–116.

Ward, H.B. and Whipple, G.C., 1959. *Freshwater Biology*, 2nd Edn. John Wiley and Sons, New York.

Chapter 27

Diversity of Zooplankton in Nagathibelagalu Tank, Bhadravathi, Karnataka

☆ *H.A. Sayeswara, K.L. Naik, T. Vasantha Naik, D.B. Sumanthrappa, E.N. Jeevan, S.G. Dhananjaya and H.M. Ashashree*

ABSTRACT

The Zooplankton composition of Nagathibelagala tank was studied for a period of six months from April to September 2011. Zooplankton assume a great ecological significance in aquatic ecosystem as they play vital role in food web of the food chain, nutrient recycling, and in transfer of organic matter from primary producers to secondary consumers like fishes. The zooplankton determines the quantum of fish stock. The failure of fishery resources is attributed to the reduced zooplankton population. A total of 19 species belonging to 16 genera of zooplankton were recorded, of which cladoceran and rotifers were dominant. Relative abundance of zooplankton in tank showed maximum of cladocera (31.57 per cent) followed by rotifera (26.31 per cent), copepoda (21.05 per cent) and protozoa (21.05 per cent). The ecological status of the tank was found to be impoverished in terms of species composition.

Keywords: Diversity of zooplankton, Nagathibelagalu tank, Bhadravathi.

Introduction

Zooplanktons are ecologically and economically important heterogeneous group of tiny aquatic organisms that can move at the mercy of water currents, as they have weak power of locomotion. Zooplanktons are tiny animals found near the surface in aquatic environments. Like phytoplankton, zooplanktons are usually weak swimmers and usually just drifts along with the currents. Zooplanktons are either herbivores, feeding on phytoplankton or carnivores feeding on other zooplankton. Zooplankton assume a great ecological significance in aquatic ecosystem as they play vital role in food web of the food chain, nutrient recycling, and in transfer of organic matter from primary producers to secondary consumers like fishes (Krishnamurthy *et al.*, 1979). The zooplanktons determine the quantum of fish stock. The failure of fishery resources is attributed to the reduced zooplankton population (Stottrup, 2000). Hence zooplankton communities, based on their quality and species diversity, are used for assessing the productivity vis-à-vis fishery resource, fertility and health status of the ecosystem. The productivity of aquatic environment is directly correlated with the density of zooplankton. Further, physico-chemical factors of water are directly related to their production. The seasonal changes in zooplankton species are related to the physico-chemical and biological parameters of aquatic environment. Several investigators have studied the diversity of zooplankton of pond ecosystems (Mishra and Sexana, 2002; Fassibuddin and Kumar, 1990; Pandey *et al.*, 1994; Choudhary and Singh, 2000; Pawar *et al.*, 2003; Bhagat and Meshram, 2007; Dayananda, 2009; Purushothama *et al.*, 2011).

The present study has been carried out to estimate the zooplankton diversity of Nagathibelagalu tank. This type of a study is relevant since the Nagathibelagalu tank forms the source of water for public distribution systems.

Study Area

Nagathibelagalu tank is located at about 8 km away from Bhadravathi town, situated between 75°43'17" E latitude and 13°55'27"N longitude. The tank is a perennial one and acquires a total area of 25 acres with average depth of 9 feet. It is surrounded by Paddy and Sugarcane fields in all directions. The water is used for domestic purposes. The tank receives copious water supply from Bhadra canal. It also receives sewage water from Nagathibelagalu village. The water has undergone moderate changes in its physico-chemical properties due to ecological degradation, overflowing of water from adjacent paddy fields and other excessive human activities. The literature revealed that there is no scientific study carried out with respect to ecological characteristics of this pond. The basis of selection of Nagathibelagalu tank was that its water is used by a large population which receives adequate wastewater and periodic flooding from plains.

Materials and Methods

The study was carried out from April to September 2011. The zooplankton samples were collected by plankton net of mesh size 50 microns. Further, these samples were fixed in 4 per cent formaldehyde and identification of zooplankton was done with the help of monographs and workshop manual (Needham and Needham, 1941 and Battish, 1992).

Results and Discussion

Results of Zooplankton diversity of Nagathibelagalu tank water are given in Table 27.1 and Genus compositions of different groups are depicted in Figures 27.1–27.4.

Zooplanktons play a significant role in determining the productivity of ponds and form food for many aquatic organisms which in turn are good sources of food for water birds. A total of 19 zooplankton species representing four groups namely cladocera, rotifera, copepoda and protozoa were reported (Table 27.1). Zooplankton showed a dominant position of cladoceran members (31.57 per cent), followed by rotiferans (26.31 per cent), copepodans (21.05 per cent) and protozoas (21.05 per cent).

Table 27.1: List of Zooplankton in Nagathibelagalu Tank

Class: Cladocera

1.	*Alona purchella*	2.	*Diaphanpsoma sarsi*
3.	*Diaphanpsoma excisum*	4.	*Daphnia carinata*
5.	*Macrophrix goeldi*	6.	*Macrophrix laticornis.*

Class: Rotifera

7.	*Brachionus caudatus*	8.	*Brachionus calyciflorus*
9.	*Brachionus falcatus*	10.	*Filinia longiseta*
11.	*Rotatoria neptunia*		

Class: Copepoda

12.	*Heliodieptomus viduus*	13.	*Mesocyclops hyalinus*
14.	*Mesocyclops leuckarti*	15.	*Paracyclops fimbriatus*

Class: Protozoa

16.	*Arcella* sp.	17.	*Difflugic* sp.
18.	*Paramecium* sp.	19.	*Vorticella* sp.

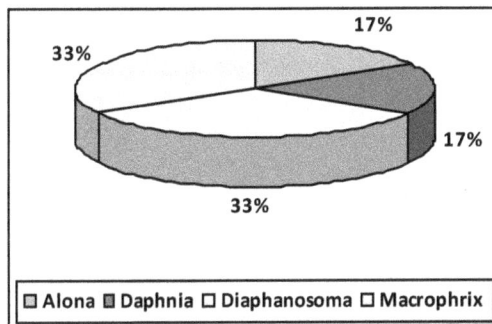

Figure 27.1: Genus Composition of Cladocera.

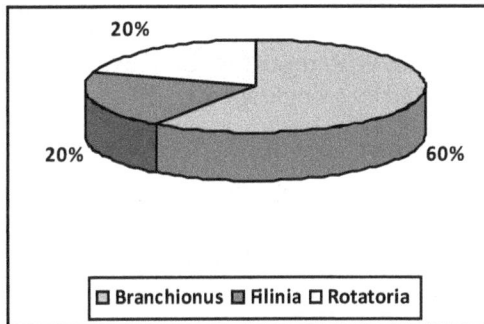

Figure 27.2: Genus Composition of Rotifera.

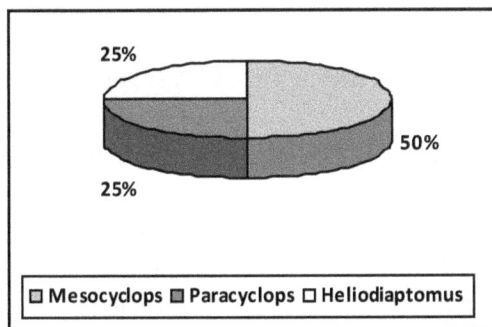

Figure 27.3: Genus Composition of Copepoda.

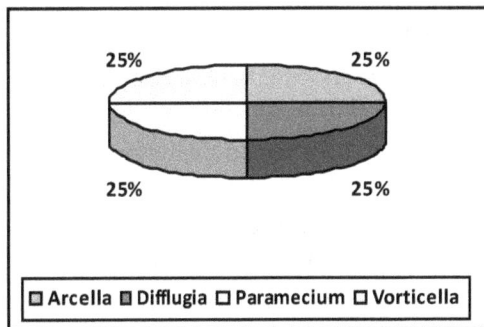

Figure 27.4: Genus Composition of Protozoa.

In the present study cladocera invariably constitute a dominant component of tank. Cladocerans are known to be abundant in water with good littoral vegetation, while ponds and lakes without vegetation have fewer cladoceran species (Idris and Fernando, 1981). Decay of this vegetation during summer may serve as food, thus maximum during that season. A total of 5 genera and 6 species of cladocera were recorded and constituting 31.57 per cent of total zooplankton population. *Macrothrix* was represented by 2 species, *Daphnia, Moina, Ceriodaphnia* and *Diaphanosoma* by a single species each.

Rotifers are the smallest animals and occur worldwide in primarily freshwater habitats. Like the other zooplankton, rotifers also form a link in the aquatic food chain. They have a rapid turnover and high metabolic rates and feed on detritus. These organisms serve as bioindicators to depict water quality and are extensively cultured for use as fish feed. Nagathibelagala tank recorded 3 genera and 5 species of rotifers constituting 26.3 per cent of total zooplankton population. In diversity, the genus *Brachionus* was represented by 3 species, *Lepadella* and *Rotatoria* by single species each.

Copepods are aquatic crustaceans, smaller relatives of the crabs and lobsters, in terms of their size, abundance and diversity of life. Well developed copepods are abundant in littoral than pelagic areas. Comparatively copepods population recorded was less which represents 4 genera and 4 species and constituting 21.05 per cent of total zooplankton population. In diversity, the genus *Heliodiaptomus, Mesocyclops, Tropocyclops* and *Neodiaptomus* represented by a single species each.

Protozoan showed minimum population density in the present study. Nagathibelagalu tank supported 4 genera and 4 species of protozoans constituting 21.05 per cent of total zooplankton population. With regard to their diversity, genus *Arcella, Difflugia, Paramecium* and *Vorticella* were represented by a single species.

Conclusion

The water samples from Nagathibelagalu tank was collected and analyzed for zooplankton composition. The ecological status of the pond was found to be impoverished in terms of species composition. A rich zooplankton fauna with 6 species of cladocerans, 5 species of rotiferans, 4 species of copepods, 4 species of protozons were reported.

References

Battish, S.K., 1992. *Freshwater Zooplankton of India*. Oxford and IBH Publ. Co., New Delhi.

Bhagat, V.B. and Meshram, C.B., 2007. Zooplankton dynamics of Ambadi dam near Akot, Ankola, Maharashtra, *J. Aqua. Biol.*, 22(1): 19–20.

Choudhary, S. and Singh, D.K. Zooplankton populations of Bosra lake at Muzaffarpur, Bihar. *Envir. and Ecol.*, 17(2): 444–448.

Dayananda, G.Y., 2009. Zooplankton diversity in certain lentic waterbodies in Mid Western Ghats of Karnataka. *Envir. and Ecol.* 271): 1430–1431.

Fassibuddin, M. and Kumar, T., 1990. Seasonal variations in physico-chemical properties and phytoplankton periodicity in a freshwater fish pond at Bhagalpur. *Envir. and Ecol.*, 8: 929–932.

Idris, B.A.H. and Fernando, C.H., 1981. Cladocera of Malaysia and Singapore with new records, redescription and remarks on some species. *Hydrobiology*. 77: 233–256.

Krishnamurthy, K.R., Santhanam, R. and Sundaraj, V., 1979. The trophies tier. *Bulleting of Marine Sciences*, p. 209–217.

Mishra, S.R. and Sexana, D.N., 2002. Seasonal abundance of Zooplankton of wastewater from the industrial complex at Birla Nagara (Gwalior), India. *Hydrobiology*, 18: 215–220.

Needham, J.M. and Needham, P.R., 1941. *A Guide to the Study of Freshwater Biology*. Comstock, Theca, New York, USA.

Pandey, B.N., Jha, A.K., Das, P.K.I. and Pandey, K., 1994. Zooplankton community in relation to certain physico-chemical factors of Kosi Swamp, Purnia, Bihar. *Environ. Ecol.*, 12: 563–569.

Pawar, S.K., Madlapure, U.R. and Pulle, J.S., 2003. Study of Zooplankton community of Sirur dam water near Mukhed in Nanded district (M.S.). *J. Aqua. Biol.*, 18: 37–40.

Purushothama, R., Sayeswara, H.A., Mahesh Anand Goudar and Harish Kumar, K., 2011. Physico-chemical profile and zooplankton community composition in Brahamana kalasi tank, Sagara, Karnataka, India. *The Ecoscan.*, 5(1 and 2): 43–48.

Stottrup, J.G., 2000. The elusive copepods: Their production and suitability in marine aquaculture. *Aquaculture Research*, 31: 703–711.

Chapter 28

Fluctuations in the Total Ribose Nucleic Acid Content in Freshwater Fish, *Channa punctatus*, under Influence of Temperature Stress

☆ *A.R. Jagtap, R.P. Mali and P.N. Chavan*

ABSTRACT

Temperature in environment affects the metabolism at which energy and material resources are taken up from the environment, transformed within an organism, which help to maintain growth and reproduction in animals. The alterations in biochemical parameters in animals helps to understand the organisms respond to environmental changes. The complex biomolecules are basic building blocks possesses organic compounds such as amino acids, nucleotides and monosaccharide's. The biochemical parameters like nucleic acids play an important role in physiological mechanism. Any alteration in nucleic acid content leads to variations in protein profile. Hence the present study is designed to determine the effect of temperature stress on the total RNA content in freshwater fish, *Channa punctatus*. The RNA content at different temperature showed fluctuations.

Keywords: Temperature stress, RNA, Channa punctatus.

Introduction

Temperature in environment affects the metabolism. Metabolism is the rate at which energy and material resources are taken up from the environment, transformed within an organism, which help to maintain growth and reproduction in animals. The metabolism is a fundamental physiological trait found in plants as well as animals. The changes in biochemical parameters in animals helps to understand the organisms respond to environmental changes (Dahlhoff, 2004). Biochemical parameters are the complex organic molecules which are the basic structural components in a living cell. The complex biomolecules are basic building blocks possesses organic compounds such as amino acids, nucleotides and monosaccharide's. The biochemical parameters like nucleic acids play an important role in physiological mechanism.

Nucleic acids have an important role in all biological activities and also regulate the biological synthesis of proteins which are structural and functional. Any alteration in nucleic acid content leads to variations in protein profile (Raj *et al.*, 1992).

The frequency of changes in the composition of biochemical constituents of any organism varies with the fluctuations of the environmental changes. Biochemical studies are good parameters which help to see the effect of temperature on biochemical composition of vital tissue of fish. Hence, attempt has been made to find out biochemical changes in tissues like Liver and Muscle of freshwater fish, *Channa punctatus.*

Materials and Methods

The freshwater fish, *Channa punctatus* was collected from the Godavari River, Nanded (Maharashtra) with the help of local fisherman for the present investigation. They were brought to the laboratory and kept in glass aquarium with continuously aerated tap water. The physico-chemical parameters of water were maintained and determined by standard methods (APHA, AWWA, WPCP, 1992). (pH-7.0–7.2, Dissolved Oxygen-7.6 – 8.0 ppm, Carbondioxide-2.02 mg/L, Salinity 0.186 gms/L, Chlorinity - 0.112 gms/L). The fishes were acclimated at room temperature for 8-10 days prior to experimentation. Fishes were feed with the small pieces of earthworms to avoid the effect of starvation. The water in the aquarium was replaced daily with fresh tap water. The feeding was stopped one day before the starting of experiment to eliminate the effects of differential diet. The snake headed fish, *Channa punctatus* were subjected to above and below room temperatures to carry out experiment. The experimental set was designed to investigate Total RNA content in tissues of freshwater fish, *Channa punctatus*. The fishes were acclimated up to 96 hrs period of exposure.

The RNA was isolated in muscle and liver of freshwater fish, *Channa punctatus*. The supernatant contains nucleotide released from RNA which was used further for the estimation of RNA. The isolated RNA was used for the further estimation by Orcinol Method (Bial, 1902 modified by Sadasivam *et al.*, 1975). The method depends on the conversion of pentose, ribose in the presence of hot acid to furfural, which then reacts with orcinol to yield a green colour. The colour formed largely depends on the concentration of hydrochloric acid, ferric chloride, orcinol reagent and the time of heating at 100° C up to maximum level. The isolated RNA was dissolved completely in the buffer solution containing 10 m M tris acetate, 1 m M EDTA buffer. To the above mixture by adding distilled water the volume was made up to 3 ml. 0.4 ml of 6.0 per cent orcinol prepared in alcohol was added, mixed and kept in boiling water bath for 20 minutes. The contents were cooled and absorbance read at 660 nm against blank sample. The amount of RNA content was calculated by using standard graph. The values were expressed in mg of RNA/gm wet wt. of tissue.

Results and Discussion

Temperature is an important ecological factor which affects the chemical composition of various tissues. There is a tight relation between the organism and its environment. Each and every organism faces the problem of environmental changes. To overcome this problem the animal tries to adapt themselves against environmental fluctuations. This process of temperature acclimation seen in poikilothermic animals adjust their metabolic rate in order to maintain physiological activities at a more nearly constant level over a range of environmental temperatures. Thus, they attain a measure of independence of temperature (Prosser and Brown, 1961).

Ribonucleic Acid (RNA) is a biologically important molecule that consists of a long chain of nucleotide units. A nucleic acid is a macromolecule composed of chains of monomeric nucleotides. These molecules carry genetic information or form structures within cells. Nucleic acids are nitrogen containing compounds of high molecular complexes and are known to be association with proteins in the cell. They play an important role in the metabolism of protein synthesis. The RNA content found in tissues is correlated with the protein synthetic activities. Nucleic acid content is considered as an index of capacity of an organism for protein synthesis (Mali *et al.*, 2010).

Physiological responses of the snake headed fish *Channa punctatus* to temperature change, found to be more fluctuable in cold temperatures compared to warm temperature stress. *Channa punctatus* was acclimated for 24 to 96 hrs of continuous exposure at 15, 20, 30 and 35° C. The values were compared with control set. The values were represented in plotted in Figures 28.1 and 28.2; expressed in mg/gm wet wt. of tissue. The freshwater fish *Channa punctatus* exposed to cold and warm temperature showed remarkable changes in ribose nucleic acid content. Under cold temperature stress the fish muscle and liver showed increase in RNA content up to 96 hrs period of exposure. As the temperature increases the RNA content in muscle and liver was found to be decreased up to 96 hrs period of exposure.

The biochemical change in tissues of goldfish was studied by Das, (1967) under cold and warm temperature stress. He reported that RNA content in liver of fish and muscle revealed an argumentation

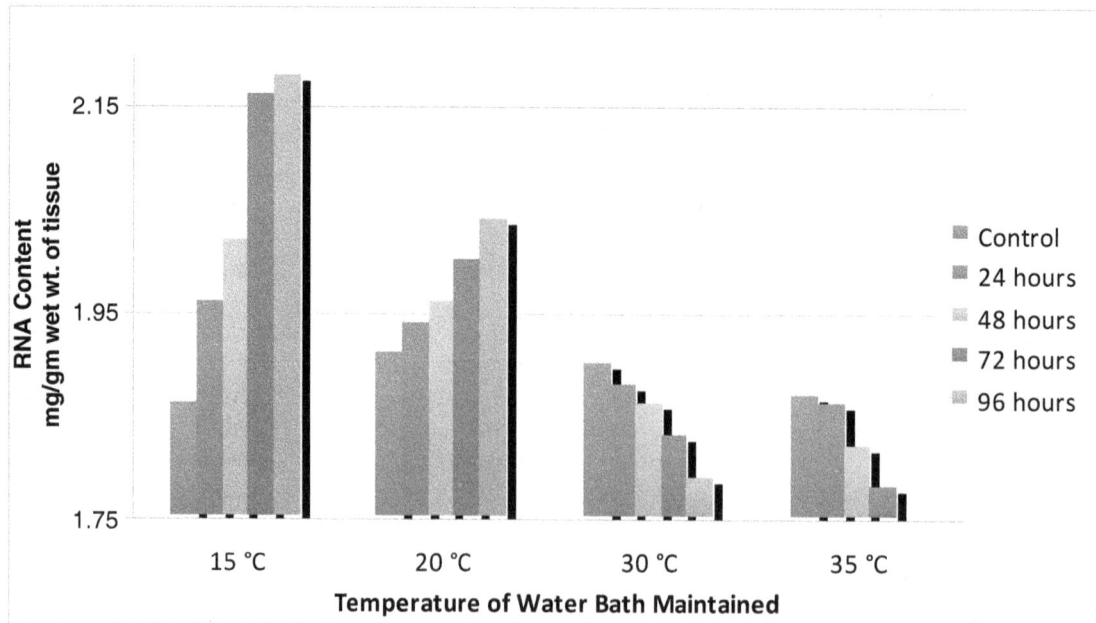

**Figure 28.1: Effect of Temperature on RNA Content in Muscle of Freshwater Fish,
Channa punctatus at Cold and Warm Temperature Stress.**

(Each Value is Mean of Six Observations ± S. D.)

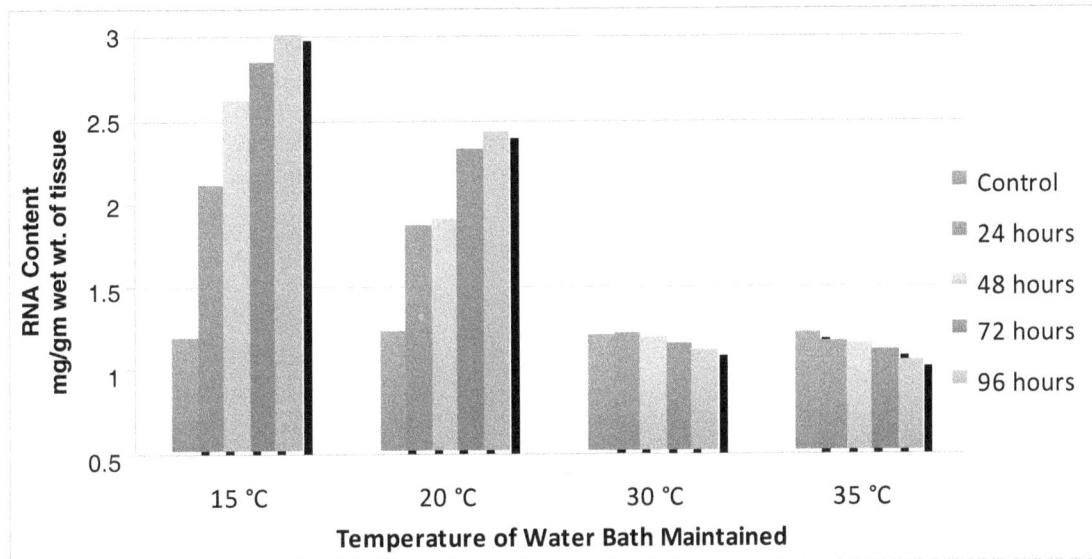

Figure 28.2: Effect of Temperature on RNA Content in Liver of Freshwater Fish,
***Channa punctatus* at Cold and Warm Temperature Stress.**

(Each Value is mean of six observations ± S. D.)

during cold acclimation. Haschemeyer *et al.* (1984) recorded data on the role of ribose nucleic acid in synthesis of proteins and controlled production of protein in toad fish, *Opsanus tau* under temperature stress.

The effect of temperature acclimation on tissues of goldfish acclimated under cold and warm condition was recorded by Das (1967). The RNA content in fish liver and muscle revealed heightening during cold acclimation. Currie *et al.* (2000) also reported same observations on rainbow trout *Oncorhynchus mykiss*. An increase in hsp 70 mRNA level in blood, brain, heart, liver, red and white muscle of rainbow trout *Oncorhynchus mykiss* was found at temperature stressed fish. Currie *et al.* (1997) found that increase in RNA level in blood of rainbow trout indicates fish is in stress condition. Therefore, tissue of fish must be considered as to right way to evaluate the stress by molecular means. The quantity of RNA present beneath the cell was highly variable and reflects the activity of protein synthesis reported by Khan *et al.* (1991) and Tripathi *et al.* (2003).

Effect of temperature under starvation and salinity influence on the RNA: DNA ratio, growth of fish, enzymatic activities were reported on rainbow trout *Salmo gairdneri* Richardson by Jurss *et al.* (1987). An increase in RNA content and RNA: DNA ratio was increased at cold temperature stress as compared to normal rainbow trout fish, *Salmo gairdneri*. Similar results were reported by Kent, *et al.* (1992). The rate of protein synthesis is directly correlated with the growth rate in animals and RNA: Protein concentration. The protein synthesis rate relative to concentration of RNA showed enhanced activity after meal Houlihan *et al.* (1989). The cryfish *Onconectes virilis* showed increased level of RNA during normal and premolt stage (Gorell *et al.*, 2004).

Jayaprakas *et al.* (1996) was found increased RNA content in tissues (Hepatopancreas and muscle) of prawn *Penaeus indicus*. The enhanced rate of RNA correlated with the rise in protein synthesis rate and growth of crustacean. Marsh *et al.* (2001) reported the correlation between the RNA and protein synthesis rate while the total amount of RNA content found to be higher amounts in Antarctic species by Fraser *et al.* (2007) and Pace *et al.* (2010). They suggests that increased rate of protein synthesis directly correlates with the available synthetic machinery. The rate of protein synthesis may be controlled by the available RNA amount that could counteract the assumed low rates of biosynthesis which occur when the animals acclimated under cold condition. The protein synthesis rate might be regulated by the high peptide chain elongation rates from ribosome's translating protein from mRNA transcripts.

References

APHA, AWWA and WPCP, 1992. *Standard Methods for Examination of Water*, 18th Edn. American Public Health Associated, Washington.

Bial, M., 1902. *Den. Med. Woch.*, 28: 58–63.

Currie, S. and Tufts, B., 1997. Synthesis of stress protein 70 (Hsp70) in rainbow trout (*Oncorhynchus mykiss*) red blood cells. *J. Exp. Biol.*, 200: 607–614.

Currie, S., Moyes, C. D. and Tufts, B.L., 2000. The effects of heat-shock and acclimation temperature on hsp70 and hsp30 mRNA expression in rainbow trout: *In vivo* and *in vitro* comparisons. *J. Fish Biol.*, 56: 398–408.

Dahlhoff, E.P., 2004. Biochemical indicators of stress and metabolism: applications for marine ecological studies. *Annual Review of Physiology*, 66: 183–207.

Das, A.B., 1967. Biochemical changes in tissues of goldfish acclimated to high and low temperatures. II. Synthesis of protein and RNA of subcellular fractions and tissues composition. *Comp. Biochem. Physiol.*, 21: 469–485.

Fraser, K.P.P. and Rogers, A.D. 2007. Protein metabolism in marine animals: The underlying mechanism of growth. *Adv. Mar. Biol.*, 52: 267–362.

Gorell, T.A. and Gilbert, L.L., 2004. Protein and RNA synthesis in the pre molt crayfish, Orconectes virilis. *J. Comp. Physiol. Neuroethology, Sensory, Neural and Behavioral Physiology*, 73: 345–356.

Haschemeyer, Audrey E.V. and Roger Persell, 1984. Comparative analysis of leucine transport in temperate fish liver *in vivo*. *Journal of Comparative Biochemistry and Physiology Part A: Physiology*, 78(2): 381–385.

Houlihan, D.F., Hall, S.J. and Gray, C., 1989. Effect of ration on protein turnover in cod. *Journal of Aquaculture*, 79: 103–110.

Jayaprakas, V. and Sambhu, C., 1996. Growth response of white prawn, *Penaeus indicus* to dietary L. carnitine. *J. Asian Fisheries Science*, 9: 209–219.

Jurss, K., Bittrof, Th., Vokler, Th. and Wacke, R., 1987. Effect of temperature, food deprivation and certain enzyme activities in rainbow trout (*Salmo gairdneri* Richardson). *Journal of Comparative Biochemistry and Physiology*, 87B: 241–253.

Jeffery, Kent C., Prosser, Ladd and Graham, Clenn, 1992. Alterations in liver composition of channel catfish (*Ictalurus punctatus*) during seasonal acclimatization. *Journal of Physiological Zoology*, 65(50): 667–684.

Khan, M.A. and Jafri, A.K., 1991. Protein and nucleic acid concentration in the muscle of catfish *Clarias batrachus* at different dietary protein level. *Asian Fisheries Science*, 4: 75–84.

Mali, R.P. and Kadam, M.S. 2010. Comparative fluctuation in RNA contents of *Barytelphusa guerini* after exposure to zinc and cadmium sulphate toxicity. *Journal of Eco-physiology Occupational Health*, 10: 173–176.

Marsh, A.G., Maxson, R.E. and Manahan, D.T., 2001: High macromolecular synthesis with low metabolic cost in Antarctic sea urchin embryos. *Science,* 291: 1950–1952.

Pace, D.A., Maxon, R. and Manahan, D.T., 2010. Ribosomal analysis of rapid rates of protein synthesis in the Antarctic sea urchin *Sterechinus neumayeri. Journal of Biol. Bull.,* 218: 48–60.

Prosser, C.L. and Brown, F.A., 1961 *Comparative Animal Physiology*. W.B. Saunders, Philadelphia.

Raj, D.S. and Selvarajan, V.R., 1992. Influence of quinalphos a organophosphrous pesticide, on the biochemical constituents of the tissues of fish, *Oreochromis mossambicus. Journal of Environmental Biology*, 13: 181–185.

Sadasivam, S., Radhashanmugasundaram and Shanmugasundaram, E.R.B., 1975. *Arogya Journal Health Science*, 1: 125.

Tripathi, G. and Verma, P., 2003. Starvation induced impairment of metabolism in a freshwater catfish. *Z. Naturforsch*, 58C: 446–451.

Chapter 29

Study of Lipase Activity in Snake Headed Fish, *Channa punctatus* (Bloch, 1972) Under Influence of Temperature Stress

☆ *A.R. Jagtap*

ABSTRACT

Temperature is the dominant ecological factor influences of enzyme activity on all animal lives. The aquatic animals are highly sensitive to the temperature fluctuations. In present chapter, snake headed fish *Channa punctatus* was exposed for 24, 48, 72 and 96 hours under stress conditions of temperature. The notable changes in protease enzyme content in liver were observed and compared with control set. It showed that sudden asphyxia developed in animals facing stress conditions and their normal activities were affected. As a result, the lipase enzyme activity was significantly increased at as temperature increases than that of the control.

Keywords: Lipase, Liver, Temperature, Channa punctatus.

Introduction

The existence of enzymes was established in the middle of the 19th century by scientists studying the process of fermentation. Their role as catalysts of all living things followed rapidly. Enzymes were known for many years as ferments, a term derived from the Latin word for yeast. In 1878 the name enzyme, from the Greek words meaning "in yeast," was introduced; since the late 19th century it has been universally used.

Enzymes are essential to sustain life. They are widely known biological catalyst. They are mostly protein catalysts which are able to speed a chemical reaction up and they are not expended during the reaction. Digestive enzymes are secreted along the gastrointestinal tract and break down the food in the body so that the nutrients can be absorbed. They are present in the food which has a great importance having plenty of raw foods in the diet. The enzymes help to start the process of digestion reduces the body's need to secret digestive enzymes. Proteases are a group of proteinases or proteolytic enzymes. They belong to class of hydrolases, catalyze the reaction of hydrolysis of various bonds along with water molecules and are participated in the digestion of long chain proteins into short fragments, splitting of peptide bonds within amino acid residues. Some proteases detach the terminal amino acids from the protein chain (such as aminopeptidases, carboxypeptidases); some on the internal peptide bonds of protein molecules (trypsin, pepsin, chymotrypsin, papain and elastase). These enzymes are involved in the various metabolic activities (Slenzka *et al.*, 1994; Zavodszky *et al.*, 1998).

Shaklee (1977) recorded that the differences in environmental temperature leads to significant changes in the levels of activity of different enzymes were observed in one or more tissues. Enzymes in a single metabolic pathway (*e.g.*, glycolysis) usually exhibited parallel changes in a given tissue. Enzymes in different pathways, on the other hand, frequently exhibited changes in opposite directions, one group of enzymes increased in the cold while the other group decreased, indicating that major metabolic reorganizations were occurring. The patterns of change among tissues were distinctly different; therefore changes observed in one tissue cannot readily be generalized to other tissues.

Temperature is a major factor of influence of enzyme activity (Kuzmina *et al.*, 1996). Digestive glands are the main site of extra and intracellular digestion; they typically store large amounts of sugars, proteins and lipids. Aquatic environment contains the largest pool of diversified genetic material and, hence represents an enormous potential for different sources of enzymes. Freshwater fishes were regularly sampled for physiological measurements. Therefore the present investigation focused study on changes in composition of digestive enzyme *i.e.* lipase under cold and warm temperature stress.

Materials and Methods

The freshwater fish, *Channa punctatus* were collected from the river Godavari of Nanded (Maharashtra). They were kept in glass aquarium for 8-10 days prior to experimentation. Digestive enzyme *i.e.* lipase in freshwater fish *Channa punctatus* estimated procedure given by Nagabhushnam *et al.* (1981). The estimation was carried out on liver of snake headed fish, *Channa punctatus*. The enzyme activity of lipase was estimated by following procedure:

Preparation of Tissue Extract

The freshwater fish, *Channa punctatus* dissected and 2 gm digestive gland *viz.* liver and stomach were separated. The tissues were homogenized in mortar and pestle to prepare 2 per cent extract. Half extract boiled to inactivate the digestive enzymes. The set was treated as control. The remaining half quantity was used for experimental set. The tissue extract prepared used for further estimation for lipase activity.

Estimation of Lipase Activity

To the known volume of 2 per cent tissue extract an equal volume of 5 per cent amyl-acetate was added in a conical flask and mixed thoroughly; a thin layer of toluene also added. The extract solution was plugged with cotton and incubated at room temperature for 24 hours. This serves as experimental extract. After boiling the extract of control set the same procedure was applied up to incubation period. After 24 hours, the toluene layer was removed from control and experimental set. Then, 20 ml of experimental extract taken in a conical flask and 3-4 drops of phenolphthalein indicator was used and titrated against N/20 sodium hydroxide solution. The end point was recorded when the colourless solution turns pink. The same procedure was carried out for control extract. The difference between two readings gives the amount of lipase activity. The lipase activity expressed in terms of mg/gm wet wt. of tissue/hr.

Results and Discussion

The results of lipase activity liver of snake headed freshwater fish *Channa punctatus* in relation to thermal stress (acclimation at cold and warm conditions) have been represented in the graph plotted along with control; the values were expressed in mg/gm wet wt. of tissue/hr.

Table 29.1: Effect of Temperature on Lipase Activity in *Channa punctatus* at Cold and Warm Temperature Stress.

Sl.No.	Temperature of Water Bath Maintained	Period of Exposure	Liver (mg/gm wet wt. of tissue/hr)
1.	15° C ± 1° C	CONTROL SET (26°C ± 1°C)	2.280 ±0.62
		24 Hrs	2.321 ±1.65
		48 Hrs	2.262 ±0.86
		72 Hrs	2.161 ±0.04
		96 Hrs	2.042 ±0.54
2.	20° C ± 1° C	CONTROL SET (26°C ± 1°C)	2.281 ±0.33
		24 Hrs	2.243 ±1.70
		48 Hrs	2.211 ±0.32
		72 Hrs	2.182 ±0.68
		96 Hrs	2.142 ±1.34
3.	30° C ± 1° C	CONTROL SET (26°C ± 1°C)	2.280 ±0.25
		24 Hrs	2.281 ±0.12
		48 Hrs	2.298 ±0.14
		72 Hrs	2.304 ±1.01
		96 Hrs	2.312 ±0.84
4.	35° C ± 1° C	CONTROL SET (26°C ± 1°C)	2.276 ±0.34
		24 Hrs	2.362 ±0.12
		48 Hrs	2.461 ±0.14
		72 Hrs	2.640 ±1.01
		96 Hrs	2.844 ±0.84

Each Value is Mean of Six Observations ± S. D.

The freshwater fish, *Channa punctatus* exposed to cold temperature stress at 15 ± 1° C showed remarkable changes in lipase activity in liver. The lipase activity was suddenly decreased up to 96 hrs period of exposure as compared to control set. The lipase activity in control set was 2.280 mg/gm wet wt. of tissue/hr. The values obtained for lipase activity in liver at 24 hrs, 48 hrs, 72 hrs and 96 hrs period of exposure were found to be 2.321, 2.262, 2.161 and 2.042 mg/gm wet wt. of tissue/hr

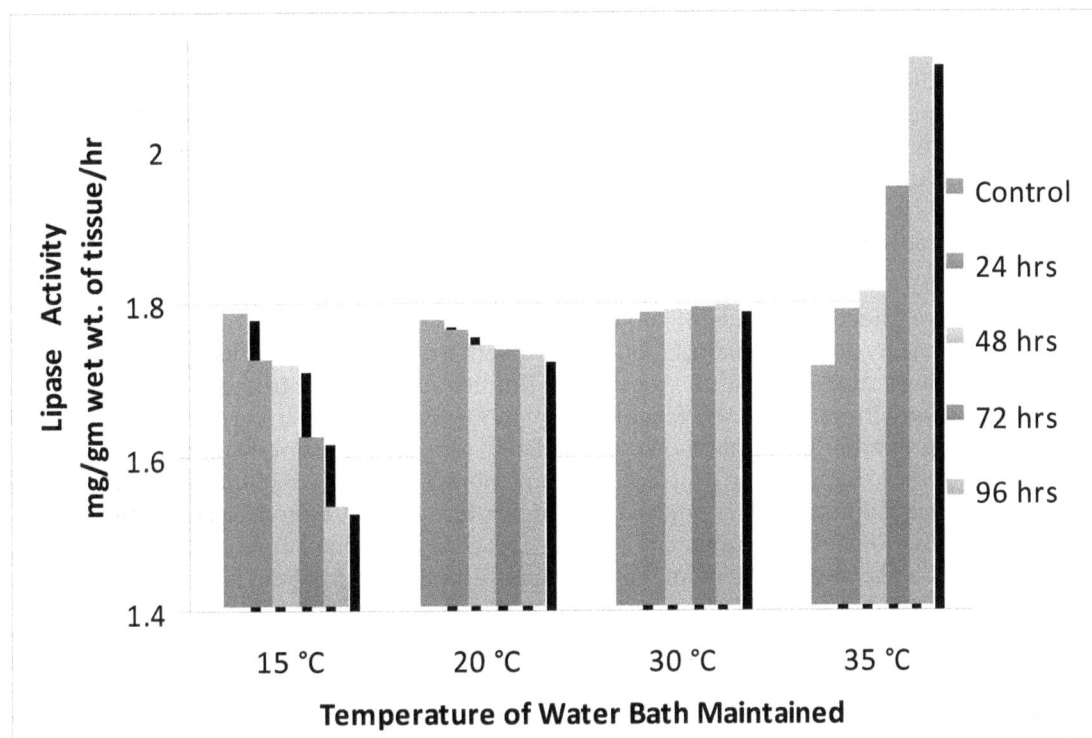

Figure 29.1: Effect of Temperature on Lipase Activity in Liver of Freshwater Fish,
***Channa punctatus* at Cold and Warm Temperature Stress.**

respectively. The recorded lipase activity in liver for control set for 20 °C was decreased at 20 ± 1 °C. The values obtained for 24 hrs, 48 hrs, 72 hrs and 96 hrs were found to be 2.243, 2.211, 2.182 and 2.142 and for control set was 2.281 mg/gm wet wt. of tissue/hr respectively. At 30 ± 1° C lipase activity found in liver was slowly increasing at this temperature stress. The values obtained were for 24 hrs, 48 hrs, 72 hrs and 96 hrs were 2.281, 2.298, 2.304 and 2.312 mg/gm wet wt. of tissue/hr respectively. At 30 ± 1° C the lipase activity in liver in control set was 2.280 mg/gm wet wt. of tissue/hr. Temperature has profound effect on the lipase activity in liver of *Channa punctatus* at 35 °C temperature stress. The fishes exposed to 35 °C temperature stress showed increase in lipase activity as compared to control set (2.276 mg/gm wet wt. of tissue/hr). The values up to 96 hrs of period of exposure were 2.362, 2.461, 2.640 and 2.844 mg/gm wet wt. of tissue/hr respectively.

The impact of climatic variations on aquatic communities has been well-documented (Bakun, 1996). Warming of the atmosphere affects all stages of life of aquatic populations directly or indirectly. Critical temperatures for different fish species have been defined by the onset of anaerobic metabolism due to the mismatch of oxygen demand and oxygen supply (Portner *et al.*, 1998). Exposure of animals to temperatures above or below the limits of critical temperatures leads to death if thermal acclimation is not prevalent (Sommer *et al.*, 1997). A regulatory enzyme must be capable of both efficiently catalyzing a metabolic transformation and of varying its rate of catalysis in response to changes in the cell's need for the product(s) of the pathway. Changes in the external environment of the organism, such as the ambient temperature and oxygen content, may also affect the chemistry of the cell. Therefore, modulations in the activities of regulatory enzymes must occur according to the limitations imposed by the changing external environment.

The metabolism of lipids is under the control of adipokinetic and hyperglycemic hormones. Lipids taken as nutrients are degraded by lipase enzymes to diglycerides and monoglycerides for absorption after which they converted to phospholipids and stored in hepatopancreas in crabs (liver in fishes) (Mulford and Villena, 2000). The present investigation lipase activity was more at warm temperature stress in liver of *Channa punctatus* while under cold acclimation the fishes showed decrease in lipase activity as temperature decreases. The increased lipase activity reveals that degradation of lipids.

Temperature affects all levels of biological organization, is a crucial determinant of the biogeography and physiological characteristics of poikilotherms. It alters the velocity of chemical and enzymatic reactions, rates of diffusion, membrane fluidity and protein structure (Hochachka and Somero, 2002). The thermal sensitivity of membrane processes is due to the strong effect of temperature on the physical properties of membrane lipids, which in turn have a major influence on associated proteins. A decrease in temperature usually reduces membrane fluidity, which can lead to membrane dysfunction. Poikilotherms usually counteract this temperature effect by remodelling membrane lipids, a process known as homeoviscous adaptation (HVA), *via* changes in phospholipid head groups, fatty acid composition and cholesterol content that compensate for the effect of temperature on membrane structure (Hazel, 1995).

The digestion rate and amount of enzymes produced decrease with decreasing temperature (Vonk *et al.*, 1984; Jobling 1995). Several workers reported the information on different teleost species (Ugolev *et al.*, 1983; Hazel 1993; Kuzmina *et al.*, 1996) demonstrates a different effect of temperature on digestive enzyme performance of warm water and coldwater species. Several researchers report digestive enzyme optima at temperatures not encountered in nature (45-60 °C) (Kuzmina *et al.*, 1991; Pyeun, *et al.*, 1991), while other studies confirm thermal inactivation (50 per cent) of fish digestive enzymes at 35-55 °C (60min) (Dimes *et al.*, 1994).

References

Bakun, A., 1996. *Patterns in the Ocean Processes and Marine Population Dynamics*. California Sea Grant College System, La Jolla, pp. 323.

Dimes, L., Garcia-Carreno, F. and Haard, N., 1994. Estimation of protein digestibility–3. Studies on the digestive enzymes from the pyloric caeca of rainbow trout and salmon. *Comparative Biochemistry and Physiology*, 109A: 349–360.

Hazel, J.R., 1993. Thermal biology. In: *The Physiology of Fishesm* (Ed.) D. Evans. pp. 427–467. CRC Press, London, UK, 592pp. (Chapter14).

Hazel, J.R., 1995. Thermal adaptation in biological membranes: Is homeoviscous adaptation the explanation? *Annu. Rev. Physiol.*, 57: 19–42.

Hochachka, P.W. and Somero, G.N., 2002. *Biochemical Adaptation: Mechanism and Process in Physiological Evolution*. Oxford University Press, New York, pp. 466.

Jobling, M., 1995. Digestion and absorption. In: *Environmental Biology of Fish*, (Ed.) M. Jobling. Chapman and Hall, London, UK, 6: 176–210.

Kuzmina, V.V. and Kuzmina, Y.G., 1991. Characteristics of some digestive tract enzymes in the sterlet, *Acipenser ruthenus*. *Journal of Ichthyology*, 31: 120–129.

Kuzmina, V.V., Golovanova, L.L. and Izvekova, G.I., 1996. Influence of temperature and season on some characteristics of intestinal mucosa carbohdrases in six freshwater fishes. *Comparative Biochemistry and Physiology*, 113B: 255–260.

Mulford and Villena, 2000. Cell cultures from crustaceans shrimp, crabs and crayfish. In: *Aquatic Invertebrates Cell Culture*, (Eds.) C. Mothersil and B. Austin. Springer, Chichester, pp. 63–96.

Nagabhushnam, R., Awad, V. R. and Sarojini, R., 1981. *A Practical Handbook on 'Laboratory Exercises in Animal Physiology'*, University Leadership Project in Biology (Zoology) Publication, Sponsored by U.G.C., pp. 18–19.

Portner, H.O., Hardewig, I., Sartoris, F. J. and Van Dijk, P.L.M., 1998. In: *Cold Ocean Physiology*. pp. 88–120, (Eds.) H.O. Portner and R. Playle. Cambridge University Press, UK.

Pyeun, J., Cho, D. and Heu, M., 1991. Comparative studies on the enzymatic properties of trypsins from cat-shark and mackerel. 1. Purifications and reaction conditions of the trypsins. *Bulletin of Korean Fisheries Society*, 24: 273–288.

Shaklee, J. B., Christiansen, J. A, Sidell, B. D. Prosser, C. L. and Whitt, G. S., 1977. Molecular aspects of temperature acclimation in fish: contributions of changes in enzyme activities and isozyme patterns to metabolic reorganization in the green sunfish. *J. Exp. Zool*, 201: 1–20.

Slenzka, K., Appel, R. and Rahmann, H., 1994. Development and altered gravity dependent changes in glucose 6–phosphate dehydrogenase activity in the brain of cichlid fish, *Oreochromis mossambicus*, *International Journal of Neurochemistry*, 26: 579–585.

Sommer, A., Klein, B. and Portner, H.O., 1997. Temperature induced anaerobiosis in two populations of the polychaete worm *Arenicola marina* (L). *J. Comp. Physiol* (B). 167: 25–35.

Ugolev, A.M., Egorova, V., Kuzmina, V.V. and Grudskov, A., 1983. Comparative molecular characterization of membrane digestion in fish and mammals. *Comparative Biochemistry and Physiology* 76B: 627–635.

Vonk, H.J. and Western, J.R.H., 1984. *Comparative Biochemistry and Physiology of Enzymatic Digestion*. Academic Press, New York, NY, USA, pp. 501.

Zavodszky, P., Kardos, J., Svingor, A. and Petsko, G.A., 1998. Adjustment of conformational flexibility is a key event in the thermal adaptation of proteins. *Proc Natl Acad Sci USA*, 95: 7406–7411.

Chapter 30

Protease Activity in Liver of Snake Headed Fish, *Channa punctatus* (Bloch, 1972) on Exposure to Temperature Stress

☆ *A.R. Jagtap and R.P. Mali*

ABSTRACT

Temperature is the dominant ecological factor influences of enzyme activity on all animal lives. The aquatic animals are highly sensitive to the temperature fluctuations. In present chapter, snake headed fish *Channa punctatus* was exposed for 24, 48, 72 and 96 hours under stress conditions of temperature. The notable changes in protease enzyme content in liver were observed and compared with control set. It showed that sudden asphyxia developed in animals facing stress conditions and their normal activities were affected. As a result, the protease enzyme activity was significantly increased at as temperature increases than that of the control.

Keywords: Protease, Liver, Temperature, Channa punctatus.

Introduction

Enzymes are essential to sustain life. They are widely known biological catalyst. They are mostly protein catalysts which are able to speed a chemical reaction up and they are not expended during the reaction. Digestive enzymes are secreted along the gastrointestinal tract and break down the food in the body so that the nutrients can be absorbed. They are present in the food which has a great importance having plenty of raw foods in the diet. The enzymes help to start the process of digestion reduces the body's need to secret digestive enzymes. Proteases are a group of proteinases or proteolytic enzymes.

They belong to class of hydrolases, catalyze the reaction of hydrolysis of various bonds along with water molecules and are participated in the digestion of long chain proteins into short fragments, splitting of peptide bonds within amino acid residues. Some proteases detach the terminal amino acids from the protein chain (such as aminopeptidases, carboxypeptidases); some on the internal peptide bonds of protein molecules (trypsin, pepsin, chymotrypsin, papain, elastase). These enzymes are involved in the various metabolic activities (Slenzka *et al.*, 1994; Zavodszky *et al.*, 1998).

Temperature has been considered as a measure of the molecular motion (Lehinger, 1964). There is usually a limit beyond for biochemical processes, damage tissue and damage of tissues. The activity of enzymes also depends on the environmental temperature, substrate concentration. The change in temperature also changes protein synthetic activities, degradation of proteins (Prosser *et al.*, 1952). The low and high temperature affects the physiological activities of living organisms. Enzyme cause an increase in the ratio of the reaction, while not being consumed in the reaction. The enzymes are affected by many factors *viz.* temperature, pH and environmental stress condition. The activity of enzymes changed with change in temperature. As temperature inclined the rate of chemical reaction increases because of increase in motion of molecules causes increase in interactions between an enzyme and its substrate (Campbell *et al.*, 1999; Morgan and Judith, 2000). The effect of temperature leads to regulatory sensitivity, catalytic potential and structural stability to maintain enzymatic properties (Somero, 1975).

Temperature is a major factor of influence of enzyme activity (Kuzmina *et al.*, 1996). Digestive glands are the main site of extra and intracellular digestion; they typically store large amounts of sugars, proteins and lipids. Aquatic environment contains the largest pool of diversified genetic material and, hence represents an enormous potential for different sources of enzymes. Freshwater fishes were regularly sampled for physiological measurements. Therefore the present investigation focused study on changes in composition of digestive enzyme *i.e.* protease under cold and warm temperature stress.

Material and Methods

The freshwater fish, *Channa punctatus* were collected from the Godavari River, Nanded (Maharashtra). They were kept in glass aquarium for 8-10 days prior to experimentation. Digestive enzyme *i.e.* protease was estimated by Nagabhushnam *et al.* (1981). The tissue was homogenized in mortar and pestle to prepare 2 per cent extract. Half extract boiled to inactivate the digestive enzymes (control). The remaining half quantity was used for experimental set. The equal volume of 1 per cent peptone was added as a substrate in homogenate and the contents were mixed thoroughly. A thin film of toluene was added and incubated for 24 hrs. This serves as experimental extract. After incubation by removing the extract 20 ml experimental extract treated with 10 per cent formaldehyde by adding 3-4 drops of phenolphthalein as an indicator. The mixture titrated against N/10 sodium hydroxide solution. The end point was recorded when the colourless solution turns pink. The same procedure applied for control extract. The difference between two readings gives the amount of protease activity. The protease activity expressed in terms of mg/gm wet wt. of tissue/hr.

Results

The results of protease activity in liver of snake headed fish *Channa punctatus* in relation to thermal stress (acclimation at cold and warm conditions) have been represented in the graphs plotted along with control.

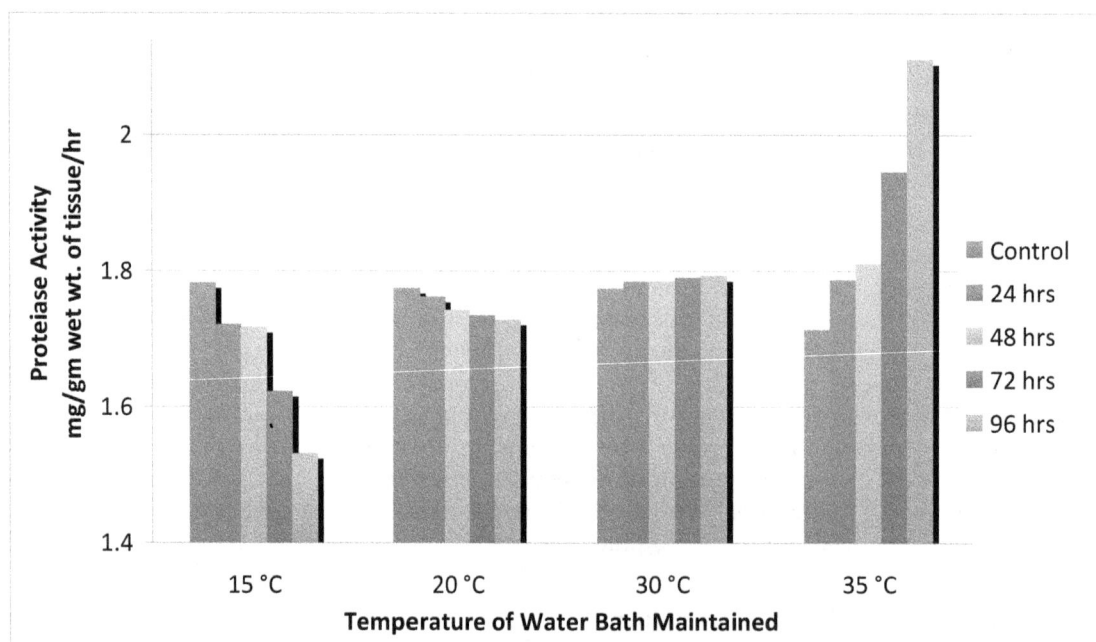

Figure 30.1: Effect of Temperature on Protease Activity in Liver of Freshwater Fish,
***Channa punctatus* at Cold and Warm Temperature Stress.**
(Each Value is Mean of Six Observations ± S. D.)

The freshwater fish, *Channa punctatus* exposed to cold temperature stress at 15 ± 1° C showed remarkable changes. The protease activity in liver of fish *Channa punctatus* in control set was ranged between 1.782 mg/gm wet wt. of tissue/hr. The level of protease activity in liver of fish subjected to cold condition was found to be decreased as compared to normal conditions as the day's progress. The obtained values at 24 hrs, 48 hrs, 72 hrs and 96 hrs period of exposure were found to be 1.721, 1.716, 1.622 and 1.531 mg/gm wet wt. of tissue/hr respectively. At 20 °C protease activity in liver of *Channa punctatus* for control set was 1.774 mg/gm wet wt. of tissue/hr. The decreasing trend was also found at 20 °C temperature stress. The values obtained for 24 hrs, 48 hrs, 72 hrs and 96 hrs were found to be 1.761, 1.742, 1.734 and 1.727 mg/gm wet wt. of tissue/hr respectively. The noticeable increase in protease activity was observed at 30 ± 1° C. The trend continues also for 35 ± 1° C temperature stress. The maximum increase in protease activity was at 35 ± 1° C as compared to 30 ± 1° C. The protease activity in liver for control set at 30 ± 1° C and 35 ± 1° C were 1.773 and 1.712 mg/gm wet wt. of tissue/hr respectively. At 30 ± 1° C the protease activity in liver for 24 hrs, 48 hrs, 72 hrs and 96 hrs period of exposure were found to be 1.784, 1.785, 1.789 and 1.792 mg/gm wet wt. of tissue/hr respectively. At 35 ± 1° C protease activity up to period of 96 hrs were 1.786, 1.831, 1.944 and 2.110 mg/gm wet wt. of tissue/hr respectively.

Discussion

Enzymes reduce the activation energy required for reaction, while temperature influences the fraction of molecules with enough energy to react. A primary determinant of the inherent temperature sensitivity of any reaction is the enzyme catalytic efficiency. Enzymes which are highly efficient catalysts typically have low temperature sensitivity. There are so many factors, which can change the functioning of enzymes and thereby the temperature sensitivity of biochemical reactions that this could be considered as one of the mega problems of ectothermy. Any change in temperature may well differentially perturb a wide range of biochemical processes and integrating these effects to achieve an overall function is a huge problem for ectotherms (Hochachka,1991).

Fishes being poikilothermous animals show the changes in the digestive capabilities with the thermal fluctuations. Al-Hussain (1949 b), reviewing the study on the physiology of digestion in fishes, described correlation between food and digestive enzymes. The enzyme proteases facilitate the breakdown of proteins. The present study showed the increased activity of proteases at warm temperature stress in liver of freshwater fish, *Channa punctatus*. The increase in proteolytic activity showed breakdown of protein molecules as temperature increases. There was decline in protease activity was found in liver of fish at cold temperature stress fishes.

The digestion rate and amount of enzymes produced decrease with decreasing temperature (Jobling, 1995). Several workers reported the information on different teleost species (Hazel, 1993; Kuzmina *et al.*, 1996) demonstrates effect of temperature on digestive enzyme performance of warm water and coldwater species. Several researchers report digestive enzyme optima at temperatures not encountered in nature (45-60 °C) (Pyeun, *et al.*, 1991), while other studies confirm thermal inactivation (50 per cent) of fish digestive enzymes at 35-55 °C (60 min) (Dimes, *et al.*, 1994).

References

Al Hussain, 1949b. On the functional morphology of the alimentary tract of some fishes in relation to their feeding habits. *Qurt. J. Micro. Sci. London*, 90: 328.

Campbell, Neil, Reece, J. and Mitchell, L., 1999. *Biology*, 5th Edn. Menlo Park, CA. Benjamin Cummings.

Dimes L., Garcia-Carreno, F. and Haard, N., 1994. Estimation of protein digestibility–3. Studies on the digestive enzymes from the pyloric caeca of rainbow trout and salmon. *Comparative Biochemistry and Physiology*, 109A: 349–360.

Hazel, J.R., 1993. Thermal biology. In: *The Physiology of Fishes*, (Ed.) D. Evans. pp. 427–467. CRC Press, London, UK, 592pp.

Hochachka, P.W., 1991. Temperature: The ectothermy option. In: *Biochemistry and Molecular Biology of Fishes*, (Eds.) P.W. Hochachka and T.P. Ommsen. Amsterdam: Elsevier Science Publishers B.V., 1: 313–322.

Jobling, M., 1995. Digestion and absorption. In: *Environmental Biology of Fishes*, (Ed.) M. Jobling. Chapman and Hall, London, UK, pp. 176–210.

Kuzmina, V.V., Golovanova, L.L. and Izvekova, G.I., 1996. Influence of temperature and season on some characteristics of intestinal mucosa carbohydrases in six freshwater fishes. *Comparative Biochemistry and Physiology* 113B: 255–260.

Lehninger, A., 1964. *Bioenergetics*. Benjamin Inc. pp. 1–10.

Morgan and Judith, 2000. *General Biology Laboratory Bio III*. Menlo Park, CA. Pearson Custom Publishing.

Nagabhushnam, R., Awad, V. R. and Sarojini, R., 1981. *A Practical Handbook on 'Laboratory Exercises in Animal Physiology'*, University Leadership Project in Biology (Zoology) Publication, Sponsored by U.G.C., pp. 18–19.

Prosser, C.L., Bishop, D., Brown, F., Jahn, T. and Wulff, V.J., 1952. *Comparative Animal Physiology*. Saunders Company, pp. 341–381.

Pyeun, J., Cho, D. and Heu, M., 1991. Comparative studies on the enzymatic properties of trypsins from cat-shark and mackerel. 1. Purifications and reaction conditions of the trypsins. *Bulletin of Korean Fisheries Society*, 24: 273–288.

Slenzka, K., Appel, R. and Rahmann, H., 1994. Development and altered gravity dependent changes in glucose 6-phosphate dehydrogenase activity in the brain of cichlid fish, *Oreochromis mossambicus*. *International Journal of Neurochemistry*, 26: 579–585.

Somero, G. N., 1975. Temperature as a selective factor in protein evolution: the adaptational strategy of compromise. *J. Exp. Zool.* 194: 175–188.

Zavodszky P., Kardos, J., Svingor, A. and Petsko, G.A., 1998. Adjustment of conformational flexibility is a key event in the thermal adaptation of proteins. *Proc. Natl. Acad. Sci., USA*, 95: 7406–7411.

Chapter 31

Histochemistry of Pancreas in Freshwater Fish, *Clarias batrachus*

☆ *B. Laxma Reddy and G. Benarjee*

ABSTRACT

In freshwater teleost, *Clarias batrachus* pancreas is dark-reddish in colour and second largest gland associated with alimentary canal. The pancreas has typical polyhedral cells with exocrine and endocrine parts. The endocrine part is formed by the islets of langerhans. The exocrine portion secretes pancreatic juice which is essential for the digestion of carbohydrates, fats and proteins and the endocrine portion secretes hormones which are essential for the control of carbohydrate metabolism. The present chapter deals with the histochemical nature of pancreas.

Keywords: Clarias batrachus, Pancreas, Histochemistry.

Introduction

The pancreas is the second largest gland associated with alimentary tract. It exists as an easily definable compact gland and opens into intestine and it is of dark-reddish, diffused nature. The structural modifications of the alimentary canal has greatly influenced, the nature of the pancreas particularly in the location, distribution and cytoarchitecture. In freshwater teleosts pancreas may be extrahepatic or intrahepatic. An exocrine portion secretes pancreatic juice which is essential for the digestion of carbohydrates, fats and proteins and the endocrine portion secretes hormones which are essential for the control of carbohydrate metabolism. In the teleost fishes, the presence of pancreas was first described by Weber (1827) and Legouis (1873). Others who have contributed to the knowledge of teleostean pancreas in fishes were Shyam Sundari *et al.* (1982), Saritha Singh (1983), Bhatt (1984), Marconi *et al.* (1984), Reifel (1988), Plantikov *et al.* (1990), Lozano *et al.* (1991). The intensive investigation dealing with histological and histochemical aspects of pancreas has not been carried out much.

Hence an attempt was made to study the histology and histochemistry of the pancreas in freshwater cat fish, *Clarias batrachus*.

Material and Methods

The fish *Clarias batrachus* for the present investigation were collected from fishermen at nearby from water tanks and were sacrificed by cervical dislocation. The pancreas was removed and cut into small pieces then they preserved in various fixatives *i.e.*, Bouin's, Susa, Carnoy and Zenker's fluid. After the routine histological preparation of the tissue the sections of 5 micron thickness were cut on rotatory microtome. To study the normal histology of the pancreas Heidenhains Azan (Gurr, 1962) stain was used. Different kinds of histochemical procedures were also followed to elucidate the chemical nature of pancreas. The presence of carbohydrates, proteins, lipids and nucleic acids was determined by using the technique mentioned by Pearse (1968), Gomorie (1952), McManus and Mowry (1960), Lilliee (1965), Humason (1967) and Bancroft (1975).

Results and Discussion

In the present chapter, pancreas has been observed as a prominent digestive gland associated with many activities. The pancreas has typical polyhedral cells with exocrine and endocrine parts. The endocrine part is formed by the islets of langerhans which are seen distributed as isolated patches taking deep stain, scattered in the exocrine pancreas. But the pancreas of a single largest islets of langerhans in the pancreas of *Gambusia affinis* was reported by Bullock (1967). The islets of langerhans contain peripheral alpha cells and central beta cells. The alpha cells differ from beta cells by their larger ovoidal nuclei, while beta cells have spherical smaller nuclei. These findings show a similarly with those reported by Bucke (1971), Williams (1974) and Shyam Sundari *et al.* (1982). Sivadas (1964) reported the presence of only beta cells in *Tilapia mosaambica*. Bucke (1971) reported alpha cells in the centre and beta cells in the periphery in *Esox lucius*. Reifel (1988) described the endocrine cells in the gastrointestinal tract of *Stomachless teleostean fish*. Plantikov (1990) observed the fine structure of pancreatic exocrine cells in tissue of yellow and silver eel, *Anguilla anguilla*. Lozano *et al.* (1991) studied pancreatic endocrine cells in a sea boss *Dicentrachus labrax*.

The histochemical tests have shown that a similar nature in pancreas exists as that of liver. Fairly good amounts of carbohydrates, proteins, lipids and moderate amount of DNA and RNA were found in pancreatic cells. Little quantities of nucleic acids were also demonstrated. As there is a variability in the nature and distribution of the pancreas the collection of pancreatic tissue becomes difficult and as such the complete chemical nature of this important digestive gland is not known. However, Saxena (1965) indicated that pancreas is rich in enzymes mostly zymogens, which serve in digestion of proteins, carbohydrates and fats. The sections of pancreas when subjected to PAS reactions, produced a majenta colour indicated the presence of carbohydrates. With PAS after saliva digestion the pancreatic cells showed a little positivity and thus suggesting a low concentration of glycogen. The complete removal of PAS reaction after acetylation and restoration of the same after deacetylation exhibited the occurrence of 1:2 glycol groups. Schiff's reagent alone has given a mild magenta colour to the section showing the presence of free aldehydes. But with alcian blue at 1.0 pH and 2.5 pH a negative response occurred and thus exocrine and endocrine parts revealed as intense positivity with PAS suggesting the occurrence of large quantities of carbohydrates. When PAS technique was used in combination with alcian blue, the exocrine part revealed the presence of weakly acidic mucosubstances while the endocrine part revealed the presence of neutral mucopolysaccharides. The presence of rich quantities of basic proteins were noticed by the intense positive reaction to mercuric bromophenol blue which is indicative of the proteinous nature of the pancreas. The presence of mild positivity with millon's reaction and P-DMAB nitrite method elucidated the presence of little quantities of tyrosine and tryptophan respectively. Presence of disulphides were indicated when stained with $KMnO_4$/AB staining. With ninhydrin/Schiff method exhibited the occurrence of protein bound amino groups. When stained with ferric ferricyanide the presence of sulfydryl groups of proteins were noticed. Presence of glycoproteins were indicated when the sections of pancreas were stained with congored. Similarly the concentration of protein bound amino groups was more in the endocrine region than exocrine region.

Table 31.1: Histochemical Tests Applied for Pancreas

Histochemical Tests	Results	Histochemical Tests	Results
Periodic Acid/Schiff (PAS)	+	PAS/Saliva	+
Schiff's without oxidation	+	Acetylation/PAS	−
Deacetylation/PAS	+	Alcian blue 1.0 pH	−
Alcian blue 2.5 pH	−	Alcian blue 1.0 pH/PAS	+
Alcian blue 2.5 pH/PAS	+	Alcian blue/Aldehyde fuchsin	+
Mercuric Bromophenol blue	+	Ninhydrin/Schiff	+
Ferric ferricyanide	+	Congored	+
p-DMAB nitrite	+	Feulgen $KMnO_4$/Alcian blue	+
Millon's reaction	+	Copper pthalocyanin	+
Sudan black 'B'	+	Pyronin-Y	+
Methyl green/Pyronin-Y	+	Feulgen Reaction	+

++: Moderately positive; +: Possitive; −: Negative.

The occurrence of lipids and phospholipids was observed in the pancreatic tissue when stained with sudan black 'B' and copperpthalocyanin respectively. The presence of DNA and RNA was also conformed with the positive result of the sections of the pancreas to feulgen's reaction and pyronin-Y technique (Table 31.1).

Acknowledgements

Authors thank Head, Department of Zoology, Kakatiya University for providing necessary laboratory facilities.

References

Bancroft, J. D., 1975. *Histochemical Techniques*. Butterworths, London and Boston.

Gurr, E., 1974. *Staining Animal Tissue: Practical Theoretical*. Leonard Hill (Books), Ltd, pp. 631.

Pearse, A.G.E., 1968. *Histochemistry, Theoretical and Applied*, 2nd Edn. Littlle Brown and Company, Boston, MSS.

Humason, G., 1967. *Animal Tissue Techniques*. W.H. Freeman and Co., San Francisco and London.

McManus, J.F.A. and Mowry, R.F., 1990. *Stainig Methods: Histological and Histochemical*. Paul B. Boeber Inc. Medical Division of Harper and Bros., New York.

Lilliee, R.D., 1965. *Histopathologic Technique and Practical Histochemistry*, 3rd Edn. McGraw Hill, New York.

Saxena, A.B., 1965. The morphology and enzymology of the digestive organs in *Rasbora daniconius*. *Proc. of the Zool. Seminar*, Vikram Univ., p. 111–118.

Weber, E.H., 1827. The exocrine pancreatic tissue of seven teleosts. *S. Afr. J. Med. Sci. Biol. Suppl.*, 11: 79–86.

Legoius, P., 1873. Recherches Sur les tubes deweber et., surles pancreas des poisons osseux. *Anm. Sci. Nat. Zool.*, 17: 1–107.

Lozano, M.T., Ayala, A. Gracia, Abad, M. and Agutleiro, B., 1991. Pancreatic endocrine cells in a sea bass. II. Immunocytochemical study of insulin and somatostalin peptides. *Gen. Comp. Endocrinol.*, 81(2): 198–206.

Marconi, Strip, Antonio Carios and Ferri, Sylvio, 1984. Electronic microscopic study of cytoplasm inclusions in acinar cells of teleost (*Pinelodus maculatus*) pancreas. *Zool. Anz.*, 212(1/2): 117–121.

Plantikov, H., Kennizt, P., Putzke, H.P., Sporman, H. and Letko, G., 1990. Fine structure of pancreatic exocrine cells in tissue of yellow and silver cell cells (*Anguilla anguilla* L.). *Zool. Anz.*, 225(1/2): 55–62.

Reifel, C.W., 1988. Endocrine cells in the gastrointestinal tract of stomachless teleostean fish. *Anatanz.*, 167(4): 259–263.

Saritha singh 1983. Comparative studies of pancreas in some teleostean fishes. *Mastya*.9–10: 125–128.

Shyam Sundari, K., and Raj Kumari. V. J. V. and Hanumantha Rao, K., 1982. Observations on the panc reas of the marine Lizard fish, *Saurida tumbil* (Bloch). *J. Fish. Biol.*, 21: 449–454.

Bhatt, S. D,1984. Light microscopic changes in the Islets of Langerhans *in C. batrachus* (Linn.) under certain experimental conditions. *Matsya*, 9–10: 1–7.

Bucke, D., 1971. The anatomy and histology of the alimentary tract of the carnivorous fish the pike, *Esox lucius* L. *J. Fish. Biol.*, 3: 421–431.

Williams, R. H., 1974. The pancreas. In: *Textbook of Endocrinology*, (Ed.) R.H. Willis. W.B. Saunders, Co., London, pp. 613–802.

Chapter 32

Effect of Sublethal Concentrations of Insecticide Hitcel on Oxygen Consumption Rates of Snail *Lymnaea accuminata* (Lamarck)

☆ *A.N. Lonkar*

ABSTRACT

Snails *Lymnaea accuminata*, were exposed to 3.6 mg/l (1/10th of 96 hr LC50) and 7.2 mg/l(1/5th of 96hr LC50) of insecticide Hitcel. Initially at 24 hours, at both the concentrations Increase in oxygen consumption rate is noted. At 48 hours stage, decrease in oxygen consumption rate is noticed at both the sublethal concentrations. At 72 hours stage in both the sublethal concentrations reduction in oxygen consumption is noted. At the end of 96 hours, again in both the concentration decrease in oxygen consumption prevailed. These experimental snails upon transfer to freshwater for 24 hours, showed recovery in oxygen consumption rate. Oxygen consumption rate are discussed with respect to sublethal concentration and time of exposure to toxicant Hitcel.

Keywords: Hitcel, Oxygen consumption, Lymnaea accuminata.

Introduction

Change in the oxygen consumption rates serves as the one of the indicators of environmental stress. Effect of insecticides on the oxygen consumption of mollusc has been studied by Agarwal (1978) he observed effects of Endrin on certain freshwater gastropods, Rao and Mane (1978) noted effects of Malathion on survival and respiration of *Mytilus gallanoprovincialis*. Moorthy *et al.* (1984) reported changes in the respiration and ionic constituents in tissues of fresh water mussel exposed to

Methyl parathion, Thosar and Lonkar (1994) studied effect of Metasystox on the oxygen consumption of *Vivipara bengalensis* and Rohankar and Kulkarni (2005) reported alteration in oxygen consumption in freshwater snail *Bellamya bengalensis* during pesticide exposure. Lonkar (2012) reported changes in oxygen consumption rates of mollusc *Indoplanorbis exustus* exposed to sublethal concentrations of insecticide Tricel.

Less information is available on oxygen consumption of snail *Lymnaea accuminata* exposed to insecticide. Therefore the present investigation is attempted.

Material and Methods

The specimens of snail, *Lymnaea accuminata* were collected locally and were acclimated to the laboratory condition for 7 days in the glass aquaria filled with chlorine free tap water. Physico-chemical parameters of chlorine free tap water showed following ranges, pH 7.1 to 7.4, dissolved oxygen 7.2 to 7.6 mg/l, free CO_2 Nil, Total hardness (as $CaCO_3$) 161-173 ppm. Alkalinity 142-160 mg/l, Temperature 27 to 29 °C.

For finding 96 hr LC 50, static bioassay experiments were set by using the toxicant Hitcel (Profenfos 40 per cent + Cypermethrine 4 per cent EC, Manufactured by Excel Crop Care Limited Jogeshwari west Mumbai Maharashtra, India) insecticide compound. Initially bioassay experiments were set with a wide range of toxicant and finally with closer ranges. Various concentrations were prepared by dilution method. Cleaned similar sized and preweighed Snails were exposed to 3.6 mg/l (1/10th of 96 hrLC50) and 7.2 mg/l (1/5th of 96 hr LC50) concentration for 24, 48, 72 and 96 hours. The toxicant solutions were renewed after every 24 hours. "A closed chamber" method was used for the measurement of oxygen consumption of snails. The oxygen content was determined by Winkler method at the end of 24, 48, 72 and 96 hours. The recovery rates were determined by transferring the experimental animals in toxicant free water. Oxygen consumption was calculated in terms of mg/hr/gram body weight of snail. Respiratory response values were found after calculating the percent normal oxygen consumption. In each experiment about 10-15 animal were used.

Results

Snail *Indoplanorbis exustus* of size 1.4. cm and average weight 0.3.7 gms. Were exposed to two sublethal concentration 3.6 mg/l (1/10th of 96 hr LC 50) and 7.2 mg/l (1/5th of 96 hr LC 50) of Hitcel. Initially after 24 houra, at both the concentrations an increase in oxygen consumption rate is noticed (109.40 per cent and 107.11 per cent) at lower and higher concentration respectively. At 48 hours stage decrease in oxygen consumption rate is recorded at lower concentration (Oxygen consumption rate is 98.40 per cent), while at higher concentration oxygen consumption rate is found to be decreased to 94.30 per cent. At 72 hours stage in both the sublethal concentrations, reduction in oxygen consumption rate is noticed the values being 90.30 per cent and 88.12 per cent for lower and higher concentration respectively. At the end of 96 hours, again in both the concentrations, decrease in oxygen consumption rate prevailed. At lower concentration the oxygen consumption rate is 88.50 per cent and for higher concentration it is 82.42 per cent. Upon transfer to freshwater for 24 hours these experimental snails showed recovery in oxygen consumption rate (104.32 per cent and 101.18 per cent) in snails previously exposed to lower and higher concentrations respectively (Table 32.1 and Figure 32.1).

Discussion

In the present investigation at 24 hours stage an increase in oxygen consumption rate is noticed in the snail *Lymnaea accuminata* exposed to both sublethal concentration. At 48 hours decrease in

Table 32.1: Changes in Oxygen Consumption Rate of *Lymnaea accuminata* Exposed to different Concentration of Hitcel

Concentration of Hitcel in mg/l	Exposure Period in Hours				Recovery in Tap Water	Oxygen Consumption
	24	48	72	96		
Normal snail	0.0412	0.0307	0.0430	0.0336	0.0352	Rate mg/hr/gm body weight
	100 per cent	100 per cent	100 per cent	100 per cent	100 per cent	Taken as 100 per cent
3.6 mg/l Hitcel	0.04507	0.03020	0.03829	0.02973	0.03672	Rate mg/hr/gm body weight
	109.40 per cent	98.40 per cent	90.30 per cent	88.50 per cent	104.32 per cent	Percent of normal
7.2 mg/l Hitcel	0.04412	0.02895	0.03784	0.02769	0.03561	Rate mg/hr/gm body weight
	107.11 per cent	94.30 per cent	88.12 per cent	82.42 per cent	101.18 per cent	Percent of normal

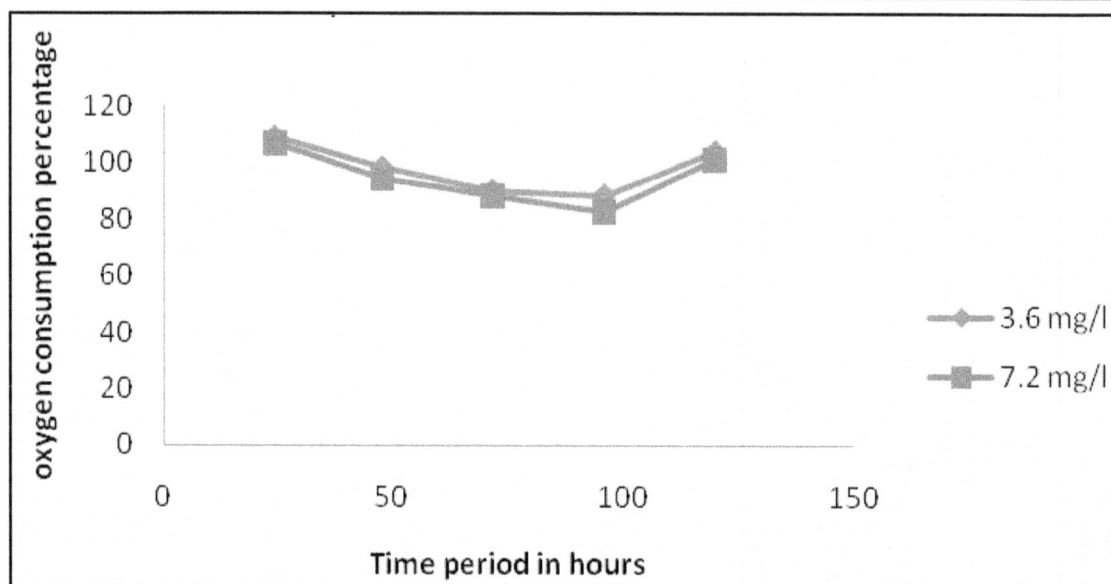

**Figure 32.1: Changes in Oxygen Consumption Rate of *Lymnaea accuminata*
Exposed to different Sublethal Concentration of Hitcel.**

oxygen consumption was noted at lower and higher concentration, at 72 hours stage the decrease in oxygen consumption is found at both the concentration. At 96 hours exposure at both the concentration decrease in oxygen consumption is noted. Rao and Mane (1978) noticed an initial stimulation in oxygen consumption rate of *Mytilus gallanoprovincialis* exposed to low sublathal concentration of Malathion. Hanumante *et al.* (1980) noticed an increase in respiratory rate for 96 hours in the pulmonate, *Onchidium vessiculatum* exposed to sublathal concentration of DDT. Moorthy *et al.* (1984) noticed an initial elavation in oxygen consumption of freshwater mussel *Lamellidens marginalis* followed by decrease in oxygen consumption. (Mane *et al.*, 1984) studied the effect of Cythion - Malathion (0.081 ppm) on oxygen consumption in three freshwater bivalve for 96 hours and reported that, the rate of oxygen consumption increased initially in *Lamellidens carianus* (Lea) and *L. marginalis* (Lamarck), whereas mortality occured in *Indonia caeruleus*. But when *I. caeruleus* were exposed to lower concentration (0.004 ppm and 0.012 ppm) of Cythion - Malathion, the rate of oxygen consumption increased as compared to control. They have attributed this increase in oxygen consumption in all three bivalves to their behaviour resulting in excessive muscular activity. Thosar *et al.* (2001) noticed an increased in the oxygen consumption of snail *L. accuminata* at 24, 72 and 96 hours exposed to 3.4 mg/l concentration of Metasystox. They also found that when these snail were exposed to higher sublathal concentration (6.8 mg/l), the Oxygen consumption rate is increased at all exposure periods *i.e.* 24, 48, 72 and 96 hours. Rohankar and Kulkarni (2005) reported the alteration in oxygen consumption of freshwater snail *Bellamya bengalensis.* exposed to Phosphomidon. The snails was exposed to lethal concentration (0.135mg/l) and two sublethal concentrations (0.045mg/l and 0.0675mg/l) of Phosphamidon an acute and chronic treatment at normal room temperature. They reported significantly rise in oxygen consumption of snail by 1.14 per cent at lethal concentration, 7.6 per cent at higher sublethal concentration,10.88 per cent at higher sublethal concentration at the end of 24 hours exposure. Lonkar

(2012) reported initial increase in oxygen consumption at 24 hours exposure in *Indoplanorbis exustus* exposed to sublethal concentrations 1.4 mg/l and 2.8 mg/l of insecticide tricel.

Agrawal (1978) reported decrease in oxygen consumption of snail *L. accuminata, L. luteola* and *V. bengalensis* exposed to sublathal concentration of organ chlorine insecticide Endrine. He inferred that, the death of snail is due to great reduction in the oxygen consumption. Mahendra and Agrawal(1981) also noticed reduction in the oxygen consumption of snail *Lymnea acuminate* exposed to 10 mg/l and 20 mg/l of Trichlorphon for 24 and 48 hours. Thosar and Lonkar(1994) noted reduction in oxygen consumption rate in snail *V. bengalensis* exposed to two sublathal concentration (1.30 mg/l and 2.60 mg/l) of organophosphorus insecticide Metasystox. Thosar *et al.* (2000) studied respiratory response of the snail *Vivipara bengalensis* exposed to sublethal concentration of the insecticide Fenval (1.85 mg/l and 3.70 mg/l). Fall in oxygen consumption rate was noted at 3.7mg/l concentration. Thosar *et al.* (2001) exposed the snail *Lymnaea accuminata* to Metasystox and reported decrease in oxygen consumption rate at 48 hours exposure at lower sublethal concentration.

Experimental snails *Lymnaea accuminata* after transfer to toxicant free, chlorine free tap water for 24 hour, showed recovery this may be due to observation that the 24 hours time is sufficient for the gills to heal up.

References

Agarwal, H. P., 1978. some observations on the toxic effect of Endrin on certain freshwater gastropods. *Sci. and Cult.*, 44(8): 375–377.

Hanumante, M. M., Deshpande, V. A. and Nagabhushnam, R., 1980. Effect of DDT on the oxygen consumption of normal, pharmacologically treated marine pulmonate, *Onchidium vessiculatum. Indian J. Expt. Biol.*, 18: 753–754.

Lonkar, A. N., 2012. Changes in oxygen consumption rates of Mollusc *Indoplanorbis exustus* (Deshyasa) exposed to sublethal concentrations of insecticide Tricel. *Vidarbha Journal of Science*, 7 (1–2) 13–16.

Mahendra, V. K. and Agrawal, R.A., 1981. Changes in carbohydrate metabolism in various organs of snail Lymnea accuminata following exposure to Trichlorphon. *Acta Pharmacol. Toxicol.*, 48(5): 377–381.

Mane, U. H., Akarte, S.R. and Mulay, D.V., 1984. Effect of Cythion-Malathion on respiration of three freshwater bivalve molluscs from Godavari river near Paithan. *J. Environ Biol.*, 1 (2): 71–80.

Moorthy, K. S., Reddy, B. K., Swamy, K. S. and Chetty, C.S., 1984. Changes in the respiration and ionic constituents in tissues of freshwater mussel exposed to Methyl parathion. *Toxicol. Lett.*, p. 287–291.

Rao, M. B. and Mane, V. H., 1978. Effect of Malathion on survival and respiration of black sea mussel, *Mytilus gallanoprovincialis. J. Hydrobiologia*, 14(6): 100–104.

Rohankar, P. H. and Kulkarni, K. M., 2005. Alteration in oxygen consumption in freshwater snail *Bellamya bengalensis* (Lamarck) during pesticide exposure. *Indian J. Environment and Ecoplanning*, 10(1): 45–48.

Thosar, M. R. and Lonkar, A. N., 1994. Effect of Metasystox on the oxygen consumption of *Vivipara bangalensis* (Lamarck). *Proc. of Nat. Symp. on Eco-environment Impact and Organism Response*, pp. 6.1 – 6.4, Amravati.

Thosar, M.R., Huilgol, N.V. and Lonkar, A. N., 2000. Respiratory response of snail *Vivipara bengalansis* (Lamarck) exposed to sublathal concentration of Fenval. *Proc.87ᵗʰ session of the Indian Sci. Cong.,* Pune. Abst. No.2 Page 2.

Thosar, M.R., Huilgol, N.V. and Lonkar, A. N., 2001. Effect of Metasystox on the respiratory responce of the snail *Lymnea accuminata. J. Aq. Biol.,* 16(2).

Chapter 33

Socio-economic Study of Fishermen Community with Special Reference to Jawahar Reservoir of Nipani from Chikodi Tahsil of Belgaum District, Karnataka

☆ *L.P. Lanka and S.A. Khabade*

ABSTRACT

In present investigation it was observed that the fishermen community has improved their socio-economic status and obtained bread and butter due to fishery activity from Jawahar reservoir. Cultivated fishes *i.e.* major carps, local fishes, crabs, prawns play an important role in socio-economic development of the fishermen community.

Keywords: *Socio-economic study, Fishermen community, Jawahar reservoir, Nipani.*

Introduction

Jawahar reservoir is located in Nipani city of Chikodi Tahsil from Belgaum (Karnataka). The reservoir is man-made in nature and it is a habitat for variety of planktons and animal forms. The catchment area is about 27 hectares. The water storage capacity of the reservoir is about 70 million cubic ft.

This water reservoir is constructed during 1952 for the improvement to water supply system in Nipani City. The value of sub-work was about 2.06 crores. The water of reservoir is used only for drinking and not for agriculture and industrial use. Nipani city utilizes about 12 to 13 lakh gallons water per day from this reservoir. (Municipal Corporation Nipani 2010).

From pre-historic period fishes have been used as protein-rich diet for human beings. The popularity of India began taking active interest in the development of fisheries in 1944 but till 1947 Indian fisheries followed ancient mode of practice.

Nowadays, people are aware about the fisheries sector and started the development of fisheries. More and more people are working in fisheries sector for bread and butter.

Men who know the art of fishing belong to a particular cast and we call them as fishermen. In India; there was a fishermen population of about 7,145,000 (Govt. of India 1988). Day to day this population went on increasing and they are engaged in full time and part time fishing operations.

The fishery potential was studied by Angadi (1986). The socio-economic study of fishermen community from Tasgaon Tahsil of Sangli District was carried out by Sathe *et al.* (2000). The socio-ethnographic profile of fishermen community from Saurashtra was studied by Joshi (1996).

Till today the socio-economic aspects of fishery dependents from Nipani reservoir of Chikodi Tahsil of Karnataka was not studied so the present work was carried out during the year 2009-2011.

Material and Methods

The socio-economic status of fishery dependents was studied by questionnaire method. Morphometry of lakes, according to Welch (1952) pertains to branch of limnology which deals with the measurement of significant morphological features of any basin and its included water mass. These significant features include the shore line, area, depth, slope and volume etc.

Then morphometric features of Jawahar reservoir were studied by collecting data from Nipani Municipal Council.

Results and Discussion

The developmental activities generally bring the changes in society's economic status. To assess the probable impact of fishery development the socio-economic aspects of the people related to fishery activities have been studied.

According to Rao *et al.* (1998), fish is a valuable source of protein and occupies a significant position in the socio-economical fabric of South-Asian countries.

The socio-economic status of fishermen society was studied by studying religion and caste, educational status, number of earning family members, age of the fishermen and average income from fishery activity etc. The results of the present study are incorporated in the Tables 33.1–33.5.

Table 33.1: Religion and Caste of Fishermen

Sl.No.	Religion	Caste	No. of Fisherman	Percentage of Fisherman
1	Hindu	a. Koli	6	22.22 per cent
		b. Bhoi	9	33.33 per cent
		c. Khatik	5	18.51 per cent
		d. Bhangi	3	11.11 per cent
2	Muslim	–	4	14.81 per cent
		Total	**27**	**99.98 per cent**

Among 27 fishermen, 23 fishermen were Hindu by religion while 4 fishermen were Muslim by religion. It means 85.17 per cent fishermen were Hindu and 14.81 per cent were Muslims.

In present chapter it has been found that Bhoi and Koli caste say that the fishing is their inheritance business or occupation. Other fishermen communities engaged willingly in fishing sector. Bhoi cast was found dominant in fishing business.

Table 33.2: Educational Level of Fishermen

Number of Illiterate Fishermen	Number of Literate Fishermen	Total Number of Illiterate and Literate fishermen	Per cent of Illiteracy	Per cent of Literacy
0	27	27	0 per cent	100 per cent

All the fishermen engaged in fishing business were found literate. Their educational qualification was about 4th to 12th standard.

Table 33.3: Number of Earning and Non-earning Members of Fishermen Communities

Number of Earning Members	Number of Non-earning Members	Total Number of Households
27	23	50

In Nipani about 27 fishermen were working in fisheries sector. Other than fishing they work in other industries also. Their family members help them in traditional business but not working anywhere. Their wives help them in selling fishes. Children are engaged in school and colleges.

Table 33.4: Working Age of Male and Female Fishermen

Working Age of Male Fishermen	Working Age of Female Fishermen
35 to 55 yrs.	27 to 45 yrs

Table 33.5: Rates of Fishes during Regular Fishing Season and during Odd-Season

Name of the Fish Market	Name of the Fishes	Rate of Fishes during Fishing Season (Dec–May)	Rate of fishes during Odd Season (Aug–Nov)	Rate of Big Sized Fishes during Fishing Season and Odd Season in Rs.	Average Rate of Fishes during Fishing and Odd Season (Aug–May) in Rs.
Nipani fish Market	Catla Rohu Mrigal Cyprinus	90–100 Rs/kg	70–80 Rs/kg	110–120 Rs/kg	90–100 Rs/kg (mean-95 Rs/Kg)
	Tilapia	50–60 Rs/kg	40–50 Rs/kg	—	45–50 Rs/kg (mean-50 Rs/kg)
	Local fish varieties	30–40 Rs/kg	25–30 Rs/kg	—	27–35 Rs/kg (mean-31Rs/Kg)

Wives of fishermen not directly engage in fishing operation but after fishing they help them for selling the fishes in market. The working age of male fishermen is about 35 to 55 yrs of age and the female is about 27 to 45 yrs.

In present investigation it is found that fishes of Jawahar Reservoir were sold by fishermen in Nipani market. During fishing season the cost of the fishes obtained was 90 to 100 Rs/kg for Catla, Rohu, Mrigal and Cyprinus. While during odd season the cost obtained was 70 to 80 Rs/kg for the same fishes. The above big sized major carps gives cost of 110 to 120 Rs/kg. The average rate of major carps during fishing and odd season ranges between 90 to 100 Rs/kg. The average cost of the Tilapia was 45 to 55 Rs/kg during and odd season. The average cost of local fish variety obtained was 27 to 35 Rs/kg during fishing and odd season.

Other than above fishes some other fishes caught were *Hypophthalmichthys molitrix* (Silver carps), *Mastacembelus armatus* (Eel or wam). The cost of theses fishes ranges between 120 to 150 Rs/kg. The number of these fishes obtained was negligible

The crustaceans obtained were crabs (Paratelphusa) and small prawns (*Macrobrachium kistnensis*). The cost of crabs in Nipani market range between 25 to 30 Rs/kg. The cost of prawns range between 100 to 150 Rs/kg. Economics of fishery is given in the Table 6.

Table 33.6: Economics of Fishery of Jawahar Reservoir during 2011

Sl.No.	Item	Requirement/Rate (Rs)	Total Cost (Rs)
1.	**Recurring cost**		
	1) Fingerlings	50,000 Nos/₹500 per 1000 Nos.	₹25,000/-
	2) Lease amount to municipal corporation	2000 ₹/yr	₹2000/-
	3) Transportation	3200 ₹/yr	₹3200/-
			Total—Rs 32,000/-
2.	**Income**		
	1) Fish production	a. 2000 kg fish – (major carps)/ ₹95 per kg	₹1,90,300/-
		b. 500 kg fish (local fishes)/ ₹50 per kg	₹25,000/-
		c. 100 kg other fishes per ₹31 per kg	₹3,100/-
			Total—₹218,100/-
	2) Minus recurring cost	——	₹32,000/-
	3) Total profit	——	₹1,86,100/-

In present chapter it has been found that total recurring cost is about 32000/- in one year. The fish production of the same year was around 2,18,100/-. The total profit obtained in the year was about 1,86,100/-.

Thus, fishing business gives bread and butter to the fishermen community and it is a profitable business if scientifically carried out.

Rainfall is important factor in the development of fisheries in the region. Rainfall received during the year 2011 is reported in the Table 33.7.

Table 33.7: Rainfall Received during the Year 2011 in mm

Date	Jan	Feb	March	April	May	June	July	Aug	Sept	Oct	Nov	Dec
1					22.7		3.7	5.3	13.9			
2							5.7	7.9	12.0	10.5		
3							4.1	9.2	18.3			
4						17.4	2.1	10.7	11.0			
5						8.2		6.0	1.5			
6								4.1	1.9			
7								6.6	2.6			
8							1.0	3.2	21.2	16.8		
9							6.2	6.6	0.6	7.0		
10						18.0	1.8	2.8		4.9		
11				6.3		20.5	3.6	1.6				
12						11.1				2.0		
13						2.5	18.9	4.1		24.8		
14						4.4	10.1		3.0			
15						2.2	1.5			8.3		
16							4.9		1.0	13.1		
17						22.4	24.0	1.3				
18						5.6	21.5	11.4	3.5			
19							19.0	3.0	1.3			
20						1.5	7.6			10.3		
21							8.8	3.5				
22												
23						3.2	9.5					
24		26.5		1.5			16.7					
25												
26						11.7		4.9				
27						3.5		2.9				
28						11.7		11.0				
29						3.5	2.0	18.5				
30							2.0	14.5				
31							1.0	8.1				
Total	0.0	26.5	0.0	7.8	22.7	147.4	175.7	147.2	91.8	97.7	0.0	0.0
Prog. Total	0.0	26.5	26.5	34.3	57.0	204.4	380.1	527.3	619.1	716.8	716.8	716.8
Rainy Days		1		1	1	14	15	20	8	8		
P.R. days	0	1	1	2	3	17	32	52	60	68	68	68

Source: University of Agricultural Sciences, Dharwad Agricultural Research Station, Nipani.

References

Angadi S.M., 1986. Hydrobiological status of Rajaram Tank. *M.Phil. Thesis,* Shivaji University, Kolhapur.

Government of India, 1988. An appraisal of the marine fishery resources of the Indian Exclusive Economic Zone. Fishery survey of India, Bombay.

Joshi, M.V., 1966. *Economics of Fisheries*. APH Publishing Corporation, New Delhi.

Municipal Corporation Nipani, 2010. Morphometric data of Jawahar Reservoir, Nipani.

Rao, L.M., Rao, G.V. and Sivani, G., 1998. Hydrobiology and ichthyofauna of Mehadrigedda Stream of Vishakhapatnam, Andhra Pradesh. *J. Aqua. Biol.,* 13(1 and 2): 25–28.

Sathe, S.S., Khabade, S.A. and Hujare, M.S., 2000. Studies on wetlands of Tasgaon Tahsil and its importance in relation to fisheries and agricultural productivity. Project repot submitted to University Grants Commission, Western Regional Office, Pune.

Chapter 34

A Revision of Journals in Aquatic Sciences

☆ *Patricio De los Ríos Escalante*

ABSTRACT

The literature about aquatic sciences journals involves multidisciplinary topics from biology, chemistry, physics, and earth sciences. Some of the papers about aquatic sciences are included in topics such as natural sciences that involves mainly biology and its respective branches, whereas it can be included as aquatic sciences journals that involves all disciplines related with aquatic environment, such as aquaculture, fisheries, limnology and oceanography. Also, recently some journals with agricultural directions are including papers on aquatic environment with the aim of involve literature on aquatic topics, mainly related with aquaculture and environmental aquatic protection, and some journals with silviculture orientation includes topics of aquatic environments associated to ecosystems services related to native forests. It discussed some topics about potential trends in aquatic sciences journals in according to specialized information sources.

Keywords: Aquatic sciences, Basic sciences, Applied sciences.

The literature about information sciences has a notable increase that include an ordered in structured systems that include editorial process and inclusion of journals in ordered ranking classification process, that include quality of the journals in according to exigent quality characteristics such as current problems exposed in the literature, statistical management, coherence in related contents, coherence in the exposed contents and directions of the journal and periodicity of the journal (Rojas and Rivera, 2011; DOAJ, 2012; SCIELO, 2012).

In according to Rojas and Rivera (2011), there are the following international topics, that include journals from all continents:

DOAJ (Directory of Open Access Journals): It is managed by Lund University library system, and it include all open access journals for scientists and professors.

SCOPUS: It was created in 2004 by Elsevier, it includes at least 27000 journals (16500 peer review, and 1200 with open access), more of the contents are published out of United States of America, and 21 per cent are published in different to English languages.

WEB OF SCIENCE AND WEB OF KNOWLEDGE: ISI (Institute for Scientific Information): It was adquired in 2002 by Thompson Reuters, the first Web of Science included 10000 prestigious scientific journals, and Web of Knowledge include social sciences, arts and human sciences journals.

As free access index information it is available SCIMAGO Journal Ranking, that is developed by SCIMAGO laboratory and powered by SCOPUS (SCIMAGO, 2007), it includes journals ranking by continent, country and discipline areas, and these are separated in quartiles being the quartile 1 the best group, and the last group is the quartile 4. It used the information based in "H index", total documents each year and the last three years, total references cited, total cites in three years, citable documents, cites per document in the last two years and references by document.

If it is considered under the Aquatic sciences subject category in the first journal included is Fish and Fisheries from United Kingdom with Scimago Journal Rank indicator of 3.295, whereas the last is Phuket Marine Biological Center from Thailand with Scimago Journal Rank indicator of 0.001. If we search under Oceanography, the first journal is Paleoceanography from United States of America with Scimago Journal Rank indicator of 2.668, whereas the last is Phuket Marine Biological Center from Thailand with Scimago Journal Rank indicator of 0.001. If it is search under subject Water Science and Technology the first journal is Water Research from United Kingdom with Scimago Journal Rank indicator of 2.446, whereas the last is Water Science and Technology from United Kingdom with a Scimago Journal Rank of 0.001. It it is search under Animal Science and Zoology subject, the first journal is Journal of Animal Ecology from United Kingdom with a Scimago Journal Rank of 2.741, whereas the last journal is Wild Life Society Bulletin from United States of America with a Scimago Journal Rank of 0.001. It itsearch under the criteria of Plant Science the first journal is Annual Review of Aquatic Science with Scimago Journal Rank of 14.740, whereas the last journal is ScientaAgraria from Brazil with Scimago Journal Rank of 0.001. If it is search under Ecology, the first journal is Annual Review of Ecology, Evolution and Systematics from United States of America, with a Scimago Journal Rank of 8.821, whereas the last is journal is Wild Life Society Bulletin from United States of America with a Scimago Journal Rank of 0.001. Finally if it is search under Ecology, Evolution and Systematics the first journal is Trends in Ecology and Evolution from United Kingdom with Scimago Journal Rank of 8.702, whereas the last journal is Marine Biodiversity Records from United Kingdom with Scimago Journal Rank of 0.001.

If it is search by country in India, and as Aquatic Science as subject, India is in 13th location with 420 published documents the 2011, with an H index of 53. Whereas in oceanography subject India is located in 13th location with 282 published documents with H index of 49. If it is search under Water Science and Technology, India is in 8th location with 685 documents and H index of 64. It is search under Animal Science and Ecology, India is located in 11th location with 842 published and H index of 35. If it is search under Plant Science, India is in third site with 2016 documents and H index of 86. If it is search under ecology, India is in the tenth site with 582 published documents with H index of 56. If it is search under ecology, evolution and systematics, India is located in 12th location with 523 published documents with H index of 48.

If we search under Aquatic sciences, in India there are two journals Indian Journal of Aquatic Sciences with Scimago Journal Rank of 0.131, and International Journal of Ocean and Oceanography Scimago Journal Rank of 0.117. If it is search under Oceanography, the first Indian journal is Indian Journal of Marine Sciences with 0.267 and the second journal is International Journal of Ocean and Oceanography Scimago Journal Rank of 0.117. In water science and technology there are not Indian journals. In Animal Science and Technology, the first journal is Indian Journal of Animal Sciences with Scimago Journal Rank of 0.255, whereas the last journal is Journal of Advances Zoology with 0.102. For Plant Science, the first journal is Tropical Ecology with Scimago Journal Rank of 0.380, and the last is Plant Science with Scimago Journal Rank of 0.000. If we search under Ecology, the first journal is Pestology with Scimago Journal Rank of 0.188, and the last is Conservation and Society with Scimago Journal Rank of 0.000. Finally, if it is search under Ecology, Evolution and Systematics, the first journal is Bioscience, Biotechnology Research Asia with Scimago Journal Rank of 0.141, and the last journal is Conservation and Society with Scimago Journal Rank of 0.000.

References

DOAJ (2010) Selection Criteria. Visited February 13, 2013:

http://www. doaj. org/doaj?func=loadTempl and templ=forPublishers and uiLanguage=en

Rojas, M. A. and S. Rivera, 2011. Guía de Buenas prácticas para revistas académicas de acceso directo. O. N. G. Derechos Digitales, Santiago de Chile. 26 p

SCIELO Chile, (2013). Criterios SCIELO Chile (In Spanish, in English mean: Journal Evaluation).

www. scielo. cl/criterios/es.

SCImago. (2007). SJR — SCImago Journal and Country Rank. Visited February 13, 2013, from http://www. scimagojr. com

Chapter 35

A Biogeographic View Point of Comparison between Patagonian and Indian Crustacean Zooplankton

☆ *Patricio De los Ríos Escalante*

ABSTRACT

The crustacean zooplankton communities in Patagonia are characterized by the calanoid dominance and the presence of many endemic species that were originated from common ancestor from Gondwana, when Antarctica, Australia, India and South America. Also, it is possible found some widespread species from North America such as Daphnidscladocerans. Nevertheless, in spite of it, there are not shared species with Indian inland waters. The literature described that in Indian inland waters there are very few endemic species and many species shared with African and another Asian zones. Then in this scenario, the zooplankton identification and study for Indian inland waters requires different directions in comparison to Patagonian inland waters.

Keywords: Cladocerans, Copepods, Inland waters, Dispersion, Biogeography.

The inland water species in different regions are characterized by determined patterns due biogeographical causes that generate the presence of endemic and cosmopolite species (Boxshall and Defaye, 2008). The inland freshwater crustaceans are characterized by the presence of two main groups, Branchiopoda and Copepoda that include a wide species groups (Battish, 1992). In southern hemisphere, the main dominant crustacean zooplankton are calanoids copepods mainly of the genus *Boeckella* and *Calamoecia*, that have a similar origin from Gondwana, and they are widespread in Antarctica, Australia, South America and Subantarctic islands (Bayly, 1992a), whereas these genus are not reported for African and Indian inland waters (Boxshall and Defaye, 2008).

A Biogeographical View Point as Cause of Species Presence

During Jurasic period when existed Gondwana continent, it supposed the presence of a common ancestor of calanoids copepods, and during last period of continental derive it produced an speciation process that generate the presence of species of *Boeckella* genus in South America, that are different to species reported from Australia and New Zealand (Bayly, 1993), and a few species shared with sub-Antarctic islands (Table 35.1; Menu-Marque *et al.*, 2000). There are cladocerans that are endemic and cosmopolite species (Table 35.1; De los Ríos-Escalante, 2010).

Table 35.1: Crustacean Species Reported for Patagonian Lakes

	Kind of Habitat
Copepoda, Calanoida	
Boeckellabrasiliensis (Lubbock, 1855)	Ponds
B. brevicaudata (Brady, 1875)	Ponds
B. gracilipes (Daday, 1902)	Small and large lakes and ponds
B. meteoris (Kiefer, 1928)	Ponds
B. michaelseni (Mrázek, 1901)	Small and large lakes and ponds
B. poopoensis (Marsh, 1906)	Ponds
B. poppei (Mrázek, 1901)	Ponds
Parabroteassarsi (Mrázek, 1901)	Large lakes and Ponds
Copepoda, Cyclopoida	
Acantocyclops vernalis (Fisher, 1853)	Small lakes
Mesocyclops longisetus (Thiebaud, 1914)	Small and large lakes and Ponds
Tropocylops prasinus (Fisher, 1960)	Large lakes
Microcyclops sp.	Large lakes
Branchiopoda, Anostraca	
Branchinecta gaini (Daday, 1902)	Ponds
B. granulosa (Daday, 1902).	Ponds
B. vuriloche (Cohen, 1985)	Ponds
Branchiopoda, Cladocera	
Daphnia ambigua (Scourfield, 1967)	Small lakes and Ponds
D. dadayana (Paggi, 1999)	Ponds
D. obtusa (Kurz, 1874)	Ponds
D. pulex (Scourfield, 1877)	Small lakes and ponds
Ceriodaphnia dubia (Richard, 1894)	Small lakes and ponds
Neobosmina chilensis (Daday, 1902)	Small and large lakes and ponds
Chydurus sphaericus (O. F. Müller, 1785)	Lakes and ponds

In South America the southern zone between approximately 38-54° S, is known as Patagonia, that is characterized by the presence of semiarid plains in oriental zones, and perennial forests in western zones, with different kinds of inland water bodies (Quiros and Drago, 1998). The literature about crustacean zooplankton described the presence of a zooplankton assemblages with few species richness, that is directly related with chlorophyll concentration and simultaneously inversely related with conductivity (Soto and De los Ríos, 2006; De los Ríos-Escalante, 2010). For species determination, it is possible use for cladocerans the descriptions of Araya and Zú iga (1985), for calanoids identification the descriptions of Bayly (1992a, b); for cyclopoids it is possible use the references of Araya and Zú iga (1985) and Reid (1985).

The literature described the presence of large, deep and oliogotrophic lakes with very few species richness, that are two or three species in according to its oligotrophy, normally it is possible found *Boeckellagracilipes* or *Boeckellamichaelseni*, *Tropocyclopsprasinusmeridionalis* and *Neobosminachilensis* (De los Ríos-Escalante, 2010). Whereas for small lakes with fishes, that are normally mesotrophic, it is possible found the same species in comparison to large lakes, and large bodies cladocerans such as *Daphniapulex*, *D. ambigua* and *Ceriodaphniadubia*, and another cyclopoids such as *Mesocyclopsaraucanus* (Soto and De los Ríos, 2006; De los Ríos-Escalante, 2010). Whereas for small ponds without fishes, it is possible found large bodies crustacean zooplankton, where the main species reported are calanoids such as *Boeckellapoppei*, and *Parabroteassarsi*, and large cladocerans such as *Daphnia dadayana*, and the species reported for large and small lakes (Soto and De los Ríos, 2006; De los Ríos-Escalante, 2010).

It is possible found large Anostracans of *Branchinecta* genus (De los Ríos-Escalante, 2010), many of these species are endemic (Belk and Brték 1995), there are very scarce studies about these genus, nevertheless it would be possible that this genus inhabits in low conductivity and oligotrophic ephemeral pools (De los Ríos-Escalante, 2010).

In Indian inland waters it was found the presence of very few endemic species, for cladocerans (Raghunathan and Kumar, 2003), similar situation would happen with copepods (Boxshall and Defaye, 2008). It was found species from Africa, Europa and Asia, in according to Bosxhall and Defaye (2008), it is possible found a kind of Indian Raft, due the presence of dispersion zone that include India and Africa. The literature for crustacean zooplankton indicate the presence of two main books for identification zooplankton species, these are the descriptions of authors such as Fernando, Sharma and Kumar that were published before 1990, perhaps the most recent references for identify Indian inland water cladocerans are Battish (1992); Murugan *et al.* (1998). It suggest the standardization of zooplankton identification procedures for Indian crustacean zooplankton, considering the marked differences in zooplankton communities for both regions.

References

Araya, J.M. and Zú iga, L.R., 1985. Manual taxonómico del zooplancton lacustre de Chile. *Boln. Limnológico, Universidad Austral de Chile*, 8: 1–110.

Battish, S.K., 1992. *Freshwater Zooplankton of India*. Oxford and IBH Publishing, New Delhi.

Bayly, I.A.E., 1992a. The non–marine Centropagidae (Copepoda, Calanoida) of the world. Guides to the identification of the microinvertebrates of the continental waters of the world, 2: 1–30. (SPB Academic Publishers, Amsterdam).

Bayly, I.A.E., 1992b. Fusion of the genera Boeckella and Pseudoboeckella and a revision of their species from South America and subantarctic islands. *Revista Chil. Hist. Nat.*, 65: 17–63.

Bayly, I.A.E., 1993. The fauna of athalassic saline waters in Australia and the Altiplano of South America: comparison and historical perspectives. *Hydrobiologia*, 267: 225–231.

Boxshall, G.A. and Defaye, D., 2008. Global diversity of copepods (Crustacea, Copepoda) in freshwater. *Hydrobiologia*, 595: 195–207.

Belk, D. and Brték, J., 1995. A check list of the Anostraca. *Hydrobiologia*, 298: 315–353.

De los Ríos-Escalante, P., 2010. Crustacean zooplankton communities in Chilean inland waters. *Crustaceana Monographs*, 12: 1–109.

Murugan, N., Murugavel, P. and Kodarkar, M.S., 1998. Cladocera: The biology classification, identification and ecology. *Ass. Aq. Biol.*, 5: 1–55.

Quiros, R. and Drago, E., 1999. The environmental state of Argentinean lakes: An overview. *Lake Reserv. Res. and Manag.*, 4: 55–64.

Raghunathan, M.B. and Kumar, R.S., 2003. Checklist of Indian cladocera (Crustacea). *Zoos Print J.*, 18(8): 1108–1182.

Reid, J., 1985. Chave de identifiç o e lista de referencias para as speciescontinentais sudamericanas de vida livre da ordem Cyclopoida (Crustacea, Copepoda). *Bolm. Zool. Univ. de S o Paulo*, 9: 17–143.

Soto, D. and Ríos, P. De los, 2006. Trophic status and conductivity as regulators of daphnids dominance and zooplankton assemblages in lakes and ponds of Torres del Paine National Park. *Biologia, Bratislava*, 61: 541–546.

Previous Volumes–Contents

— Volume 1 —

2007, xvi+194p., figs., tabls., ind., 25 cm   Rs. 950

ISBN 81-7035-483-8

— Volume 2 —

2008, xvi+143p., col. plts., figs., tabls., ind., 25 cm   Rs. 750

ISBN 81-7035-559-5

— Volume 3 —

2010, xiv+176p., col. plts., figs., tabls., ind., 25 cm   Rs. 800

ISBN 978-81-7035-633-2

— Volume 4 —

2010, xvii+182p., figs., tabls., ind., 25 cm   Rs. 750

ISBN 978-81-7035-657-8

— Volume 5 —

2011, xviii+231p., col. plts., tabls., figs., ind., 25 cm   Rs. 1200

ISBN 978-81-7035-697-4

— Volume 6 —

2012, xv+345p., col. plts., figs., tabls., ind., 25 cm   Rs. 1800

ISBN 978-81-7035-782-7

— Volume 7 —

2013, xiv+239p., 25 cm   Rs. 1300

ISBN 978-81-7035-820-6

Index